Lecture Notes in Operations Research

Lecture Notes in Operations Research is an interdisciplinary book series which provides a platform for the cutting-edge research and developments in both operations research and operations management field. The purview of this series is global, encompassing all nations and areas of the world.

It comprises for instance, mathematical optimization, mathematical modeling, statistical analysis, queueing theory and other stochastic-process models, Markov decision processes, econometric methods, data envelopment analysis, decision analysis, supply chain management, transportation logistics, process design, operations strategy, facilities planning, production planning and inventory control.

LNOR publishes edited conference proceedings, contributed volumes that present firsthand information on the latest research results and pioneering innovations as well as new perspectives on classical fields. The target audience of LNOR consists of students, researchers as well as industry professionals.

Ali Emrouznejad · Emmanuel Thanassoulis · Mehdi Toloo
Editors

Advances in the Theory and Applications of Performance Measurement and Management

Proceedings of DEA45—International Conference on Data Envelopment Analysis

Springer

Editors
Ali Emrouznejad
Surrey Business School
University of Surrey
Guildford, UK

Emmanuel Thanassoulis
Aston Business School
Aston University
Birmingham, UK

Mehdi Toloo
Surrey Business School
Guildford, UK

ISSN 2731-040X ISSN 2731-0418 (electronic)
Lecture Notes in Operations Research
ISBN 978-3-031-61596-2 ISBN 978-3-031-61597-9 (eBook)
https://doi.org/10.1007/978-3-031-61597-9

This Springer imprint is published by the registered company Springer Nature Switzerland AG
The registered company address is: Gewerbestrasse 11, 6330 Cham, Switzerland

Preface

Data Envelopment Analysis (DEA) has emerged as a powerful tool in the realm of performance assessment, efficiency measurement, and decision-making across various sectors. These proceedings include selected papers presented at DEA45, a conference organized by the Centre for Business Analytics in Practice at the University of Surrey in September 2024. DEA45 was convened not only as a platform for showcasing the latest developments in DEA, but also to commemorate the 45th Anniversary of the seminal paper by Charnes, Cooper, and Rhodes,[1] which introduced DEA for the first time in 1978. As we reflect on the remarkable journey of DEA over the past four and a half decades, DEA45 provided an invaluable opportunity for scholars, practitioners, and enthusiasts to celebrate the enduring impact of Charnes, Cooper, and Rhodes' pioneering work and to chart the future trajectory of DEA research and applications.

In this comprehensive volume, we bring together a collection of 24 papers that delve into diverse aspects of DEA, showcasing its versatility and applicability in addressing real-world challenges. The chapters in this book span a wide spectrum of topics, ranging from methodological advancements to practical applications in different industries and contexts. Each chapter offers distinct perspectives, methodologies, and empirical discoveries, enriching the diverse landscape of DEA literature.

This volume begins with a tribute to Prof. Jon A. Chilingerian, a pioneer in the field whose contributions have significantly shaped the landscape of DEA research. From foundational concepts to cutting-edge methodologies, these proceedings cover a breadth of topics, including inference in dynamic, nonparametric models of production, Bayesian approaches for bias correction, and applications in various sectors such as theatrical firms, ports, and thermal energy generation plants. Exploring the intersection of DEA with other analytical techniques, such as fuzzy logic, path-based models, and intuitionistic fuzzy models, we uncover new avenues for enhancing decision-making processes and gaining robust insights. Moreover, we examine the

[1] Charnes A., Cooper W. W. and Rhodes E. L. "Measuring the efficiency of decision making units", European Journal of Operational Research, 2 (6) (1978).

impact of DEA in diverse domains, including banking, education, healthcare, and environmental sustainability. In the face of the unprecedented challenges posed by events such as the COVID-19 pandemic, DEA provides valuable insights into evaluating and enhancing resource allocation and performance within regional health systems.

Furthermore, this volume sheds light on the global economic landscape, offering comprehensive analyses of economic performance over time and space through the lens of DEA and the Malmquist index. As we navigate through the intricacies of efficiency measurement and performance evaluation, we also explore the practical implications of DEA in policy formulation and decision support. From identifying best practices in educational funding to informing policy decisions in public basic education and electric power transmission, DEA serves as an invaluable tool for informed decision-making and policy design. Lastly, we introduce DEA-Viz, a software tool designed to facilitate the visualization of DEA problems, enhancing the accessibility and usability of this powerful analytical framework.

In assembling this collection of papers, we aim to provide researchers, practitioners, and policymakers with a comprehensive resource that not only showcases the state-of-the-art in DEA research but also inspires future advancements and applications. We hope that this volume serves as a catalyst for further exploration and innovation in the ever-evolving field of Data Envelopment Analysis.

We extend our heartfelt gratitude to all the authors whose dedication, expertise, and scholarly contributions have made this volume possible. Additionally, we extend our sincere thanks to all the reviewers whose invaluable feedback and constructive critiques have enhanced the quality and rigor of each chapter. We would also like to express special thanks to Springer and Springer Production, especially Jialin Yan and Christian Rauscher, who supported the publication of this volume in the Springer series of Lecture Notes in Operational Research.

Finally, it is essential to clarify that the responsibility for the content, accuracy, and interpretations presented in each chapter lies solely with the respective authors. As editors, we facilitate the compilation and organization of the material, but we do not assume any responsibility for errors, inaccuracies, or omissions that may be present within individual chapters. While every effort has been made to ensure the quality and integrity of the content, readers should exercise their judgment and discretion when interpreting and utilizing the information provided.

Guildford, UK
Birmingham, UK
Guildford, UK
March 2024

Ali Emrouznejad
Emmanuel Thanassoulis
Mehdi Toloo

Contents

Remembering Jon A. Chilingerian

Mitchell P. V. Glavin

Abstract Jon Arsen Chilingerian passed away in May 2023, and this article is a salute to him and his career. Jon was a prominent scholar in healthcare management and noted for his many contributions to the use of data envelopment analysis in research on provision of health services. Jon was also a passionate and engaging instructor who had a profound impact on students around the globe.

Keywords Chilingerian · Healthcare · Data envelopment analysis · DEA

1 Jon A. Chilingerian, 1951–2023

May 2023 brought the most unwelcome news that Jon Chilingerian had passed away. His death was a great loss for the field of healthcare management and leaves a very big hole in so many lives in so many places around the globe. Jon was Chair for my Ph.D. dissertation at the Heller School, Brandeis University. Jon taught me the methods of data envelopment analysis which I incorporated into my doctoral research. I have been blessed to have Jon be my mentor, research collaborator, and friend for over 30 years. Many DEA 45 attendees can relate similar tales about knowing and working with Jon.

Jon was generous and compassionate while brimming with intellectual curiosity. He invested that energy and passion into his teaching and scholarly research career. In truth, it was more like 3 or 4 careers as he packed so much into one lifetime.

Here is an excerpt from the announcement on Jon's passing released in May 2023 by the Heller School:

> Professor Jon Chilingerian was a force of nature. He led innovation at the frontiers of health care and management with unmatched passion and insight.

M. P. V. Glavin (✉)
Chair, Department of Healthcare Management, Meehan School of Business, Stonehill College, Easton, MA 02357, USA
e-mail: mglavin@stonehill.edu

A. Emrouznejad et al. (eds.), *Advances in the Theory and Applications of Performance Measurement and Management*, Lecture Notes in Operations Research,
https://doi.org/10.1007/978-3-031-61597-9_1

> At the Heller School for Social Policy and Management, Jon added "Management" and an accredited MBA to our school. He did so first with the launch in 1995 of a joint MD-MBA program with Tufts University's medical school, one of the first such programs in the nation.
>
> Stuart Altman brought Jon to the Heller School, and was a wonderful colleague and dear friend. In 1998 what is now the Heller Social Impact MBA was launched with Jon's leadership, and in 2016 Jon led the launch of the Executive MBA (EMBA) for physicians. The EMBA program has [educated] more than 200 physician leaders advancing the frontiers of health care in the United States and worldwide. When the current cohort of physicians learned of Jon's passing, they commented, "Jon taught us to be adaptable, so we honor him even as we have to adjust to this terrible news."
>
> Jon was a national leader in working with medical societies to advance professional development for physicians, including work with the American College of Surgeons, the Thoracic Surgery Foundation for Research and Education, the European Health Leader's Program, and the Advanced Health Policy and Leadership Academy in partnership with the Hanley Center in Maine [1].

At Brandeis, Jon also served as co-director and then director for more than 20 years of the doctoral training grant in health services research funded by the U.S. Agency for Healthcare Research and Quality.

Jon would add on to his responsibilities at Brandeis by completing a sabbatical year in France at INSEAD in 1997–98 and becoming a key contributor there for over 20 years in their Healthcare Management Initiative, European Health Leaders Program, Middle East Health Leaders Program, and much more. He also managed to fit in stints as a visiting scholar in residence with the Aston School of Management, Aston University and as a guest instructor for the University of Alabama–Birmingham.

Jon published his scholarly work in a wide array of journals, such as *Annals of Operational Research, Medical Care, European Journal of Operational Research, Production and Operations Management, Health Services Research, Health Care Management Review, Medical Care Research and Review, Journal of Medical Systems, Health Services Management Research, Journal of Operations Management, Journal of Healthcare Management, Frontiers in Public Health,* and the *Journal of Health Politics, Policy, and Law.* A significant portion of his research work and publications involved application of data envelopment analysis to health care where Jon was a trailblazer and innovator. His most notable contribution is likely his application of DEA to the performance of individual physicians, going beyond healthcare facility-level applications. He also authored or co-authored chapters included in several books on data envelopment analysis, such as *Data Envelopment Analysis: Theory, Methodology, and Applications* and *Handbook on DEA* [2, 3].

Jon received many awards over the course of his professional life. Here is a sampling:

- Best paper award, Academy of Management 1989 Annual Meeting, Health Care Management Division, Academy of Management
- Best paper award, 1993, American Marketing Association, Academy for Health Services Marketing

- "Teaching Award MD-MBA Program," 2008, Tufts School of Medicine, Tufts University
- Myron D. Fottler Exceptional Service Award, 2010, Health Care Management Division, Academy of Management
- Recipient of "2018 Top Professor of the Year" designation, International Association of Top Professionals
- Recognition of Excellence, American College of Surgeons, Association for Surgical Education
- Recipient of 2018 Albert Nelson Marquis Distinguished Humanitarian & Lifetime Achievement Award, October 2018
- "Certificate of Honor" award, Minister of Health and Prevention, United Arab Emirates, Clinical Governance Symposium, November 2019
- MD-MBA Excellence in teaching award, 2022, Tufts School of Medicine, Tufts University
- Heller Teaching Excellence Award, 2023, Heller School, Brandeis University
- 2023 Professor of the Year Award from Business Leaders and Professions' Top 100 Registry.

I think the teaching awards are the ones that Jon treasured most. Many tributes to Jon from former students are viewable here: www.kudoboard.com/boards/MYRQduVQ

In early 2007 I received an unexpected call from Jon. He persuaded me to fly out of Logan Airport during a snowstorm to conduct a site visit at a pharmaceutical plant in Ireland. Our work generated a published case study that was the basis for my research presentation when I applied to join the faculty at Stonehill College. I am still there today as Chair of the Department of Healthcare Management. I am just one of many, many people that learned so much from him and are so indebted to him. Best of all, he made everything so much fun.

2 Other Voices

2.1 Dr. Stuart Altman, Brandeis University

Jon Chilingerian was a truly remarkable person. I had the rare privilege of working closely with Jon as his supervisor, his colleague and finally as my boss. First let us start at the beginning. The Heller School at Brandeis University had started an ill-defined Master's Program in Human Services Management in the late 1970's. It combined some aspects of an MBA, a Social Policy program and a sprinkling of social work. While it was quite successful in attracting many first-rate students, few outside of Heller knew what a MHSM was and it lacked a clear focus. As the dean of the Heller School, I was responsible for making this program work. Fortunately, I met Jon. After much arm twisting, I convinced him to come to Heller to direct the program. With his Ph.D. degree from the Sloan School at MIT and a stint at

Boston City Hospital, Jon restructured the program into a focused MBA in Non-Profit Management. This restructuring did not happen overnight. But his perseverance and understanding of what a first-rate management program required helped create one of the most respected educational offerings in the country with such a focus.

As time moved on both he and I stepped down from our administrative responsibilities and became colleagues and friends. We talked often about what was needed at Heller and more importantly what was needed in our U.S. healthcare system. In particular, that our healthcare system lacked a focus on "efficiency". We frequently discussed his enthusiasm for using Data Envelopment Analysis or DEA for helping senior healthcare managers restructure their organizations to provided needed healthcare more efficiently. His enthusiasm and analytic abilities convinced many senior healthcare leaders, clinicians, and academics that the U.S. healthcare system needed to change. That brings us to the next phase of our relationship.

Jon was convinced that if our healthcare system was going to change it needed a new group of physician managers to lead the change. He gave frequent lectures around the country and around the world on why this change was necessary and the type of education that was needed to facilitate such a change. He convinced a private educational foundation in the state of Maine to develop a multi-session program to educate all the physicians in the State on the problems facing the healthcare system and what could be done to help solve these problems. Working with the American College of Surgeons, Jon developed an annual week-long certificate program at Heller with the same goals as in Maine. Several medical school programs have also adopted segments of his program. But Jon was not done. He kept coming back to the idea that the country needed hundreds, if not thousands, of physician leaders, trained in the new management science of healthcare efficiency. Again, with his dogged perseverance he convinced the leadership at Heller and Brandeis to create a new type of degree, an Executive MBA (EMBA), which would allow working physicians to obtain an MBA with half the courses taught in-person and half online over two years. In that effort, I was an enthusiastic supporter and willing faculty member. The program has now graduated 189 students via 6 cohorts with a seventh cohort in progress.

This country needs these new managers desperately. With millions of Americans reaching older ages and needing more healthcare services and with healthcare already consuming almost 20% of the country's annual income, we must find a way to provide these needed services more efficiently.

It is hard to imagine anyone other than Jon Chilingerian accomplishing so much in such a short period of time. He left us much too soon.

2.2 *Dr. Joe Zhu, Worcester Polytechnic Institute*

Data Envelopment Analysis (DEA) is a widely utilized methodology across various fields. Over the last four decades since its inception, DEA has garnered immense popularity, with over ten thousand publications that span across more than 150 journals, indicating the widespread interest in the subject. However, when it comes to DEA applications in healthcare research and practice, one name invariably stands out: Jon Chilingerian. Jon is undeniably a pioneer in the domain of DEA applications in healthcare management. His innovative work has earned international recognition and played a pivotal role in shaping the research trajectories for both DEA researchers and practitioners alike.

Being local in the Metro-Boston area, I had the honor of meeting Jon on multiple occasions to discuss DEA-related projects and ideas. Jon is an individual with energy and a wealth of ideas when it comes to promoting DEA in practical applications. Given the highly mathematical nature of DEA methodology, Jon expressed the idea of publishing books aimed at introducing DEA to practitioners and managers. Jon's research has generated real results that continue to support the use of DEA in healthcare. As both an educator and researcher, Jon derives immense satisfaction from witnessing these ideas translated into practical use, creating a positive and meaningful impact on society.

2.3 *Dr. Stephen Chick, INSEAD*

Jon was an essential and driving force in the launch and development of INSEAD's Healthcare Management Initiative and in directing and teaching in the J&J partner programme at INSEAD and the European Health Leadership Programme. Jon and I worked tightly on additional programs, the Middle East Health Leadership Programme and Innovating Health for Tomorrow, and he was active in many other Exec Ed programs for decades, both teaching and directing. He was a wonderful human being, was passionate about the learning process and inspired others to have a real impact on health systems, not only at Brandeis but though his tireless efforts and collaborations with the INSEAD Healthcare Management Initiative and in our partnership with the J&J Corporate Citizenship Trust. He has impacted healthcare leaders in INSEAD programs that number in the thousands in many countries around the world. In recent years, he had asked me to co-edit a special issue of *Frontiers in Public Health* called 'The Science and Art of Value in Healthcare: Measures, Voices and Methods' [4]. This issue speaks volumes about Jon: his passion for both scientific justification, leadership, and impact in practice, bringing together different viewpoints in the pursuit of more effective health systems, management, and leadership.

Jon's passing is a huge loss. I am proud to have learned from him, to have worked with him, and to have gotten to know him as a friend. I miss him very much. I know that many others in the INSEAD community feel the same way. I send condolences to his family and share the grief with others that have had the privilege to interact with him too.

2.4 Dr. Regina Herzlinger, Harvard University

Jon was a vibrant, positive, pleasant, effective presence. With his intellectual and personal networking skills, he brought the subject of health care management to so many students who otherwise would not have access to it. The world of health care management and I will miss him dearly.

I list some of his many accomplishments below; but impressive as they are, they are overshadowed by the sheer presence of his personality.

- Directing the U.S Agency for Healthcare Research and Quality training grant program in health services research for doctoral students at the Heller School at Brandeis University for over 20 years
- Launching and leading the MBA program and the Master's in Management program at the Heller School
- Directing the Brandeis Component of the MD-MBA program operated in conjunction with Tufts University Medical School
- Directing executive education programs at the Heller School for physician groups such as the American College of Surgeons and the Massachusetts Medical Society
- Launching and leading the Executive MBA program for physicians at the Heller School
- Serving as Program Director for the European Health Leaders program at INSEAD in France that trained nearly 800 clinical leaders from 35 countries.

2.5 Dr. H. David Sherman, Northeastern University

I so appreciated and will miss my friend and colleague Jon Arsen Chilingerian. Jon was an accomplished and creative seeker of new insights in health management and Data Envelopment Analysis; he was also an innovative, creative, kind and inspiring educator. I met Jon when I joined the Sloan MIT faculty where he was a PhD student. He enrolled in a course I taught on nonprofits and then became my research assistant. This is where I introduced Jon to DEA and my work on managing productivity of hospitals and financial institutions. Jon was captivated by the potential of DEA, and we decided to try applying it to physicians. Any illness can be treated by different physicians with many different practice patterns, so DEA offered one new way to identify the best practice patterns. Our first try was with cardiac bypass patients, and we were both excited to find that we could identify different quality and efficiency results even in a small number of surgeons and even identify the ones that one might want to avoid. This began an informal partnership working on applying DEA to physicians that continued for several years. Jon included DEA research in his work thereafter, including papers in the DEA 45 conference in Surrey. In this way, I would like to believe that I had a positive impact on Jon's professional life. During this time, we became good friends; our families enjoyed times together, even a musicale with Jon on the Mandolin (I think). My daughter, Amanda, recalled dining with the

Chilingerian family at Karoun, the Armenian restaurant in Newton, Massachusetts (owned by another of Jon's friends and memorable also as a place where dancing accompanies dessert.)

Just as I hope introducing Jon to DEA and its applications enhanced his professional life, Jon also had a significant impact on my life and my family's life. Jon introduced me to INSEAD (the business school in France) after his sabbatical at INSEAD. He opened the way for me to visit (3 times), teach, and publish with INSEAD colleagues as well as teaching in their J&J program. Equally important on a personal level, this enabled my family to move to Paris during my sabbatical at INSEAD; the positive impact and pleasure was immeasurable. Adding to our professional and personal association, Jon and I became initial managers representing Brandeis and Northeastern for the Tufts School of Medicine MD-MBA program. Jon and I really enjoyed working with the medical school deans, the faculty at Brandeis and Northeastern that taught in the program, and wonderful medical students.

Rather than restate Jon's accomplishments which others have explained, I offer a few other observations related to DEA. Jon's energy and enthusiasm for exploring DEA seemed unending. He studied and absorbed the complex issues underlying DEA and became involved in developing the theory and the applications. He particularly appreciated meeting Bill Cooper, one of the DEA founders, and he very much enjoyed being able to work with Regi Herzlinger and benefit from her encouragement and insights. I would add that from my time working with Jon, I know he admired Stuart Altman and fully returned the appreciation expressed by Stuart.

Jon Chilingerian was a kind person who cared very much about his family and directed his professional energy into trying to improve the health care system. I believe that the health care system was enhanced with Jon's work and with his influence over numerous students in a range of health fields.

3 In Conclusion

I wish to express my deepest appreciation to Dr. Ali Emrouznejad for allowing me to share some thoughts here on Jon Chilingerian. While we have indeed been cheated out of having Jon join us for the DEA 45 conference at the University of Surrey, he has left us a powerful legacy through his research, teaching, and collegiality. I also want to note that he is co-editor of a forthcoming book *The New Science of Medicine & Management: A Comprehensive, Case-Based Guide for Clinical Leaders* [5]. This book was generated out of Jon's work as Director of the Executive MBA program for physicians at Brandeis University. I encourage readers to explore Jon's thoughts on the emerging science of medicine and management and his identification of the key role to be played in improving the organization and delivery of health services by physicians who are versed in management theory and tools.

I endeavor to do as much for my students as Jon did for me and so many others, but he set a dizzyingly high standard that is impossible to match. God bless you, Jon, and thank you for everything!

Photo credit Heller School, Brandeis University

References

1. Anonymous. (2023, May 11). *The Heller community remembers and Honors Professor Jon Chilingerian.* https://heller.brandeis.edu/news/items/releases/2023/remembering-jon-chilingerian.html
2. Chilingerian, J. (1995). Exploring why some physicians are more efficient: taking DEA into the hospital. In A. Charnes, W. Cooper, A. Lewin, L. Seiford (Eds.), *Data envelopment analysis: Theory, methodology, and applications.* Kluwer Publishing.
3. Chilingerian, A., & Sherman, D. (2004). Health applications of DEA. In W. Cooper, L. Seiford, J. Zhu (Eds.), *Handbook on DEA* (1st ed.). Springer Press.
4. Chilingerian, J., Lorenz, L., True, G., Chick, S., & Tompkins, C., (special issue Eds.). (2022). The science and art of value in healthcare: Measures, voices and methods. *Frontiers in Public Health, 10.*
5. Chilingerian, J., Shobeiri, S., & Talamini, M., (Eds.). (2023). *The new science of medicine & management: A comprehensive, case-based guide for clinical leaders* (1st edn.). Springer Nature.

Inference in Dynamic, Nonparametric Models of Production for General Technologies

Léopold Simar and Paul W. Wilson

Abstract Nonparametric envelopment estimators are often used to estimate the attainable set and its efficient boundary, and to assess efficiency and changes in productivity. Kneip et al. [11] provide asymptotic results that can be used to make inference about expected changes in productivity measured by Malmquist indices and about the sources of productivity changes. These results require convexity of the attainable set, but in a number of situations this assumption is questionable. Recently, Kneip et al. [12] extend these results to allow for possibly non-convex technologies where the DEA estimators are known to be inconsistent. This paper summarizes these results, and explains how researchers should choose the appropriate method in a particular application.

Keywords Nonparametric production frontiers · DEA · FDH · Malmquist Indices

1 Introduction

Nonparametric envelopment estimators are often used to estimate the attainable set and its efficient boundary, and to assess efficiency and changes in productivity. Kneip et al. [9] provide asymptotic theory enabling inference about expected efficiency, and Kneip et al. [10] provide asymptotic results that can be used to test convexity versus non-convexity of the production set as well as constant versus variable returns to scale (when the production set is convex) using these estimators. In addition, Kneip et al. [11] and Simar and Wilson [20] provide asymptotic results that can be used to make inference about expected changes in productivity measured by Malmquist indices and about the sources of productivity changes, provided the underlying production sets

L. Simar
Institut de Statistique, Biostatistique et Sciences Actuarielles (ISBA), LIDAM, UCLouvain, Voie du Roman Pays 20, B 1348 Louvain-la-Neuve, Belgium
e-mail: leopold.simar@uclouvain.be

P. W. Wilson (✉)
Department of Economics and School of Computing, Clemson University, Clemson, SC, USA
e-mail: pww@clemson.edu

A. Emrouznejad et al. (eds.), *Advances in the Theory and Applications of Performance Measurement and Management*, Lecture Notes in Operations Research,
https://doi.org/10.1007/978-3-031-61597-9_2

are convex. But in a number of situations, convexity of the production set is questionable, and some studies (e.g., O'Loughlin and Wilson [14]) have rejected convexity using the test of Kneip et al. [10]. The results have recently been extended by Kneip et al. [12] to allow for possibly non-convex technologies, where the DEA estimators are known to be inconsistent. Kneip et al. [12] establish properties of the conical FDH estimator, a nonparametric envelopment estimator of distance to the boundary of the cone spanned by a possibly non-convex attainable set. These new results are then extended to make inference about Malmquist indices and their various components in non-convex settings. Recent results are summarized below, providing guidelines for empirical researchers that should be helpful for choosing appropriate estimators and for making inferences.

1.1 Basic Notation

A general nonparametric model of production is described as follows (e.g., see Shephard [17]). Denote by $x \in \mathbb{R}_+^p$ the vector of input quantities that are used to produce a vector $y \in \mathbb{R}_+^q$ of output quantities. The set of feasible combinations of input and output quantities (i.e., the production set) is given by

$$\Psi = \{(x, y) \in \mathbb{R}_+^{p+q} \mid x \text{ can produce } y\}. \tag{1}$$

The efficient frontier of Ψ is given by

$$\Psi^\partial = \{(x, y) \mid (\gamma^{-1}x, \gamma y) \notin \Psi, \text{ for any } \gamma > 1\}. \tag{2}$$

The following assumptions on Ψ are typically made:

- Ψ is closed;
- no free lunches exist, i.e., $(x, y) \notin \Psi$ if $x = 0$ and $y \geq 0,\ y \neq 0$; and
- both inputs and outputs are freely disposable, i.e., for all $(x, y) \in \Psi$, if $x' \geq x,\ y' \leq y$ then $(x', y') \in \Psi$.

In addition, Ψ is sometimes (but not always) assumed to be convex (here, we do not assume Ψ is convex). Some studies assume constant returns to scale (CRS) so that Ψ is such that if $(x, y) \in \Psi$, then $(ax, ay) \in \Psi$ for all $a > 0$.

The efficiency of a firm represented by a point $(x, y) \in \Psi$ is measured by its distance to the efficient frontier Ψ^∂. Several such measures have appeared in the literature, including

- the radial, input-oriented measure $\theta(x, y \mid \Psi) = \inf\{\theta \mid (\theta x, y) \in \Psi\} \leq 1$;
- the radial, output-oriented measure $\lambda(x, y \mid \Psi) = \sup\{\lambda \mid (x, \lambda y) \in \Psi\} \geq 1$;

- the hyperbolic measure $\gamma(x, y \mid \Psi) = \inf\{\gamma \mid (\gamma x, \gamma^{-1} y) \in \Psi\} \leq 1$; and
- the directional distance measure $\beta(x, y; d_x, d_y \mid \Psi) = \sup\{\beta \mid (x - \beta d_x, y + \beta d_y) \in \Psi\} \geq 0$, where $d_x \geq 0$ and $d_y \geq 0$ are direction vectors.

See Simar and Wilson [19] for a review and detailed references.

Of particular interest for purposes of the Malmquist Productivity Index (MPI) is the cone $\mathcal{C}(\Psi)$ spanned by Ψ, defined by

$$\mathcal{C}(\Psi) = \{(\tilde{x}, \tilde{y}) \mid \tilde{x} = ax, \tilde{y} = ay, \forall a \in \mathbb{R}_+ \text{ and } \forall(x, y) \in \Psi\}. \tag{3}$$

Since Ψ is not necessarily convex, $\mathcal{C}(\Psi)$ is not necessarily convex. In the particular case of CRS, we have $\mathcal{C}(\Psi) = \Psi$, for both convex and non-convex cases, but in general $\Psi \subseteq \mathcal{C}(\Psi)$.

In any realistic situation, Ψ is unknown, and hence $\mathcal{C}(\Psi)$ is also unknown. Consequently, both Ψ and $\mathcal{C}(\Psi)$, as well as the efficiency measures described above, must be estimated from a sample $\mathcal{X}_n = \{(X_i, Y_i)\}_{i=1}^n$ of observed input-output combinations. Typically, the observations in $\mathcal{X}_n$ are assumed to be identically and independently distributed (iid).

2 Nonparametric Estimators

2.1 A Data Generating Process (DGP)

The DGP is *the model*, i.e., the set of assumptions, that specify how the random sample $\mathcal{X}_n$ is generated. The DGP is determined by specifying the joint density $f_{XY}(x, y)$ of the random vectors (X, Y) with support on Ψ, or equivalently by specifying the probability

$$H_{XY}(x, y) = \Pr(X \leq x, Y \geq y) \tag{4}$$

of a production plan (x, y) begin dominated. Under Free Disposability and regardless of whether Ψ is convex, Ψ is the support of H_{XY} and hence

$$\Psi = \{(x, y) \mid H_{XY}(x, y) > 0\}. \tag{5}$$

Then the efficiency measure for input-oriented case can be written as

$$\theta(x, y \mid \Psi) = \inf\{\theta \mid H_{XY}(\theta x, y) > 0\}. \tag{6}$$

In a nonparametric model, no particular structure is imposed on $H_{XY}(x, y)$ and a natural estimator is provided by its empirical analog $\widehat{H}_{XY}(x, y) = n^{-1} \sum_{i=1}^n \boldsymbol{I}(X_i \leq x, Y_i \geq y)$, where $\boldsymbol{I}(\cdot)$ is the indicator function. This provides the basis of various envelopment estimators of Ψ.

2.2 Popular Envelopment Estimators

Under free disposability assumption of Ψ, without assuming convexity, Deprins et al. [4] propose the Free Disposal Hull (FDH) estimator obtained by applying the plug-in principle:

$$\begin{aligned}\widehat{\Psi}_{FDH}(\mathcal{X}_n) &= \{(x, y) \mid \widehat{H}_{XY}(x, y) > 0\} \\ &= \left\{(x, y) \in \mathbb{R}_+^{p+q} \mid y \leq Y_i,\ x \geq X_i,\ (X_i, Y_i) \in \mathcal{X}_n\right\}. \end{aligned} \tag{7}$$

Imposing convexity of Ψ leads to the Data Envelopment Analysis (DEA) estimator

$$\begin{aligned}\widehat{\Psi}_{DEA}(\mathcal{X}_n) = \{(x, y) \in \mathbb{R}_+^{p+q} \mid y \leq \sum_{i=1}^{n} \eta_i Y_i; x \geq \sum_{i=1}^{n} \eta_i X_i \text{ for } (\eta_1, \ldots, \eta_n) \\ \text{such that } \sum_{i=1}^{n} \eta_i = 1; \eta_i \geq 0, i = 1, \ldots, n\} \end{aligned} \tag{8}$$

which is the convex hull of $\widehat{\Psi}_{FDH}(\mathcal{X}_n)$ (see Farrell [7], and Charnes et al. [3]). Neither $\widehat{\Psi}_{FDH}(\mathcal{X}_n)$ nor $\widehat{\Psi}_{DEA}(\mathcal{X}_n)$ impose constant returns to scale.

2.3 Cone Spanned by Ψ

In general, $\Psi \subseteq \mathcal{C}(\Psi)$, and $\Psi = \mathcal{C}(\Psi)$ if and only if CRS holds; otherwise, $\Psi \subset \mathcal{C}(\Psi)$. This is true regardless of whether Ψ is convex. If Ψ is convex, then $\mathcal{C}(\Psi)$ is the conical hull of Ψ and $\mathcal{C}(\Psi)$ is convex. Alternatively, if Ψ is *not* convex, then $\mathcal{C}(\Psi)$ is not necessarily convex. Note that CRS may (or may not) hold regardless of whether Ψ is convex.

Under free disposability, but without assuming convexity of Ψ, $\mathcal{C}(\Psi)$ can be estimated by the *conical* FDH (CFDH) estimator

$$\begin{aligned}\widehat{\Psi}_{CFDH}(\mathcal{X}_n) := \mathcal{C}(\widehat{\Psi}_{FDH}(\mathcal{X}_n)) = \big\{(x, y) \in \mathbb{R}_+^{p+q} \mid (x, y) = (a\widetilde{x}, a\widetilde{y}) \text{ for some} \\ a \geq 0 \text{ and } (\widetilde{x}, \widetilde{y}) \in \widehat{\Psi}_{FDH}(\mathcal{X}_n)\big\}. \end{aligned} \tag{9}$$

If in addition CRS holds, then $\widehat{\Psi}_{CFDH}$ also estimates Ψ. When both free disposability and convexity of Ψ are assumed, either $\widehat{\Psi}_{CFDH}(\mathcal{X}_n)$ or the *conical* DEA (CDEA) estimator

$$\begin{aligned}\widehat{\Psi}_{CDEA}(\mathcal{X}_n) := \mathcal{C}(\widehat{\Psi}_{DEA}(\mathcal{X}_n)) := \big\{(x, y) \in \mathbb{R}_+^{p+q} \mid y \leq \sum_{i=1}^{n} \eta_i Y_i; x \geq \sum_{i=1}^{n} \eta_i X_i \\ \text{with } \eta_i \geq 0\ \forall\, i = 1, \ldots, n\big\} \end{aligned} \tag{10}$$

can be used to estimate $\mathcal{C}(\Psi)$. If in addition CRS holds as well as free disposability and convexity, then both $\widehat{\Psi}_{CFDH}(\mathcal{X}_n)$ and $\widehat{\Psi}_{CDEA}(\mathcal{X}_n)$ estimate Ψ. But if CRS does not hold, then $\widehat{\Psi}_{CDEA}(\mathcal{X}_n)$ estimates $\mathcal{C}(\Psi) \neq \Psi$ provided Ψ is convex, and $\widehat{\Psi}_{CFDH}(\mathcal{X}_n)$ estimates $\mathcal{C}(\Psi)$ regardless of whether Ψ is convex.

2.4 *Efficiency Estimators*

Results for the input-oriented case are given below, but similar results hold for the output orientation as well as both hyperbolic and directional distances. Efficiency estimates are obtained by "plugging in" the the appropriate estimator of Ψ, depending on the assumptions one makes about Ψ.

If only free disposability (but not convexity) is assumed, then the appropriate estimator of Ψ is $\widehat{\Psi}_{FDH}(\mathcal{X}_n)$, and hence an estimate distance to Ψ^{∂} in the input direction is given by

$$\widehat{\theta}_{FDH}(x, y) = \min_{i \in D(x,y)} \left[\max_{j=1,\ldots,p} \left(\frac{X_{ij}}{x_j} \right) \right], \tag{11}$$

where $D(x, y) = \{i \mid X_i \leq x, Y_i \geq y\}$ is the set of observations dominating (x, y). For estimating Malmquist indices, estimates of distance to the boundary of $\mathcal{C}(\Psi)$ are needed, and if only free disposability is assumed these distances are estimated by distance to the boundary of $\mathcal{C}(\Psi_{FDH}(\mathcal{X}_n)$ given by

$$\widehat{\theta}_{CFDH}(x, y) = \min_{i=1,\ldots,n} \frac{\max_{j=1,\ldots,p} \frac{X_{ij}}{x_j}}{\min_{j=1,\ldots,q} \frac{Y_{ij}}{y_j}} \tag{12}$$

as shown by Kneip et al. [12]. Note that if CRS holds, then $\widehat{\theta}_{CFDH}(x, y)$ also estimates efficiency.

Under both free disposability and convexity of Ψ, efficiency is estimated by $\widehat{\theta}_{DEA}(x, y) = \min\{\theta \mid (\theta x, y) \in \widehat{\Psi}_{DEA}(\mathcal{X}_n)\}$. For estimating Malmquist indices, estimates of distance to the boundary of $\mathcal{C}(\widehat{\Psi}_{DEA})$ are needed, and these are provided by $\widehat{\theta}_{CDEA}(x, y) = \min\{\theta \mid (\theta x, y) \in \mathcal{C}(\widehat{\Psi}_{DEA}(\mathcal{X}_n))\}$. Once again, if CRS holds, then $\widehat{\theta}_{CDEA}(x, y)$ also estimates efficiency. Both the DEA and CDEA estimators $\widehat{\theta}_{DEA}(x, y)$ and $\widehat{\theta}_{CDEA}(x, y)$ can be computed by solving familiar linear programs (e.g., see Kneip et al. [11]).

2.5 *Statistical Properties*

The properties of the estimators discussed above depend crucially on assumptions made about the DGP. In all cases, the asymptotic properties rest on mild regularity

Table 1 Values of κ for the various distance estimators

Estimator	CRS, CONV	$\overline{\text{CRS}}$, CONV	CRS, $\overline{\text{CONV}}$	$\overline{\text{CRS}}$, $\overline{\text{CONV}}$
FDH	$1/(p+q)$	$1/(p+q)$	$1/(p+q)$	$1/(p+q)$
CFDH	$1/(p+q-1)$	$1/(p+q-1/2)$	$1/(p+q-1)$	$1/(p+q-1/2)$
DEA	$2/(p+q)$	$2/(p+q+1)$	—	—
CDEA	$2/(p+q)$	$2/(p+q+1)$	—	—

"CRS" and "$\overline{\text{CRS}}$" denote that CRS holds or does not hold, respectively
"CONV" and "$\overline{\text{CONV}}$" denote that convexity of Ψ holds or does not hold, respectively

assumptions, e.g., smoothness of boundaries and of f_{XY} near the relevant boundary. The results assume that $f_{XY}(x, y) > 0$ on the boundary of Ψ. See Park et al. [15, 16], Kneip et al. [8, 11, 12] and Simar and Wilson [19] for additional details.

Under appropriate assumptions, each of the distance estimators discussed above is statistically consistent and converges to a stable, non-degenerate limiting distribution at rate n^{κ} as $n \to \infty$. For the FDH and CFDH estimators, the limiting distribution is Weibull, and for the DEA and CDEA cases (under convexity) the limiting distribution is a regular distribution depending on unknown parameters, but no closed-form representations are available when p or $q > 1$. If Ψ is non-convex, the results for the DEA and CDEA estimators do not hold; neither DEA nor CDEA estimators are consistent without convexity of Ψ. The rates of convergence n^{κ} depend on the particular case, and on the values of $p + q$. The error of estimation of each efficiency measure for a fixed point (x, y) is of order $O_p\left(n^{-\kappa}\right)$, and so the *curse of dimensionality* that is common among nonparametric estimators is present here, too. Table 1 summarizes the results, with free disposability holding in all cases.

It is important to note that with free disposability, the FDH and CFDH estimators always provides consistent estimates, regardless of whether Ψ is convex. This is not the case with the DEA and CDEA estimators, which requires convexity to be consistent. Note also that the CFDH, DEA and CDEA estimators converge at faster rates under CRS than when CRS does not hold.

Inference (i.e., confidence intervals) for measures of distance from a given point (x, y) to the boundary of Ψ or $\mathcal{C}(\Psi)$ can be obtained by bootstrap methods (see Simar and Wilson [19] for a survey and references). Ordinary Central Limit Theorems (CLTs) (e.g., the Lindeberg-Feller CLT) do not hold in general for mean efficiency or distances, but recent CLT results are available making possible inference about expected efficiency or mean efficiency over groups of firms. See Kneip et al. [9, 11, 12] for details.

As a practical matter, inference requires information about convergence rates, and Table 1 makes clear that convergence rates of the various estimators depend on whether CRS or convexity hold. Whether CRS or convexity hold are ultimately empirical questions requiring inference via hypothesis testing. Kneip et al. [10] develop a test of convexity and a test for CRS versus non-CRS for convex production sets, and Kneip et al. [12] provide a test of CRS versus non-CRS that does not require convexity of production sets.

3 Dynamic Settings

An objective in dynamic, multivariate ($p,\ q \geq 1$) settings is to measure any gains or losses of productivity by firms over time, and perhaps to analyze the source of these changes. This is achieved by defining a Malmquist Productivity Index (MPI); e.g., see Malmquist [13] and Caves et al. [2]. Various studies have decomposed the MPI into indices of changes in efficiency, technology, scale efficiency, etc.

In the literature, DEA and CDEA estimators are invariably used to estimate MPIs and component indices, but as discussed above, these estimators require Ψ to be convex. Kneip et al. [11, 12] provide the tools for providing consistent estimators of the index under various assumptions (i.e., convexity or non-convexity, CRS or non-CRS), as described in the preceding section. They also derive appropriate CLTs for making inference on averages of the MPI over the units in a entire sample.

3.1 Malmquist Productivity Indices

Consider two periods t_1 and t_2 and suppose that $p = q = 1$. The relative productivity changes (using the output orientation) is easily defined as

$$\mathcal{M} = \frac{y^{t_2}/x^{t_2}}{y^{t_1}/x^{t_1}}.$$

If productivity increases between periods t_1 and t_2, then $\mathcal{M} > 1$, and if productivity decreases, then $\mathcal{M} < 1$. If there is no change, then $\mathcal{M} = 1$. For $p,\ q \geq 1$, the idea of MPI is to use distances to the boundary of $\mathcal{C}(\Psi)$ to measure these productivity changes. It turns out that the MPI defined below is equivalent to the ratio $\mathcal{M}$ when $p = q = 1$.

In order to handle dynamic settings, denote the attainable set at time t by

$$\Psi^t = \{(x, y) \mid x \text{ can produce } y \text{ at time } t\} \tag{13}$$

and its efficient boundary by

$$\Psi^{t,\partial} = \{(x, y) \mid (x, y) \in \Psi^t, (\gamma x, \gamma^{-1} y) \notin \Psi^t\ \forall \gamma \in (0, 1)\}. \tag{14}$$

Denote the cone spanned by Ψ^t by $\mathcal{C}(\Psi^t)$.

Defining the MPI in terms of hyperbolic distances

$$\gamma(x, y \mid \mathcal{C}(\Psi^t)) = \inf \left\{\gamma \mid (\gamma x,\ \gamma^{-1} y) \in \mathcal{C}(\Psi^t)\right\} \tag{15}$$

to the boundary of $\mathcal{C}(\Psi^t)$ ensures that the MPI is well-defined for all of the observations in a given sample. Recall that under CRS, $\mathcal{C}(\Psi^t) = \Psi^t$, but in general,

$\mathcal{C}(\Psi^t) \subseteq \Psi^t$. Then the MPI for a firm moving from (x^1, y^1) in period 1 to (x^2, y^2) in period 2 is defined as

$$\mathcal{M} = \left(\frac{\gamma(x^2, y^2 \mid \mathcal{C}(\Psi^1))}{\gamma(x^1, y^1 \mid \mathcal{C}(\Psi^1))} \times \frac{\gamma(x^2, y^2 \mid \mathcal{C}(\Psi^2))}{\gamma(x^1, y^1 \mid \mathcal{C}(\Psi^2))} \right)^{1/2}. \tag{16}$$

The boundaries of $\mathcal{C}(\Psi^1)$ and of $\mathcal{C}(\Psi^2)$ serve as benchmarks against which any change in productivity is measured. The MPI in (16) is the geometric mean of the changes (the ratios) of the distances of the firm to the boundary of the cone spanned by the possibility sets at the two periods. The geometric mean is natural when considering means of ratios. All of this is valid under convexity or non-convexity of the attainable sets, and with or without the CRS assumption.

3.2 Sources of Productivity Changes

Several decompositions of MPI have been proposed in the literature; see Simar and Wilson [20] for discussion and references. To facilitate notation, let $w^t = (x^t, y^t)$, $t \in \{1, 2\}$. We focus below on the decomposition

$$\mathcal{M} = \mathcal{E} \times \mathcal{T} \times \mathcal{S}_1 \times \mathcal{S}_2, \tag{17}$$

where

$$\begin{aligned} \mathcal{E} &= \frac{\gamma(w^2 \mid \Psi^2)}{\gamma(w^1 \mid \Psi^1)} && \text{(change in efficiency)}, \\ \mathcal{T} &= \left[\frac{\gamma(w^2 \mid \Psi^1)}{\gamma(w^2 \mid \Psi^2)} \times \frac{\gamma(w^1 \mid \Psi^1)}{\gamma(w^1 \mid \Psi^2)} \right]^{1/2} && \text{(change in technology)}, \\ \mathcal{S}_1 &= \frac{\gamma(w^2 \mid \mathcal{C}(\Psi^2))/\gamma(w^2 \mid \Psi^2)}{\gamma(w^1 \mid \mathcal{C}(\Psi^1))/\gamma(w^1 \mid \Psi^1)} && \text{(change in scale efficiency)} \end{aligned} \tag{18}$$

and $\mathcal{S}_2$ may be viewed as a residual term. Simar and Wilson [20] discuss how $\mathcal{S}_2$ can be interpreted as a change in the scale or shape of the technology. Note that the true production sets Ψ^t serve as the reference sets for defining $\mathcal{E}$ and $\mathcal{T}$. The scale efficiency term $\mathcal{S}_1$ measures the differences between Ψ^t and $\mathcal{C}(\Psi^t)$. Again, note that this does not require convexity of Ψ^t nor CRS, but clearly under CRS in both periods, $\mathcal{S}_1 = \mathcal{S}_2 = 1$.

3.3 Nonparametric Estimation

Estimation requires random samples of firms for two different time periods $t \in \{1, 2\}$, i.e. n iid pairs $\{(W_i^1, W_i^2)\}_{i=1}^n$, where $W_i^t = (X_i^t, Y_i^t)$. Since there may be additional observations in each period, with no counterpart in the other period, we might consider $\mathcal{X}_{n_1}^1$ and $\mathcal{X}_{n_2}^2$, with $n_t \geq n$. The Malmquist indices for each firm $i = 1, \ldots, n$ can be written as

$$\mathcal{M}_i = \left(\frac{\gamma(X_i^2, Y_i^2 \mid \mathcal{C}(\Psi^1))}{\gamma(X_i^1, Y_i^1 \mid \mathcal{C}(\Psi^1))} \times \frac{\gamma(X_i^2, Y_i^2 \mid \mathcal{C}(\Psi^2))}{\gamma(X_i^1, Y_i^1 \mid \mathcal{C}(\Psi^2))} \right)^{1/2}. \tag{19}$$

For estimation of the Malmquist index at fixed point, for a single firm $w^t = (x^t, y^t)$, $t \in \{1, 2\}$, first simplify the notation by defining $\widehat{\gamma}_C^s(w^t) = \gamma(w^t \mid \mathcal{C}(\widehat{\Psi}^s))$ for $s, t = 1, 2$, and letting this denote either the CFDH hyperbolic distance for general (possibly non-convex) technologies, or the CDEA estimator for convex technologies, of w^t to the boundary of the cone spanned by the FDH or DEA estimator of Ψ^s (respectively) estimated from the sample $\mathcal{X}_{n_s}^s$, for $s, t = 1, 2$. To be explicit,

$$\widehat{\mathcal{M}}(w^1, w^2) = \left(\frac{\widehat{\gamma}_C^1(w^2)}{\widehat{\gamma}_C^1(w^1)} \times \frac{\widehat{\gamma}_C^2(w^2)}{\widehat{\gamma}_C^2(w^1)} \right)^{1/2}. \tag{20}$$

The properties of this estimator are based on some mild regularity assumptions (see Kneip et al. [11, 12] for details). These regularity assumptions include (i) for $t = 1, 2$, we have iid observations (X_i^t, Y_i^t), $i = 1, \ldots, n_t$; (ii) for $n = \min\{n_1, n_2\}$, $((X_i^1, Y_i^1), (X_i^2, Y_i^2))$ are iid with joint density f_{12}, with (X_i^s, Y_i^s) independent of (X_j^t, Y_j^t) for $i \neq j$; and (iii) some regularity assumptions on the densities of (X^t, Y^t), f_t, $t = 1, 2$ and on $f_{12}(X^1, Y^1, X^2, Y^2)$ ensuring that $\mathcal{M}(w^1, w^2)$ and $\log \mathcal{M}(w^1, w^2)$ are well defined. Then

$$n^{\kappa}\left(\widehat{\mathcal{M}}(w^1, w^2) - \mathcal{M}(w^1, w^2)\right) \xrightarrow{\mathcal{L}} F_{w^1, w^2}, \tag{21}$$

as $n = \min\{n_1, n_2\} \to \infty$, where F_{w^1, w^2} is a regular distribution depending on the particular estimators used and on various features of the DGP. The rate κ depends on the assumption made on Ψ^1 and Ψ^2 (see Table 1). For instance, without the CRS assumption we have $\kappa = 1/(p + q - 0.5)$ when CFDH is used, and $\kappa = 2/(p + q + 1)$ for CDEA if the convexity assumption is valid.

Inference, e.g., confidence intervals for $\mathcal{M}(w^1, w^2)$, can then conducted by using the bootstrap. One can perform the bootstrap on pairs $\{(W_i^1, W_i^2)\}_{i=1}^n$ using the subsampling techniques described by Simar and Wilson [18]. The same applies for all the components appearing in the decomposition (17), as discussed by Simar and Wilson [20].

3.4 Inference for Geometric Means

In empirical studies, researchers often compute the geometric mean of the MPI over all the units belonging to a particular group. We can also provide inference (confidence intervals or tests of hypotheses) about the productivity changes of the entire group.

To derive the results, first consider the mean $\mu_{\mathcal{M}} = E(\log \mathcal{M}_i)$ where

$$\log \mathcal{M}_i = \frac{1}{2}\left[\log \gamma_C^1(W_i^2) + \log \gamma_C^2(W_i^2) - \log \gamma_C^1(W_i^1) - \log \gamma_C^2(W_i^1)\right]. \quad (22)$$

Then define $\widehat{\mu}_{\mathcal{M},n} = n^{-1}\sum_{i=1}^{n} \log \widehat{\mathcal{M}}_i$ where

$$\log \widehat{\mathcal{M}}_i = \frac{1}{2}\left[\log \widehat{\gamma}_C^1(W_i^2) + \log \widehat{\gamma}_C^2(W_i^2) - \log \widehat{\gamma}_C^1(W_i^1) - \log \widehat{\gamma}_C^2(W_i^1)\right]. \quad (23)$$

Kneip et al. [11, 12] derive CLTs for $\mu_{\mathcal{M}}$ by establishing that

$$\sqrt{n}\left(\widehat{\mu}_{\mathcal{M},n} - \mu_{\mathcal{M}} - \mathcal{R}_n\right) \xrightarrow{\mathcal{L}} N(0, \sigma_{\mathcal{M}}^2) \quad (24)$$

as $n \leq \min\{n_1, n_2\} \to \infty$, where $\mathcal{R}_n := C_{\mathcal{M}} n^{-\kappa} + \xi_{n,\kappa}$ is the leading bias term, with $\xi_{n,\kappa} = O\left(n^{-2\kappa}\right)$ when CFDH estimators are used and $\xi_{n,\kappa} = O\left(n^{-3\kappa/2}\right)$ if CDEA estimators are used. This property ensures that $\widehat{\mu}_{\mathcal{M},n} \xrightarrow{p} \mu_{\mathcal{M}}$. In addition, a consistent estimator of the variance is given by

$$\widehat{\sigma}_{\mathcal{M}}^2 = n^{-1}\sum_{i=1}^{n}\left(\log \widehat{\mathcal{M}}_i - \widehat{\mu}_{\mathcal{M},n}\right)^2 \xrightarrow{p} \sigma_{\mathcal{M}}^2. \quad (25)$$

Using the Delta Method, it is easy to obtain the desired result for the geometric mean $M_n = \left(\prod_{i=1}^{n} \mathcal{M}_i\right)^{1/n} = \exp(\mu_{\mathcal{M}})$, which is consistently estimated by $\widehat{M}_n = \left(\prod_{i=1}^{n} \widehat{\mathcal{M}}_i\right)^{1/n} = \exp(\widehat{\mu}_{\mathcal{M},n})$. Kneip et al. [11, 12] prove that

$$\sqrt{n}\left(\widehat{M}_n - \widetilde{M}\right) \xrightarrow{\mathcal{L}} N(0, \exp(2\mu_{\mathcal{M}})\sigma_{\mathcal{M}}^2) \quad (26)$$

as $n \leq \min\{n_1, n_2\} \to \infty$, where

$$\begin{aligned}\widetilde{M} &:= \exp(\mu_{\mathcal{M}} + \mathcal{R}_n) = \exp(\mu_{\mathcal{M}})(1 + \mathcal{R}_n) + O\left(n^{-2\kappa}\right) \\ &= M_n + \exp(\mu_{\mathcal{M}})C_{\mathcal{M}} n^{-\kappa} + \xi_{n,\kappa},\end{aligned} \quad (27)$$

where $C_{\mathcal{M}}$ is a constant. The variance $\exp(2\mu_{\mathcal{M}})\sigma_{\mathcal{M}}^2$ is estimated consistently by $\exp(2\widehat{\mu}_{\mathcal{M}})\widehat{\sigma}_{\mathcal{M}}^2$. Note that if κ is too small, i.e., less than or equal to 1/2, then the leading term of the bias, $\exp(\mu_{\mathcal{M}})C_{\mathcal{M}} n^{-\kappa} = O\left(n^{-\kappa}\right)$, does not vanish when multiplied by $\sqrt{n}$ and $n \to \infty$.

Kneip et al. ([11], Sect. 4) describe in detail a procedure for estimating the leading bias term by using a generalized jackknife method on the sample of common pairs $\mathcal{W}_n = \{(W_i^1, W_i^2)\}_{i=1}^n$. This provides a consistent estimate $\widehat{B}_{n,\kappa}$ of $\exp(\mu_{\mathcal{M}})C_{\mathcal{M}}n^{-\kappa}$, with error of the same order as $\xi_{n,\kappa} = o(n^{-\kappa})$. So, when $\kappa \geq 1/2$, which occurs only in limited cases (see Table 1), the CLT result

$$\sqrt{n}\left(\widehat{M}_n - \widehat{B}_{n,\kappa} - M_n + \xi_{n,\kappa}\right) \xrightarrow{\mathcal{L}} N(0, \exp(2\mu_{\mathcal{M}})\sigma_{\mathcal{M}}^2) \tag{28}$$

as $n \to \infty$ can be used to make inference about M_n. However, if $\kappa < 1/2$, the bias correction is not enough; from (27), it is evident that the bias explodes as $n \to \infty$ when $\kappa < 1/2$. This problem requires defining the geometric mean over a smaller number $n_\kappa = \lfloor n^{2\kappa} \rfloor < n$ of randomly selected units to reduce the rate of convergence. Then as $n \to \infty$,

$$\sqrt{n_\kappa}\left(\widehat{M}_{n_\kappa} - \widehat{B}_{n,\kappa} - M_n + \xi_{n,\kappa}\right) \xrightarrow{\mathcal{L}} N(0, \exp(2\mu_{\mathcal{M}})\sigma_{\mathcal{M}}^2), \tag{29}$$

where the geometric mean over the subsample, i.e. $\widehat{M}_{n_\kappa} = \left(\prod_{i=1}^{n_\kappa} \widehat{\mathcal{M}}_i\right)^{1/n_\kappa}$ replaces $\widehat{M}_n$ in (28). Then confidence intervals for M_n can be estimated using normal quantiles and $\exp(2\widehat{\mu}_{\mathcal{M}})\widehat{\sigma}_{\mathcal{M}}^2$ as a consistent estimator of the variance.

4 Malmquist Indices in Action

The recent results of Kneip et al. [11, 12] enable inference on Malmquist indices and their component indices under quite general technologies, i.e., with our without convexity of Ψ and with or without CRS. Moreover, results are provided for testing whether Ψ is convex or whether CRS holds. The computations are easy, and can be done using commands in the *FEAR* package described by Wilson [23] for the *R* programming language. Results from intensive Monte-Carlo experiments are provided by Kneip et al. [11, 12], indicating that even with small sample sizes, the CLTs provide good approximations that are usable in practice.

The methods described here have already been used in several fields of applications. For example, Simar and Wilson [20] revisit the analysis of productivity in Swedish pharmacies by Färe et al. [5] to provide inference. Simar and Wilson [22] re-examine the analysis of productivity changes among U.S. hospitals by Burgess and Wilson [1], while Simar and Wilson [21] update the analysis of productivity changes among OECD countries by Färe et al. [6], providing inference in both cases using the methods described above. The results and methods described above will likely prove useful in many other applications in the near future.

References

1. Burgess, J. F., Jr., & Wilson, P. W. (1995). Decomposing hospital productivity changes 1985–1988: A nonparametric Malmquist approach. *Journal of Productivity Analysis, 6*, 343–363.
2. Caves, D. W., Christensen, L. R., & Diewert, W. E. (1982). The economic theory of index numbers and the measurement of input, output and productivity. *Econometrica, 50*, 1393–1414.
3. Charnes, A., Cooper, W. W., & Rhodes, E. (1978). Measuring the efficiency of decision making units. *European Journal of Operational Research, 2*, 429–444.
4. Deprins, D., Simar, L., & Tulkens, H. (1984). Measuring labor inefficiency in post offices. In M. M. P. Pestieau & H. Tulkens (Eds.), *The Performance of Public Enterprises: Concepts and Measurements* (pp. 243–267). Amsterdam: North-Holland.
5. Färe, R., Grosskopf, S., Lindgren, B., & Roos, P. (1992). Productivity changes in Swedish pharmacies 1980–1989: A non-parametric Malmquist approach. *Journal of Productivity Analysis, 3*, 85–101.
6. Färe, R., Grosskopf, S., Norris, M., & Zhang, Z. (1994). Productivity growth, technical progress, and efficiency change in industrialized countries. *American Economic Review, 84*, 66–83.
7. Farrell, M. J. (1957). The measurement of productive efficiency. *Journal of the Royal Statistical Society A, 120*, 253–281.
8. Kneip, A., Simar, L., & Wilson, P. W. (2008). Asymptotics and consistent bootstraps for DEA estimators in non-parametric frontier models. *Econometric Theory, 24*, 1663–1697.
9. Kneip, A., Simar, L., & Wilson, P. W. (2015). When bias kills the variance: Central limit theorems for DEA and FDH efficiency scores. *Econometric Theory, 31*, 394–422.
10. Kneip, A., Simar, L., & Wilson, P. W. (2016). Testing hypotheses in nonparametric models of production. *Journal of Business and Economic Statistics, 34*, 435–456.
11. Kneip, A., Simar, L., & Wilson, P. W. (2021). Inference in dynamic, nonparametric models of production: Central limit theorems for Malmquist indices. *Econometric Theory, 37*, 537–572.
12. Kneip, A., Simar, L., Wilson, P. W. (2022). Conical FDH estimators of general technologies, with applications to returns to scale and Malmquist productivity indices, IIDAM Discussion Paper ISBA 2022/24; http://hdl.handle.net/2078.1/264666.
13. Malmquist, S. (1953). Index numbers and indifference surfaces. *Trabajos de Estatística y de Investigación Operativa, 4*, 209–242.
14. O'Loughlin, C. T., & Wilson, P. W. (2021). Benchmarking the performance of U.S. municipalities. *Empirical Economics, 60*, 2665–2700.
15. Park, B. U., Simar, L., & Weiner, C. (2000). FDH efficiency scores from a stochastic point of view. *Econometric Theory, 16*, 855–877.
16. Park, B. U., Jeong, S. O., & Simar, L. (2010). Asymptotic distribution of conical-hull estimators of directional edges. *Annals of Statistics, 38*, 1320–1340.
17. Shephard, R. W. (1970). *Theory of Cost and Production Functions*. Princeton: Princeton University Press.
18. Simar, L., & Wilson, P. W. (2011). Inference by the *m* out of *n* bootstrap in nonparametric frontier models. *Journal of Productivity Analysis, 36*, 33–53.
19. Simar, L., & Wilson, P. W. (2015). Statistical approaches for non-parametric frontier models: A guided tour. *International Statistical Review, 83*, 77–110.
20. Simar, L., & Wilson, P. W. (2019). Central limit theorems and inference for sources of productivity change measured by nonparametric Malmquist indices. *European Journal of Operational Research, 277*, 756–769.
21. Simar, L., & Wilson, P. W. (2023). Another look at productivity growth in industrialized countries. *Journal of Productivity Analysis Forthcoming*.
22. Simar, L., & Wilson, P. W. (2023). New tools for evaluating the performance of health-care providers using DEA and FDH estimators. In S. Grosskopf, V. Valdmanis, V. Zelenyuk (Eds.), *The Cambridge Handbook of Productivity, Efficiency & Effectivenss in Health Care*. Cambridge, UK: Cambridge University Press. Forthcoming.
23. Wilson, P. W. (2008). FEAR: A software package for frontier efficiency analysis with R. *Socio-Economic Planning Sciences, 42*, 247–254.

Can Super-Efficiencies Improve Bias Correction? A Bayesian Data Envelopment Analysis Approach

Panagiotis D. Zervopoulos, Angelos Kanas, Ali Emrouznejad, and Philip Molyneux

Abstract It has been proven that DEA efficiencies, within the interval (0, 1], are overestimated for finite samples, while asymptotically, this bias reduces to zero. In the extant literature, the statistical inference approaches yielding the best-performing DEA estimates are the smoothed bootstrap and Bayesian DEA methods. All statistical inference techniques apply to DEA models yielding efficiencies between zero and one. This study presents a new Bayesian DEA approach that takes into account efficiencies and super-efficiencies aiming to improve bias correction. We prove that efficiencies and super-efficiencies are interrelated for finite samples. However, bias correction is statistically significant only in the case of efficiencies below one. The new Bayesian super-efficiency DEA approach yields estimates with lower mean absolute error and mean square error than the extant DEA statistical inference techniques referring only to efficiencies with right-censored distributions, where efficiencies are

The views expressed herein do not necessarily reflect the views of the Hellenic Parliamentary Budget Office, Hellenic Parliament. The usual disclaimer applies.

P. D. Zervopoulos (✉)
Department of Management, University of Sharjah, Sharjah, UAE
e-mail: pzervopoulos@sharjah.ac.ae

A. Kanas
Department of Economics, University of Piraeus, Piraeus, Greece

Hellenic Parliamentary Budget Office, Hellenic Parliament, Greece

A. Kanas
e-mail: akanas@unipi.gr

A. Emrouznejad
Surrey Business School, University of Surrey, Guilford, UK

P. Molyneux
College of Business, Abu Dhabi University, Abu Dhabi, UAE
e-mail: p.molyneux@leeds.ac.uk

Leeds University Business School, University of Leeds, Leeds, UK

A. Emrouznejad
e-mail: a.emrouznejad@surrey.ac.uk

A. Emrouznejad et al. (eds.), *Advances in the Theory and Applications of Performance Measurement and Management*, Lecture Notes in Operations Research,
https://doi.org/10.1007/978-3-031-61597-9_3

not allowed to exceed unity. Drawing on formal analysis, real-world and simulated data sets, we conclude that the new Bayesian super-efficiency DEA estimates are consistent of DEA parameters.

Keywords Data envelopment analysis · Super-efficiency · Bayesian methods · Statistical inference · Banking

1 Introduction

DEA is a non-parametric approach to constructing empirical production frontiers [1, 2]. Despite the non-parametric nature of DEA, [3, 4] provided statistical justification for DEA. In particular, [5] proved that DEA, under the conditions of monotonicity and concavity, yields consistent estimators of the production frontier. Among others, the studies of [6, 7] allowed for inference on DEA efficiency estimators.

Banker [5] was the first to highlight the overestimation of DEA efficiencies when finite samples are used. Banker [5] showed that this bias asymptotically reduces to zero. In addition, [8] identified an inverse relationship between the rate of convergence of DEA efficiency estimators and the dimensionality of the production set. Simar and Wilson [9] stated that the true efficiency of a firm is unknown.

Emphasizing DEA, there are six major approaches dealing with the sensitivity of efficiency estimators: (a) Chance Constrained DEA (CCDEA); (b) Two-stage DEA-based methods; (c) Corrected Concave Non-Parametric Least Squares (C^2NLS); (d) Stochastic Non-Smooth Envelopment of Data (StoNED); (e) Bayesian DEA; and (f) bootstrap DEA. In this study, we discuss the last two approaches, which are directly related to our bias-correction method.

A Bayesian DEA approach for CCDEA was developed by [10]. This method provides statistical inference to CCDEA efficiencies relying on assumptions about the distribution (e.g., multivariate normal) of the (posterior/observed) inputs and outputs. Also, [11] put forth a Bayesian approach drawing on a CCDEA model to estimate efficiency under uncertainty. This approach does not make any distributional assumption.

Bootstrap DEA is the most widely used method for correcting bias and constructing confidence intervals of efficiency estimators [12]. Bootstrap DEA, or smoothed bootstrap, originated from [6], combines both the virtues and limitations of bootstrap and DEA. Smoothed bootstrap relies on pseudo-data obtained from an estimated data-generating process (DGP) [13]. Kneip et al. [12, 14] developed improved smoothed bootstrap algorithms providing consistent bias-corrected estimators.

The two methods discussed above dealing with the sensitivity of DEA efficiencies refer to estimators lying within the interval (0, 1]. Andersen and Petersen [15], drawing on the work of [16], presented a super-efficiency DEA model making possible efficiency estimators exceed unity, unlike the conventional DEA models, as the firm under review is excluded from its own reference set. Super-efficiency DEA procedure is used for ranking efficient units and identifying outliers [17].

However, [17] and [18] found that super-efficiency performs unsatisfactorily in ranking efficiency units. Banker and Gifford [16], Banker and Chang [17], Banker et al. [18] suggested the removal of outliers from the sample to prevent possible distortion of efficiency measurements due to observational errors. This study identifies a relationship between estimates lying below (efficiencies) and above unity (super-efficiencies). This relationship is stronger for smaller samples and weakens as sample sizes increase.

The contribution of this work is to provide statistical inference in efficiency estimates lying below one, obtained from super-efficiency DEA models. Specifically, we develop a two-stage Bayesian DEA approach: in the first stage, we use a super-efficiency DEA model, while in the second stage, we specify consistent efficiency estimates taking into account the relationship between efficiencies and super-efficiencies, after excluding outliers. We use a novel theoretical Bayesian DEA approach, which extends the work of [7] on standard DEA models, to obtain bias-corrected efficiencies for finite samples. To the best of our knowledge, this is the first work on correcting bias of efficiencies considering super-efficiency scores.

The rest of the paper unfolds as follows. In Sect. 2, we analyze the Bayesian super-efficiency DEA model. Section 3 presents the real-world and simulated data sets used in this study. Section 4 discusses the empirical results. Section 5 concludes the study.

2 Methodology

2.1 Super-Efficiency DEA

In this study, we use [19]'s model to obtain super-efficiency estimators (θ_j). This is one of the methods in the extant literature that addresses the infeasibility problem in super-efficiency DEA models. Banker et al. [20]'s model is as follows.

$$\min\ \tau + M \times \sum_{r=1}^{s} \eta_r$$

$$s.t. \sum_{\substack{j=1\\ j\neq o}}^{n} \lambda_j x_{ij} \leq (1+\tau)x_{io} \quad i = 1, 2, \ldots, m$$

$$\sum_{\substack{j=1\\ j\neq o}}^{n} \lambda_j y_{rj} \geq (1-\eta_r)y_{ro} \quad r = 1, 2, \ldots, s$$

$$\sum_{\substack{j=1 \\ j \neq o}}^{n} \lambda_j = 1$$

$$\lambda_j \geq 0, j \neq o, \eta_r \geq 0 \tag{1}$$

where M is a user-defined large positive number (e.g., 10^5), x_{ij} and y_{rj} are the inputs and outputs of the j th unit, and λ is the intensity variable. The subscript o expresses the unit under review. In addition, $(1+\tau)$ and $(1-\eta_r)$ represent the super-efficiency from the input side and the output side, respectively.

The super-efficiency estimators are defined as $1+\tau+\frac{1}{|R|}\sum\limits_{r\in R}\frac{1}{1-\eta_r}$, where $R = \{r|\eta_r > 0\}$, and $|R|$ is the cardinality of the set R.

For the application of this Bayesian DEA method for the correction of bias of efficiencies, any super-efficiency model, e.g., [21, 22], can be used instead of the model (1).

2.2 Conventional Statistical Inference

Let $\Theta = (\theta_j)$, where $\theta_j > 0 (j = 1, 2, ..., n)$, be a random vector of independently and identically distributed (iid) elements (i.e., super-efficiency estimates) obtained from model (1). Θ is assumed to be uniformly distributed from $\theta_L (0 < \theta_L < 1)$ to $\theta_U (\theta_U > 1)$. The two parameters (i.e., θ_L and θ_U) are unknown.

Acknowledging the probability density function (PDF) of Θ as

$$f_\Theta(\theta_j|\theta_L, \theta_U) = \begin{cases} \frac{1}{\theta_U - \theta_L}, & \theta_L \leq \theta_j \leq \theta_U \\ 0, & \text{otherwise} \end{cases} \tag{2}$$

the likelihood function is expressed as follows

$$L(\theta_L, \theta_U|\Theta) \overset{\text{iid}}{=} \prod_{j=1}^{n} f_\Theta(\theta_j|\theta_L, \theta_U) = \frac{1}{(\theta_U - \theta_L)^n}, \theta_L \leq \theta_j \leq \theta_U, j = 1, 2, ..., n \tag{3}$$

By partially differentiating the likelihood function (3), we find that it is monotone increasing for θ_L and monotone decreasing for θ_U. Hence, the likelihood function (3) is maximized at $\hat{\theta}_L = \min \Theta (0 < \theta_L \leq \theta_j < 1)$ and $\hat{\theta}_U = \max \Theta (1 < \theta_j \leq \theta_U)$, which is well-known in the literature.

From the marginal PDF of the maximum likelihood estimator (MLE) $\hat{\theta}_L$ and the corresponding marginal PDF of the MLE $\hat{\theta}_U$, we obtain $f_{(\hat{\theta}_L, \hat{\theta}_U)}(t, s|\theta_L, \theta_U) \neq f_{\hat{\theta}_L}(t|\theta_L, \theta_U) \cdot f_{\hat{\theta}_U}(s|\theta_L, \theta_U)$. Hence, the MLE $\hat{\theta}_L$

and $\hat{\theta}_U$ are not independent. The expected value of $\hat{\theta}_L$ is $E_n\{\hat{\theta}_L\} = \int_{\theta_L}^{\theta_U} t \cdot f_{\hat{\theta}_L}(t|\theta_L, \theta_U)dt = \theta_L \frac{n}{n+1} + \theta_U \frac{1}{n+1} < 1$. Likewise, the expected value of $\hat{\theta}_U$ is $E_n\{\hat{\theta}_U\} = \int_{\theta_L}^{\theta_U} s \cdot f_{\hat{\theta}_U}(s|\theta_L, \theta_U)ds = \theta_U \frac{n}{n+1} + \theta_L \frac{1}{n+1}$.

The unbiased estimators of parameters θ_L and θ_U are $\tilde{\theta}_L = \frac{n\hat{\theta}_L - \hat{\theta}_U}{n-1} < \hat{\theta}_L$ and, $\tilde{\theta}_U = \frac{n\hat{\theta}_U - \hat{\theta}_L}{n-1} > \hat{\theta}_U$ respectively, which satisfy $E_n\{\tilde{\theta}_L\} = \theta_L$ and $E_n\{\tilde{\theta}_U\} = \theta_U$, respectively.

The covariance of $\hat{\theta}_L$ and $\hat{\theta}_U$ is defined as follows

$$\begin{aligned}\mathrm{cov}_n(\hat{\theta}_L, \hat{\theta}_U) &= E_n\{\hat{\theta}_L\hat{\theta}_U\} - E_n\{\hat{\theta}_L\}E_n\{\hat{\theta}_U\} \\ &= \theta_L\theta_U + \frac{(\theta_U - \theta_L)^2}{n+2} - \frac{\theta_L\theta_U(n^2+1) + (\theta_L^2 + \theta_U^2)n}{(n+1)^2}.\end{aligned}$$

Also, the covariance of $\tilde{\theta}_L$ and $\tilde{\theta}_U$ is defined as follows

$$\begin{aligned}\mathrm{cov}_n(\tilde{\theta}_L, \tilde{\theta}_U) &= E_n\{\tilde{\theta}_L\tilde{\theta}_U\} - E_n\{\tilde{\theta}_L\}E_n\{\tilde{\theta}_U\} \\ &= \frac{n^2+1}{(n-1)^2}\left(\theta_L\theta_U + \frac{(\theta_U - \theta_L)^2}{n+2}\right) \\ &\quad - \frac{n}{(n-1)^2}\left(\theta_L^2 + \theta_U^2 + \frac{2(\theta_U^2 - \theta_L^2)}{(n+1)}\right) - \theta_L\theta_U \qquad (4)\end{aligned}$$

Based on expression (4), we develop Fig. 1, where an inverse relationship between the unbiased estimators $\tilde{\theta}_L$ and $\tilde{\theta}_U$ becomes explicit. These unbiased estimators are asymptotically uncorrelated as expression (4) is asymptotically zero $\left(\lim_{n\to\infty} \mathrm{cov}_n(\tilde{\theta}_L, \tilde{\theta}_U) = 0\right)$.

2.3 Bayesian Statistical Inference

The Prior

Let the vector $\Theta = (\Theta_O, \Theta_I)$ of efficiency and super-efficiency scores where $\Theta_O = (\theta_1, ..., \theta_k)$, $\theta_L \le \Theta_O < 1$, and $\Theta_I = (\theta_{k+1}, ..., \theta_n)$, $1 < \Theta_I \le \theta_U$, respectively. In the absence of any information about the distribution of the DEA scores, we assume $\Theta_O|\theta_L$ and $\Theta_I|\theta_U$ to be uniformly distributed. The same distributional assumption is made in [7]. The uniform distribution assumption supports the "non-informative" prior argument, which does not depend on the distribution of a finite sample DEA efficiencies and super-efficiencies or the sample size.

The parameter θ_L could be beta-distributed with parameters $\gamma > 0$ and $\delta > 0$.

Assuming the unbiased estimator of θ_L is equivalent to the expected value of the prior beta distribution of θ_L (i.e. $\frac{\gamma}{\gamma+\delta}$), we find that δ is an expression of γ,

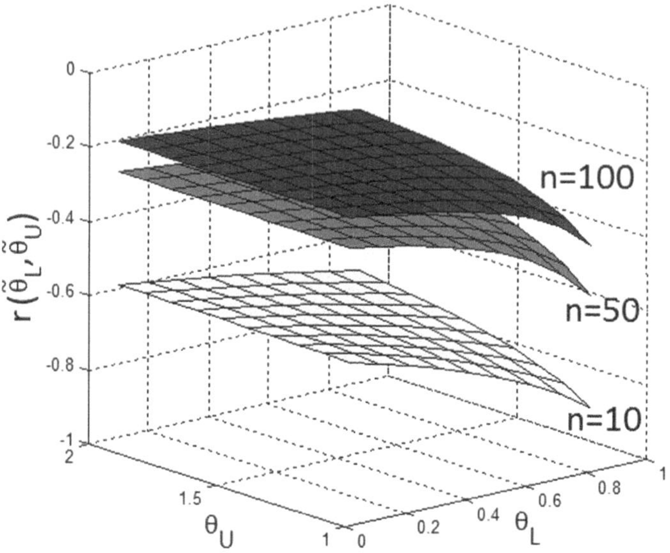

Fig. 1 Covariance of the unbiased estimators $\tilde{\theta}_L$ and $\tilde{\theta}_U$

$$\delta = \frac{(1 - \tilde{\theta}_L)\gamma}{\tilde{\theta}_L} \tag{5}$$

and the prior beta distribution is reduced to the single parameter γ.

The vector Θ_O has a joint PDF $f_{\Theta_O}(\Theta_O) = \int_0^1 f_{\Theta_O|\theta_L}(\Theta_O|\theta_L) f_{\theta_L}(\theta_L) d\theta_L = \frac{B(\gamma, \delta - k)}{B(\gamma, \delta)}$ where $\delta > k$.

Then, the parameter γ is lower bounded

$$\gamma > k \frac{\tilde{\theta}_L}{1 - \tilde{\theta}_L} \tag{6}$$

With respect to Θ_I, the parameter θ_U is assumed to be shifted gamma-distributed with parameters $\alpha > 0$ and $\beta > 0$.

Assuming the unbiased estimator is identical to the expected value $\alpha\beta + 1$ of the prior shifted gamma-distribution of θ_U, then the parameter β is an expression of α,

$$\beta = \frac{\tilde{\theta}_U - 1}{\alpha} \tag{7}$$

Hence, the prior shifted gamma distribution is reduced to the single parameter α. The vector Θ_I has a joint PDF

$$f_{\Theta_I}(\Theta_I) = \int_1^\infty f_{\Theta_I|\theta_U}(\Theta_I|\theta_U) f_{\theta_U}(\theta_U) d\theta_U = \frac{\Gamma(\alpha - n + k)}{\Gamma(\alpha)} \tag{8}$$

where $\alpha > n - k$.

Then, the parameter β is upper bounded

$$\beta < \frac{\tilde{\theta}_U - 1}{n - k}$$

The Posterior

The Bayesian PDF of $\theta_L|\Theta_O$ is

$$f_{\theta_L|\Theta_O}(\theta_L|\Theta_O) = \frac{f_{\Theta_O}(\Theta_O|\theta_L)f_{\theta_L}(\theta_L)}{f_{\Theta_O}(\Theta_O)} \tag{9}$$

The Bayesian PDF (9) becomes $\frac{1}{\mathrm{B}(\gamma,\delta-k)}\theta_L^{\gamma-1}(1-\theta_L)^{(\delta-k)-1}$

Which refers to a posterior beta distribution with parameters γ and $\delta - k$.

The posterior beta distribution shifts the corresponding prior beta distribution, with parameters γ and δ, to the right, which is justified by the following expected values.

$$E_k\{\theta_L|\Theta_O\} > E_k\{\theta_L\} \; as \; \frac{\gamma}{\gamma + \delta - k} > \frac{\gamma}{\gamma + \delta}$$

The posterior shifted gamma distribution shifts the corresponding prior distribution to the left, which justifies the underestimation of the DEA efficiencies that are greater than one. This underestimation is explained as follows.

$$E_{n-k}\{\theta_U|\Theta_I\} < E_{n-k}\{\theta_U\} \; as \; (\alpha - n + k)\beta + 1 < \alpha\beta + 1$$

2.4 Bias Correction

Let a correction parameter $\xi_L = \tilde{\theta}_L/\hat{\theta}_L$ where

$$\xi_L < 1. \tag{10}$$

Elaborating on expressions (5), (6), and (10), we estimate parameters γ and δ as follows.

$$\hat{\gamma} = k\tilde{\theta}_L/(1 - \xi_L) \tag{11}$$

and

$$\hat{\delta} = \frac{(1-\tilde{\theta}_L)\hat{\gamma}}{\tilde{\theta}_L} \tag{12}$$

Two random data sets of size k are generated, where k expresses the number of units assigned efficiencies lower than unity by the DEA program (1). The first randomly generated data set is drawn from a prior beta distribution of θ_L with parameters $\hat{\gamma}$ and $\hat{\delta}$ (see expressions (11) and (12)). The second randomly generated data set is drawn from a posterior beta distribution $\theta_L|\Theta_O$ with parameters $\hat{\gamma}$ and $\hat{\delta}-k$. The prior to posterior distribution $\Psi_L = \frac{\theta_L}{\theta_L|\Theta_O}$ is fitted by a gamma distribution with parameters $z_O > 0$ and $e_O > 0$. The MLE of these two parameters (i.e., $\hat{z}_O$ and $\hat{e}_O$) are obtained from the MATLAB function *gamfit*. The goodness-of-fit is calculated using the Wilcoxon rank sum test for equal medians.

The corrected estimator is $\vartheta_p = \theta_p \, Gamma\,(\hat{z}_O, \hat{e}_O)$, $p = 1, ..., k(p \subset j)$ with confidence interval $\overline{\vartheta}_p - \frac{s(\vartheta_p)}{\sqrt{w}}\text{cv} \le \vartheta_p \le \overline{\vartheta}_p + \frac{s(\vartheta_p)}{\sqrt{w}}\text{cv}$, where w expresses the number of Monte Carlo iterations and cv denotes the critical value of t-distribution with $w-1$ degrees of freedom. Moreover, p expresses the set of units with efficiencies below unity.

With respect to the DEA estimators that are greater than unity, we define a correction parameter $\xi_U = \tilde{\theta}_U/\hat{\theta}_U$ that satisfies $\xi_U > 1$.

We already know that $\alpha > n-k$ (8) and $\beta = \frac{\tilde{\theta}_u - 1}{\alpha}$ (7). To estimate the two parameters of the shifted gamma distribution, we introduce $\frac{(\tilde{\theta}_U - 1)}{(\tilde{\theta}_U - \hat{\theta}_U)} > 1$ in (8), which leads to $\hat{\alpha} = \frac{(n-k)(\tilde{\theta}_U - 1)}{(\tilde{\theta}_U - \hat{\theta}_U)}$ and $\hat{\beta} = \frac{\tilde{\theta}_U - 1}{\hat{\alpha}}$.

Similar to the correction process followed for the DEA estimates lying within the interval (0,1), for the estimates exceeding one, we generate two random data sets for both the prior shifted gamma distribution with parameters $\hat{\alpha}$ and $\hat{\beta}$, and the posterior shifted gamma distribution with parameters $\hat{\alpha} - n + k$ and $\hat{\beta}$. The prior to posterior distribution $\Psi_U = \frac{\theta_U}{\theta_U|\Theta_I}$ is fitted by a gamma distribution with parameters $z_I > 0$ and $e_I > 0$.

The corrected estimator is $\vartheta_q = \theta_q \, Gamma\,(\hat{z}_I, \hat{e}_I), q = 1, ..., n-k(q \subset j)$, where q denotes the set of units with efficiencies above unity.

We should note that the bias-corrected super-efficiencies are not significantly different from those obtained from the program (1).

3 Data Set and Selection of Variables

In this study, we used three real-world and five simulated data sets to test the performance of the Bayesian super-efficiency DEA approach in correcting the bias of efficiencies. The three real-world data sets consist of 50, 100, and 200 banks. In practice, the second data set (i.e., 100 banks) includes 50 new banks in addition to those of the first data set. The five simulated data sets incorporate 50, 80, 100, 200, and 500 banks. All data sets (i.e., real-world and simulated) comprise three inputs (i.e., (a) Deposits and Short-term funding; (b) Equity; (c) Fixed assets) and two outputs (i.e., (a) Gross loans; (b) Other earning assets). The data come from Moody's Analytics BankFocus. The same variables and data were used in [7].

4 Empirical Results

We should note that the statistics (i.e., Mean Absolute Error—MAE and Mean Square Error—MSE) presented in Tables 1 and 2 refer to efficiencies below one. Hence, in Table 1, the MAE and MSE refer to 37, 80, and 176 banks instead of 50, 100, and 200 units after excluding those assigned efficiencies above unity.

According to Table 1, the Bayesian super-efficiency DEA approach (i.e., Bayes DEA-2) yields consistent estimates, as well as the Bayesian DEA method put forth by [7] (i.e., Bayes DEA-1). Also, the estimates of the former approach (Bayes DEA-2) perform considerably better than those of the latter one, and the efficiency estimates obtained from the DEA smoothed bootstrap (smboot) put forth by [12, 14].

Table 1 Empirical results

Sample size	Reduced sample size[a]	MAE			MSE		
		smboot	Bayes DEA-1	Bayes DEA-2	smboot	Bayes DEA-1	Bayes DEA-2
50	37	0.07595	0.09772	0.02380	0.00889	0.00988	0.00063
100	80	0.06487	0.07570	0.01140	0.00604	0.00633	0.00014
200	176	0.07092	0.03666	0.00700	0.01052	0.00184	0.00006

[a] Banks with efficiencies below one
Bayes DEA-1: [7] Bayesian DEA; Bayes DEA-2: Bayesian super-efficiency DEA

Table 2 Simulated results

Sample size	MAE		MSE	
	smboot	Bayes DEA-2	smboot	Bayes DEA-2
50	0.17584	0.11466	0.03463	0.01362
80	0.06010	**0.08825**	0.00508	**0.00857**
100	0.06128	**0.06823**	0.00541	0.00516
200	0.05768	0.04654	0.00709	0.00298
500	0.27834	0.02461	0.09197	0.00106

Bayes DEA-2: Bayesian super-efficiency DEA

To cross-check the performance of the Bayesian super-efficiency DEA approach, we use scaled samples with simulated data. Specifically, we incorporate the same inputs and outputs (see Sect. 3) after contaminating them with moderate noise, i.e., cv = 0.25 [20], into the Bayesian bias-correction technique, which was presented in Sect. 2. For each actual sample (e.g., 50, 80, etc.—see Table 2), we generate 1,000 normally distributed pseudo-samples with the same parameters as the actual sample. According to Table 2, the Bayesian super-efficiency DEA estimates perform better than those obtained from smoothed bootstrap (smboot), referring to efficiencies below one for the samples of 50, 200, and 500 units. Also, simulated data support the consistency of the new Bayesian approach's estimates, which was evident in Table 1, where real-world data was used.

5 Concluding Remarks and Future Research

The purpose of this study was to develop a Bayesian super-efficiency DEA approach for correcting the bias of efficiencies. The distributional assumptions made for developing the new method are regarded as appropriate given the uncertainty about the distribution of the true efficiencies, while the new Bayesian DEA approach yields consistent estimates.

The new bias-correction method unfolds into two stages, where a super-efficiency DEA model is initially used to obtain super-efficiencies, followed by a Bayesian method to provide statistical inference for efficiencies. The empirical results showed a gradual decrease in MAE and MSE while the sample size increases. Also, in most cases, the new bias-correction approach's estimates perform better than those of existing statistical inference techniques.

Further research is needed to test the performance of the new Bayesian super-efficiency DEA estimates when data sets of different dimensions than those used in this study are employed.

References

1. Charnes, A., Cooper, W. W., & Rhodes, E. (1978). Measuring the efficiency of decision making units. *European Journal of Operational Research, 2*, 429–444.
2. Banker, R. D., Charnes, A., & Cooper, W. W. (1984). Some models for estimating technical and scale inefficiencies in data envelopment analysis. *Management Science, 30*, 1078–1092.
3. Banker, R. D., & Maindiratta, A. (1992). Maximum likelihood estimation of monotone and concave production frontiers. *Journal of Productivity Analysis, 3*, 401–415.
4. Banker, R. D., & Natarajan, R. (2008). Evaluating contextual variables affecting productivity using data envelopment analysis. *Operations Research, 56*, 48–58.
5. Banker, R. D. (1993). Maximum likelihood, consistency and data envelopment analysis: A statistical foundation. *Management Science, 39*, 1265–1273.
6. Simar, L., & Wilson, P. W. (1998). Sensitivity analysis of efficiency scores: How to bootstrap in nonparametric frontier models. *Management Science, 44*, 49–61.

7. Zervopoulos, P. D., Triantis, K., Sklavos, S., & Kanas, A. (2023). An alternative Bayesian data envelopment analysis approach for correcting bias of efficiency estimators. *Journal of the Operational Research Society, 74*(4), 1021–1041.
8. Simar, L. (2007). How to improve the performances of DEA/FDH estimators in the presence of noise? *Journal of Productivity Analysis, 28*, 183–201.
9. Simar, L., & Wilson, P. W. (2015). Statistical approaches for non-parametric frontier models: A guided tour. *International Statistical Review, 83*, 77–110.
10. Tsionas, E. G., & Papadakis, E. N. (2010). A Bayesian approach to statistical inference in stochastic DEA. *Omega, 38*, 309–314.
11. Charles, V., Tsolas, I. E., & Gherman, T. (2018). Satisficing data envelopment analysis: A Bayesian approach for peer mining in the banking sector. *Annals of Operations Research, 269*, 81–102.
12. Kneip, A., Simar, L., & Wilson, P. W. (2008). Asymptotics and consistent bootstraps for DEA estimators in nonparametric frontier models. *Econometric Theory, 24*, 1663–1697.
13. Dyson, R. G., & Shale, E. A. (2010). Data envelopment analysis, operational research and uncertainty. *Journal of the Operational Research Society, 61*(1), 25–34.
14. Kneip, A., Simar, L., & Wilson, P. W. (2011). A computationally efficient, consistent bootstrap for inference with non-parametric DEA estimators. *Computational Economics, 38*, 483–515.
15. Andersen, P., & Petersen, N. C. (1993). A procedure for ranking efficient units in data envelopment analysis. *Management Science, 39*, 1261–1264.
16. Banker, R. D., & Gifford, J. L. (1988). *A relative efficiency model for the evaluation public health nurse productivity*. Carnegie Mellon University, Pittsburgh.
17. Banker, R. D., & Chang, H. (2006). The super-efficiency procedure for outlier identification, not for ranking efficient units. *European Journal of Operational Research, 175*, 1311–1320.
18. Banker, R. D., Chang, H., & Zhiqiang, Z. (2017). On the use of super-efficiency procedures for ranking efficient units and identifying outliers. *Annals of Operations Research, 250*, 21–35.
19. Chen, Y., & Liang, L. (2011). Super-efficiency DEA in the presence of infeasibility: One model approach. *European Journal of Operational Research, 213*, 359–360.
20. Banker, R. D., Zheng, Z., & Natarajan, R. (2010). DEA-based hypothesis tests for comparing two groups of decision making units. *European Journal of Operational Research, 206*, 231–238.
21. Chen, Y. (2005). Measuring super-efficiency in DEA in the presence of infeasibility. *European Journal of Operational Research, 161*, 545–551.
22. Lee, H. S., Chu, C. W., & Zhu, J. (2011). Super-efficiency DEA in the presence of infeasibility. *European Journal of Operational Research, 212*, 141–147.
23. Sarath, B., & Maindiratta, A. (1997). On the consistency of maximum likelihood estimation of monotone and concave production frontiers. *Journal of Productivity Analysis, 8*, 239–246.
24. Simar, L., & Wilson, P. W. (1999). Estimating and bootstrapping Malmquist indices. *European Journal of Operational Research, 115*, 459–471.
25. Zervopoulos, P. D., Sklavos, S., Kanas, A., & Cheng, G. (2019). A multi-parametric method for bias correction of DEA efficiency estimators. *Journal of the Operational Research Society, 70*(4), 655–674.
26. Li, S., Jahanshahloo, G. R., & Khodabakhshi, M. (2007). A super-efficiency model for ranking efficient units in data envelopment analysis. *Applied Mathematics and Computation, 184*, 638–648.
27. Ray, S. C. (2008). The directional distance function and measurement of super-efficiency: An application to airlines data. *Journal of the Operational Research Society, 59*, 788–797.
28. Cook, W. D., Liang, L., Zha, Y., & Zhu, J. (2009). A modified super-efficiency DEA model for infeasibility. *Journal of the Operational Research Society, 60*, 276–281.
29. Chen, J. X., Deng, M., & Gingras, S. (2011). A modified super-efficiency measure based on simultaneous input–output projection in data envelopment analysis. *Computers & Operations Research, 38*, 496–504.

Efficiency and Productivity of Italian Theatrical Firms. A Non-parametric Approach

Concetta Castiglione, Davide Infante, and Marta Zieba

Abstract This work evaluates the efficiency and productivity of a set of 126 Italian theatrical firms over the period 2006–2014, using FDH, DEA and Malmquist total factor productivity (TFP) index. We measure an output-oriented technical efficiency (TE) assuming that Italian theatres maximise their operating revenues by the given labour and capital inputs. Public subsidies at firm level, various firm's characteristics and geographical location are included as potential efficiency determinants in a second-stage Double-Bootstrap (DB) DEA. The results demonstrate that PA firms are considerably inefficient, with the FDH TE of 0.32, the naïve DEA TE score of 0.28 and average bias-corrected DEA TE score of 0.25, while the bias-corrected scale efficiency (SE) is 0.84. There are also considerable differences between different groups of theatres, with theatre production companies being the least technically and scale efficient. We also found that public support increases efficiency of Italian theatrical firms, thus excluding rent-seeking behaviours of theatre management. Moreover, younger theatrical companies, and those organised as corporations are more technically efficient than older theatres and those organised in other legal forms. Theatres in the South of Italy are less efficient than those located in the North as expected. Moreover, total factor productivity increased only slightly in the Italian theatres and the main driver for this increase was technological change as opposed to TE or SE changes.

Keywords Technical and scale efficiency · FDH vs DEA · Double-bootstrap DEA · Malmquist TFP index · Italian theatres

---

C. Castiglione · D. Infante
Department of Economics Statistics and Finance, University of Calabria, Cosenza, Italy

M. Zieba (✉)
Department of Economics, University of Limerick, Limerick, Ireland
e-mail: marta.zieba@ul.ie

A. Emrouznejad et al. (eds.), *Advances in the Theory and Applications of Performance Measurement and Management*, Lecture Notes in Operations Research,
https://doi.org/10.1007/978-3-031-61597-9_4

1 Introduction

The aim of this work is to empirically evaluate both efficiency and productivity of a set of 126 Italian theatrical firms over the period 2006–2014, using the free disposal hull (FDH) but also the conventional DEA. The impact of various efficiency determining variables is also assessed using the second-stage double bootstrap DEA and particular emphasis is placed on examining the impact of public subsidies on the efficiency of Italian theatres.

Efficiency scores were also obtained for different samples of Italian theatres: theatre production companies, permanent theatres and opera houses, in order to compare their technologies and efficiency scores. Finally, since our dataset forms a panel covering a period of 9 years, we are also able to assess how technical efficiency and overall productivity of theatrical firms will change over time by applying the Malmquist index of total factor productivity (TFP). This allows us to decompose the TFP change into 'pure' technical efficiency change, scale efficiency change and technological change.

The study is structured as follows. Section 2 provides the context of the analysis by briefly reviewing the literature on efficiency measurement in the performing arts sector. Section 3 presents the methodology and empirical approach used. The data set and variables used are presented in Sect. 4, while Sect. 5 presents the results. Section 6 presents the concluding remarks.

2 Background

The analysis of efficiency and productivity in the performing arts (PA) sector is complicated by several factors. Firstly, the output of an artistic production process can be observed and measured only indirectly [1]. Moreover, the requirement of a direct contact with an audience, the fixed output per man-hour and the non-repeatability in artistic production, are non-standard features of the production technologies which departure from the usual assumptions [2–4]. It is also well-known that PA can also suffer from the so-called 'cost disease', first identified by Baumol and Bowen [5], whereby PA firms are unable to cover their rising costs with their revenues over time as their productivity does not rise in line with other economic sectors. However, new technologies that allow people to watch performances on new devices (TV-set, tablet, smartphones), not only increase the challenges for theatrical companies, but could also create new opportunities and accelerate the use of new modes, such as broadcasting and streaming, to consume artistic performances which could lead to improved productivity in the sector.

However, empirical evidence on the productivity and efficiency of PA firms, and in particular on what determines their productive performance, is very scarce. This could be due to the difficulty of measuring outputs and inputs with the available data, but also due to the specific features of the production characteristics of PA firms which are difficult to capture with standard model applications.

Previous studies assessed production functions in the PA sector, but have not explicitly examined the technical or productive efficiency of PA firms [2]. Other studies have relied on the estimation of cost functions for PA firms and have mainly focused on finding evidence for scale economies [6]. More recently, the improved data availability and advanced measurement methods have enabled researchers to evaluate the productivity of PA firms by directly estimating their productive efficiency. Similar to this study, an output-oriented TE has been estimated parametrically in [1, 3, 7, 8]. Other studies have assumed that PA firms have control over their inputs given their artistic output level, and have estimated an input-oriented TE, understood as the ability to use minimum inputs for the same level of output [9, 10]. Only [11] have estimated the Malmquist total factor productivity index for the German public theatres. The authors observed, in line with the cost-disease hypothesis, a decrease in the productivity for German public theatres caused by a combination of increasing labour costs and very limited opportunities to benefit from technological improvements. More recently, [8] have estimated the productivity of the PA firms using large panel data for 11 EU countries over the period 2019–2017 and have found a positive TFP growth albeit at a very small rate.

With the exception of [4], all the other efficiency studies mentioned above used parametric SFA techniques to estimate efficiency and productivity of PA firms. Marco-Serrano [4] was the only study to use a non-parametric DEA as a more flexible efficiency measurement method, as it does not require specification of the production technology. In this research, however, we apply two modifications of the basic DEA model. Firstly, we apply a Free Disposal Hull (FDH) to control for a not-well behaved production technology of theatrical firms which, as discussed in [2, 3] may have increasing marginal products and thus their production isoquants might not be convex. Furthermore, as DEA has statistical limitations, as it does not control for noise, and to investigate the determinants of efficiency, we use the double-bootstrap (DB) DEA procedure which will be outlined in Sect. 3 below.

3 Methods and Empirical Approach

The standard DEA model requires the assumption of non-increasing marginal products. However, such assumption might not be appropriate for production technologies which are non-concave in some regions as may be often the case for the performing arts firms [2, 3]. Therefore, we firstly derive the non-convex Free Disposal Hull to account for a non-convex technology of Italian theatres. Furthermore, we derive the non-parametric conventional (or naïve) DEA TE and SE scores, we use them to apply a double-bootstrap DEA algorithm of [12] to examine efficiency determinants (Sect. 3.2). Both FDH and DEA are useful procedures as they do not rely on restrictive hypothesis on the Data Generating Process (DGP) as noted by [13]. The obtained naïve TE and SE scores are also used to calculate the Malmquist TFP index in Sect. 3.3.

3.1 FDH Versus DEA

According to [14], an observed production unit (x_{it}, y_{it}) defines a firm's production possibilities set $\psi(x_{it}, y_{it})$ which under the free disposability of inputs and outputs, can be written as:

$$\psi(x_{it}, y_{it}) = \left\{(x, y) \in \mathbb{R}_{+}^{p+q} | x \geq x_{it}, y \leq y_{it}\right\} \tag{1}$$

where x is the set of inputs (labour and capital), $x \in \mathbb{R}_{+}^{p}$ used to produce a set of outputs (here total deflated operating revenues from producing artistic performances), $y \in \mathbb{R}_{+}^{q}$. Given that the set ψ cannot be observed as well as the efficiency scores, it has to be estimated from a random sample of theatrical firms, here production units $\chi = \{(x_{it}, y_{it}) | i = 1,2, \ldots, n; t = 1, \ldots, T\}$.

If we are looking in the output-direction, the Farrell's measure of output-oriented inefficiency score for a theatre operating at the level (x, y) is defined as:

$$\widehat{\theta_{FDH}}(x_{it}, y_{it}) = \sup\left\{\theta | (x_{it}, \theta y_{it}) \in \widehat{\psi}\right\} \tag{2}$$

where $\widehat{\theta}_{FDH} \geq 1$ and shows a proportional increase of output each i-th theatre in year t should achieve to be considered as being output efficient (to be located on the frontier), with input quantities held constant.

The FDH in Eq. (1) is transformed into the DEA estimator with the case of variable returns to scale (VRS) technology as proposed by [15], by applying convexity assumption as follows:

$$\widehat{\psi_{DEA}} = \left\{ \begin{array}{c} (x, y) \in \mathbb{R}_{+}^{p+q} | x \geq \sum_{i=1,t=1}^{n,T} \gamma_{it} x_{it};\, y \leq \sum_{i=1,t=1}^{n,T} \gamma_{it} y_{it}, \\ for(\gamma_{it}, \ldots, \gamma_{nT}) s.t. \sum_{i=1}^{n,T} \gamma_{it} = 1;\, \gamma_{it} \geq 0, i = 1, \ldots, n; t = 1, \ldots, T \end{array} \right\} \tag{3}$$

where γ_{it} is a vector of constants with $\gamma_{it} \geq 0$. The constant returns to scale (CRS) model developed by [16] can also be applied when the equality constraint $\sum_{i=1}^{n,,T} \gamma_i = 1$ is omitted from Eq. (3). When we are looking at the output direction, the output-oriented TE for a given theatre company i in year t (x_{it}, y_{it}) can be obtained by solving a linear program in (3) and plugging $\widehat{\psi_{DEA}}$ into (4):

$$\widehat{\theta_{DEA}}(x_{it}, y_{it}) = \max\left\{\theta | (x_{it}, \theta y_{it}) \in \widehat{\psi_{DEA}}\right\} \tag{4}$$

where $\widehat{\theta}_{DEA} \geq 1$. The reciprocal of θ defines then the technical efficiency (TE) score which is bounded between 0 and 1. If TE score equals one, the firm is considered fully efficient, which suggests maximizing output from a given set of inputs. If we detect that VRS scores are higher than CRS scores, we can conclude that a PA firm is scale inefficient. Therefore, TE scores under VRS and CRS are compared to calculate scale efficiency (SE) as follows: $TE_{it}^{CRS}/TE_{it}^{VRS}$, which is bound between 0 and 1, with an SE score of 1 inferring that a theatre company is scale efficient, and scale inefficient if the index is less than 1. The literature in the PA sector is very limited with regard to the estimation of scale efficiency and there are no empirical studies on Italian theatres that would indicate whether these companies are operating at the most productive scale size, a gap this study also seeks to address.

3.2 *Double-Bootstrap DEA Algorithm*

The DB DEA allows for the robust estimation of both the bias-corrected TE scores and the parameters of efficiency determinants,[1] and it additionally re-estimates the bias-corrected TE scores to account for these efficiency determining variables. Thus, we adopt Algorithm 2 set out by Simar and Wilson [12]. First, the "naive" output-oriented CRS and VRS DEA TE scores are estimated for each theatre i in year t, by solving the linear programming problem presented in Eqs. (3) and (4). The computations are performed in terms of Farrell's output distance measure (as defined in Sect. 3.1) which can be defined as inefficiency term. Hence, the reciprocals of the DEA TE scores ($\widehat{\theta_{it}}$) for each theatre i in year t are regressed on the vector of potential efficiency determining variables (i.e. $Z_{1it}, \ldots, Z_{pit}$). The truncated regression is run twice. The first stage of the truncated regression model is given by the following equation:

$$\widehat{\theta}_{it} = f\left(Z_{it1}, Z_{it2}, \ldots, Z_{itp}\right) + \varepsilon_{it} \sim N(0, \sigma^2) \quad (5)$$

where $\widehat{\theta}_{it}$ is the output distance term and the explanatory variables $(Z_{it1}, Z_{it2}, \ldots, Z_{itp})$ are the (in)efficiency determining variables which are presented in detail in Sect. 4; and ε_{it} is the error term which is normally distributed with zero mean and constant variance. After obtaining the results from the truncated regression model given by Eq. (5), a bootstrapping procedure is applied to correct the bias in the conventional (original) DEA efficiency scores. The truncated regression model is then re-estimated to estimate the bias-corrected reciprocals of output-oriented TE scores ($\widehat{\theta_{it}^*}$). Furthermore, these scores are obtained by adjusting the output values for the ratio of original estimates ($\widehat{\theta}_{it}$) to bias-corrected estimates ($\widehat{\theta_{it}^*}$).

[1] While the conventional DEA approach only provides a point estimate of TE scores, bootstrapping allows for creating a confidence interval around the TE score.

In the next step, the algorithm re-runs the truncated regression model; this time with the bias-corrected and adjusted by efficiency determinants, reciprocals of TE scores, that is using $\widehat{\theta_{it}^*}$ as the dependent variable:

$$\widehat{\theta}^*_{it} = f\left(Z_{it1}, Z_{it2}, \ldots, Z_{itp}\right) + \varepsilon_i \sim N\left(0, \sigma^2\right) \tag{6}$$

In the final step, in the second truncated regression model given by Eq. (6), L = 2000 drawings of residuals are taken from a truncated normal distribution and, the truncated regression model is re-estimated for each drawing. Following this, a set of robust coefficients of environmental variables is estimated in the truncated regression of $\widehat{\theta}*$ on environmental variables. The lower and upper bounds for coefficients of the determinants are also obtained. The robust CRS and VRS TE scores are further used to derive equally robust SE scores according to the definition of scale efficiency as elucidated above.

Several scholars have widely employed DB DEA over the last decade, but it has also been the subject of some criticism. First, a second-stage regression of efficiency estimates on some environmental (Z) factors can be helpful only if the Z-variables do not influence the form of the efficient (optimal) frontier. Even if this condition is to some degree satisfied, as it will be shown later in the form of separability tests in Sect. 5,[2] the inefficiency estimates utilized in the second stage might be biased due to the latent variable problem in the second-stage regression, causing significant concerns [18–20]. Second, using Monte Carlo simulations, [21] demonstrated that the two-stage DEA OLS strategy outperformed Simar and Wilson's approach regarding lower mean, median absolute deviations, and more accurate coverage rates. Hence, in this study a number of robustness checks are also performed using the usual OLS regression, and the results delivered similar results with regard to determinants and they are available on request.

3.3 Malmquist TFP Index

To calculate the productivity change of Italian theatrical companies, the Malmquist total factor productivity (TFP) index is derived in line of [22]. The index can be firstly decomposed into the technical efficiency change (TEC) and technological change (TECHC) in terms of changes of output distance functions defined as $d_o(y, x) = \theta^{-1}$, and in periods s and t as follows:

$$m_o(y_s, x_s, y_t, x_t) = TEC \times TECHC = \frac{d_o^t(y_t, x_t)}{d_o^s(y_s, x_s)} \left[\frac{d_o^s(y_t, x_t)}{d_o^t(y_t, x_t)} \times \frac{d_o^s(y_s, x_s)}{d_o^t(y_s, x_s)} \right]^{0.5} \tag{7}$$

[2] Separability tests have been suggested in [17] which are also applied in this study.

where y is the output of theatre i in years t and s, and x is the vector of inputs in years t and s. Furthermore, [22] decomposed TEC further into the pure TE change under VRS technology and scale efficiency change (SEC). Using SE measure as the ratio between CRS and VRS DEA scores, we can define an output-orientated SE change as the ratio of SE measures associated with input vectors x_t and x_s, in periods t and s, measured with respect to a specific output vector, y. Here, in line with [23], we take a geometric average as an appropriate measure of SEC as follows:

$$SEC_o^{s,t}(x_s, x_t, y_s, y_t) = \left[SEC_o^s(x_s, x_t, y_s) \times SEC_o^t(x_s, x_t, y_t)\right]^{0.5} \quad (8)$$

The final TFP index can be obtained as the product of all three indexes as follows: $TFP_O^{s,t} = TEC \times TECCH \times SEC$. All the indexes and TFP index are ratio values with the scores above one indicating positive percentage change and a score below one indicating a negative percentage change from one period to another.

4 Data Set and Variables

The data sample on Italian firms for the period of 2006–2014 was collected from two sources: the Analisi Informatizzata delle Aziende (AIDA)–Computerised Analysis of Firms; and MiBACT–Ministry of Cultural Heritage and Activities and Tourism (2014). The AIDA database is carried out by Bureau van Dijk and contains detailed accounts of Italian companies following the scheme of the 4th Directive EEL. The balance sheet data includes the information on the year of incorporation, ownership, total revenues, tangible and intangible assets, and the number of employees. Following matching the AIDA with MiBACT dataset, and using only firms with non-missing values for output and input variables, we obtained an effective unbalanced panel data for a total of 126 theatres over the period 2006–2014 (656 observations). We also divide the full sample into three distinct groups of Italian theatrical firms: theatre production companies (TPCs), permanent theatres (PTs) and Operas. The descriptive statistics (mean and standard deviation) of the output and input variables for all theatrical companies and their respective categories are presented in Table 1.

The main determinant of efficiency used in the second stage DB DEA is the level of *subsidies* (in natural log-values) received by theatres (Fig. 1) which have been decreasing over time. The other firm's characteristics included: *age* of firm measured in years, the *size* and *legal form* of the firm and *location*. Here, we use the Eurostat classification of firm's sizes: '*micro*' (0–9 employees), '*small*' (10–49 employees), '*middle sized*' (50–249 employees), and '*large*' (more than 250 employees) firms. While TPCs and PTs are usually micro or small firms, Operas are mostly run as middle sized or large companies.

To control for regional differences in TE between Italian theatres, we use dummy variables for the four Italian macro territorial areas ('North-West', 'North-East', 'Centre', and 'South and Islands'). Another variable of interest is the *legal form*

Table 1 Summary statistics of output and inputs of Italian theatrical firms

Variable and description	All firms	TPCs	PTs	Operas
Output = Total deflated operating revenues in 1,000 EUR	1,895 (5,433)	684 (1,261)	1,063 (1,094)	10,222 (12,779)
Capital = Capital stock (total deflated assets) in 1,000 EUR	9834 (30,227)	609 (1,282)	1539 (1,069)	77,297 (51,359)
Labour = Number of full-time employees	66.45 (11.83)	16.34 (12.66)	35.53 (27.85)	403.89 (269.6)
No. firms	126	87	27	12
No. observations	656	421	158	77

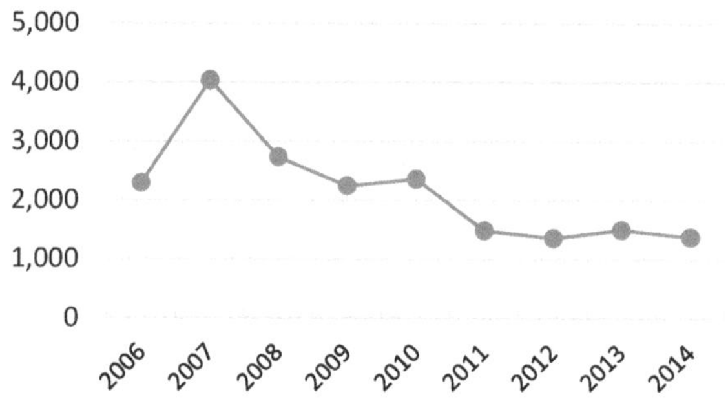

Fig. 1 Average subsidies per theatre in 1,000 EUR. *Source* AIDA database 2006–2014

which we divide into: '*cooperative companies*', '*corporations*' which include limited liability companies, and '*other legal forms*'. The foundation is the dominating legal form in the last category and it is the most common organisation form for opera houses.

5 Findings

The results of TE and SE scores are presented in Subsect. 5.1, and Subsect. 5.2 presents the estimated effects of efficiency determinants using DB DEA, while Subsect. 5.3 discusses the TFP index and its components.

5.1 TE and SE Scores

The FDH efficiency scores, the conventional (naïve) DEA and DB DEA efficiency (SE and TE) scores are presented in Table 2, for the whole sample and for three groups of Italian theatres. The FDH and DEA TE scores are the reciprocals of the estimated output distance (inefficiency) as outlined in Subsect. 3.1. The TE scores

Table 2 Obtained average TE and SE scores for Italian theatrical firms

Variable and description	All firms	TPCs	PTs	Operas
FDH efficiency (TE)	0.316	0.381	0.429	0.520
DEA efficiency (VRS TE)				
Naïve DEA	0.282*	0.291*	0.356	0.393*
Bias-corrected DB DEA	0.252	0.259	0.312	0.319
DB DEA lower bound	0.234	0.242	0.284	0.280
DB DEA upper bound	0.279	0.284	0.357	0.383
Scale efficiency (SE)				
Original DEA	0.803*	0.922*	0.937*	0.911*
Bias-corrected DB DEA	0.837	0.958	0.972	1.001
DB DEA lower bound	0.822	0.945	0.955	0.962
DB DEA upper bound	0.858	0.977	0.995	1.056

* Denotes the DEA efficiency score is outside the bootstrapped 95% confidence interval, i.e., it is significantly different from the bias-corrected efficiency score

are shown for VRS DEA technology, whilst the derived SE scores present the ratio between the CRS TE and VRS TE scores. Furthermore, the bootstrapped lower and upper bounds at 95% confidence level are presented for the bias-corrected DB DEA scores.

The FDH efficiency score equals 0.316 on overage for all theatrical firms, whereas the VRS TE score from conventional DEA equals 0.282, indicating that the theatres are between 68 and 72% technically inefficient, and they could significantly increase their output from artistic production given available inputs. From Table 2 also follows that based on naïve DEA results, all theatrical firms are also scale inefficient at 0.803 (for DB DEA SE = 0.837) and hence they could increase their SE by about 20% in order to achieve the most productive scale size. Similar results are found for the different sub-groups of theatrical firms, although some slight differences in the obtained TE and SE scores are apparent. The TPCs which constitute the greatest sample with 87 firms and 421 observations, have an average TE score of 0.381 (FDH) or 0.291 (DEA), which is lower than the average TE score found for Operas (FDH = 0.520, DEA = 0.393) or for the PTs (FDH = 0.429, DEA = 0.356). Thus, the TPCs are the least efficient firms in the sample, while PTs and Operas also operate closer to their full-scale efficiency.

From Table 2 also follows that the conventional DEA TE scores are higher than the DB DEA scores, although the difference is very small, as the bias-corrected VRS DB TE score is 0.252 on average for the whole sample. The conventional DEA TE scores lie outside the CI's limits, except for PTs, indicating that the bias is statistically significant. This finding is also confirmed by the very similar kernel density distributions for both the conventional and bias-corrected (CRS and VRS) TE scores (figures available upon request). Therefore, for consistency, we use both conventional and bias-corrected VRS DEA scores as the benchmark in the further analysis and discussion.

Table 3 Double-bootstrap VRS DEA estimates of TE determinants 2006–2014[3]

TE determinants	All firms	TPCs	PTs	Operas
Intercept	−42.962*	12.156	−80.418*	−9.348
Subsidies	−2.875*	−9.475*	1.266	−2.685
Small size	2.814	10.656	18.884*	n/a
Medium size	−10.703	7.026	11.822*	−6.891
Large size	−7.948	n/a	n/a	n/a
Age	1.046*	1.154*	0.545*	1.134*
Cooperation	25.365*	27.70	20.462*	n/a
Other legal forms	51.478*	n/a	48.510*	45.691*
North-West	1.046	−4.453	11.043*	−33.402*
North-East	0.546	−13.278	10.010*	−2.094
South-Islands	0.059*	22.455*	15.702*	5.406

* Significant at 5% of bootstrapped CIs

5.2 Estimates of Efficiency Determinants

Table 3 presents the estimated coefficients of TE determinants using DB DEA algorithm of [12] as outlined in Sect. 3.2. The estimated coefficients of the TE determinants indicate a direct effect on the output-distance (inefficiency) value, and hence an inverse effect on TE. The results are presented for the whole sample of theatrical firms but also for the three theatrical firm categories (TPCs, PTs and Operas).

The results for the full sample strongly indicate that the effect of subsidies on TE is significant and positive, but it also varies between different types of theatrical firms. For the whole sample, the effect of subsidies on TE is positive and significant, indicating that 1% increase in subsidies would reduce the inefficiency by 2.875%, and hence increase TE by approximately 11%. This finding particularly applies to the TPCs, which are also the least efficient group of theatres, with the significant subsidy coefficient of 9.475 in absolute terms. On the other hand, the effect of subsidies is not significant for PTs and opera houses. This might be explained by the fact that both groups of theatres are substantially more technically efficient than TPCs (see Table 2 above), and therefore the additional effect of subsidies on TE might disappear for these firms.

Interestingly, the age variable has always a significant and negative effect on TE in Table 3 indicating that PA firm's TE decreases with age and that younger Italian theatres are more technically efficient than older firms. The categorical variable indicating the *size* of theatrical firms has no effect on the TE scores in Table 3 for the full sample, but also specifically for Operas and TPCs. The results are only statistically

[3] In Table 3, *'large'* is the base category for the size of Operas, and *'micro'* is the base category for the size of all other firms, whilst 'corporations' is the base category for 'cooperation' and 'other legal forms', and 'centre' is the base category for location.

significant for PTs for which the coefficient is significant and positive for small firms (10–49 employees) and the medium-sized firms (50–249 employees) indicating that these companies are less technically efficient than the micro firms (0–9 employees). The legal form also matters for the technical efficiency of examined Italian theatres, with the corporations (base category) being more efficient than 'cooperations' and 'other legal forms'. These results are consistent and statistically significant for all groups of theatrical firms in Italy with the exception for TPCs for which the effect of cooperation is negative but not significant. The findings also indicate that PA firms located in the South and Islands are less technically efficient than firms localised in the northern regions of Italy.

Moreover, the separability tests of [17], which tests the null hypothesis that an environmental Z-variable influences neither the shape nor the level of the boundary of the attainable set, were applied. The results are available upon request and indicate in each case that separability conditions hold and that the potential effect of Z-variables on the production process is only through the distribution of the inefficiencies.

5.3 *Total Factor Productivity*

The Malmquist TFP index and its components are presented in Table 4. The results for the whole sample imply that total factor productivity was above one on average for the Italian theatres. Albeit, in total, TFP increased by only 1.9% - a result partly similar to that found by [8] for PA firms in 11 EU countries. Although not presented here, for the most years from 2006 until 2014, the total factor productivity also did not change substantially for the Italian theatres, as the index was slightly above one or very close to one. Furthermore, the results in Table 4 imply that technological progress (TECHC), and not technical efficiency change (TEC) or the scale efficiency change (SEC), was the main driver of productivity growth of Italian theatrical firms. In fact, TEC was lower for TPCs and Operas than for PTs, whilst in total the increase in TE was only at 2.4% for all theatrical firms (TEC = 1.024).

Table 4 TFP index and components – average 2006–2014–by the type of firm

Firm category	TFP	TEC	SEC	TECHC
Theatre production companies (TPCs)	1.014	1.029	1.017	1.030
Permanent theatres (PTs)	1.259	1.105	1.013	1.149
Opera houses	1.134	1.053	1.191	1.056
All theatrical firms	**1.019**	**1.024**	**1.005**	**1.040**

6 Conclusions

This work provides important new insights into the non-parametric evaluation of technical and scale efficiency in the Italian theatre sector. Most importantly, by applying the DB DEA model we present robust findings on the impact of various determinants of efficiency of Italian theatrical firms. Furthermore, the bias-corrected DB DEA results imply that Italian theatres are considerably inefficient with VRS TE scores at 25%. The naïve DEA TE score is slightly higher at 28% with the highest TE score obtained at 32% for the FDH which accounts for a non-convex technology. However, irrespective of the method used, the average TE scores are very low for Italian theatrical firms indicating ample room for improvement in their production efficiency. Theatres are also scale inefficient with an average naïve SE score of 0.80 and a bias-corrected SE score of 0.84.

Furthermore, by applying the two-stage DB DEA we find that public funding positively influences the TE of Italian theatrical firms. Our findings also imply that younger theatrical companies are technically more efficient. It is therefore younger companies that are experimenting with new forms of expression and introducing technical and visual innovations, which in time can lead to higher efficiency and productivity. Another interesting finding concerns the legal form of Italian theatrical firms, showing that corporations, which are usually run as limited liability companies and probably allow for greater flexibility in theatre management, are more technically efficient than other legal forms, such as cooperations or foundations. Finally, these results confirm that regional differences matter, with northern Italian regions being more technically efficient than southern ones.

The Malmquist total factor productivity index is above unity for Italian theatrical firms over the examined period, indicating that the TFP has increased over time, albeit slightly, but providing a counterargument that positive productivity change is not present in the theatre sector. In fact, our results show that the main driver of productivity is technological change. Therefore, following [24], we suggest that investments in innovation and digitalisation could play an important role in increasing the productivity of theatrical firms and Italian PA firms should be more incentivised to increase their technical efficiency, which could be achieved through more public funding.

References

1. Zieba, M. (2011). An analysis of technical efficiency and efficiency factors for Austrian and Swiss non-profit theatres. *Swiss Journal of Economics and Statistics, 147*(2), 233–274.
2. Gapinski, J. H. (1980). The production of culture. *Review of Economics and Statistics, 62*(4), 578–586.
3. Zieba, M., & Newman, C. (2013). Organisational structure and managerial efficiency: A quasi-experimental analysis of German public theatres. *Homo Oeconomicus, 29*(4), 497–534.
4. Marco-Serrano, F. (2006). Monitoring managerial efficiency in the performing arts: A regional theatres network perspective. *Annals of Operations Research, 145*(1), 167–181.

5. Baumol, W. J., & Bowen, W. G. (1965). On the performing arts: The anatomy of their economic problems. *The American Economic Review, 55*(1/2), 495–502.
6. Taalas, M. (1997). Generalised cost functions for producers of performing arts—Allocative inefficiencies and scale economies in theatres. *Journal of Cultural Economics, 21*(4), 335–353.
7. Castiglione, C., Infante, D., & Zieba, M. (2018). Technical efficiency in the Italian performing arts companies. *Small Business Economics, 51*(3), 609–638.
8. Castiglione, C., Infante, D., & Zieba, M. (2023). Public support for performing arts. Efficiency and productivity gains in eleven European countries. *Socio-Economic Planning Sciences, 85*, 1–17.
9. Last, A. K., & Wetzel, H. (2010). The efficiency of German public theaters: A stochastic frontier analysis approach. *Journal of Cultural Economics, 34*(2), 89–110.
10. Fernández-Blanco, V., Rodríguez-Álvarez, A., & Wiśniewska, A. (2019). Measuring technical efficiency and marginal costs in the performing arts: The case of the municipal theatres of Warsaw. *Journal of Cultural Economics, 43*(1), 97–119.
11. Last, A. K., & Wetzel, H. (2011). Baumol's cost disease, efficiency, and productivity in the performing arts: An analysis of German public theaters. *Journal of Cultural Economics, 35*(3), 185–201.
12. Simar, L., & Wilson, P. W. (2007). Estimation and inference in two-stage, semi-parametric models of productive efficiency. *Journal of Econometrics, 136*(1), 31–64.
13. Daraio, C., & Simar, L. (2007). Conditional nonparametric frontier models for convex and nonconvex technologies: A unifying approach. *Journal of Productivity Analysis, 28*(1–2), 13–32.
14. Farrell, M. J. (1957). The measurement of the productive efficiency. *Journal of the Royal Statistical Society, Series A, 120*(3), 253–290.
15. Banker, R. D., Charnes, A., & Cooper, W. W. (1984). Models for estimation of technical and scale inefficiencies in data envelopment analysis. *Management Science, 30*, 1078–1092.
16. Charnes, A., Cooper, E., & Rhodes, E. (1978). Measuring the efficiency of decision-making units. *European Journal of Operational Research, 2*(6), 429–444.
17. Daraio, C., Simar, L., & Wilson, W. P. (2018). Central limit theorems for conditional efficiency measures and tests of the 'separability' condition in non-parametric, two-stage models of production. *The Econometrics Journal, 21*, 170–191.
18. Simar, L., & Wilson, P. W. (2011). Two-stage DEA: Caveat emptor. *Journal of Productivity Analysis, 36*(2), 205–218.
19. Kneip, A., Simar, L., & Wilson, P. W. (2015). When bias kills the variance: Central limit theorems for DEA and FDH efficiency scores. *Econometric Theory, 31*, 394–422.
20. Moradi-Motlagh, A., & Emrouznejad, A. (2022). The origins and development of statistical approaches in non-parametric frontier models: A survey of the first two decades of scholarly literature (1998–2020). *Annals of Operations Research, 318*, 713–741.
21. Banker, R., Natarajan, R., & Zhang, D. (2019). Two-stage estimation of the impact of contextual variables in stochastic frontier production function models using data envelopment analysis: Second stage OLS versus bootstrap approaches. *European Journal of Operational Research, 278*(2), 368–384.
22. Färe, R., Grosskopf, S., Norris, M., & Zhang, Z. (1994). Productivity growth, technical progress, and efficiency change in industrialized countries. *American Economic Review, 84*(1), 66–83.
23. Coelli, T. J., Rao, D., O'Donnell, C. J., & Battese, G. E. (2005). *An introduction to efficiency and productivity analysis* (2nd ed.). Springer.
24. Cowen, T. (1996). Why I do not believe in the cost-disease. *Journal of Cultural Economics, 20*(3), 207–214.

DEA with Bootstrap, Weight Restrictions and Negative Outputs: Shortcomings of the Proposal for the Brazilian Distribution System Operators Regulation

Ana Beatriz Rodrigues Alves Branquinho Barbosa, Ana Lopes-Ahn, Bruno Vilela, Marcelo Costa, Rafael de Oliveira Gomes, and Heinz Ahn

Abstract In 2015, the Brazilian regulatory agency proposed the current methodology for determining the efficient operating costs of the Brazilian energy distribution system operators (DSOs). This methodology applies an input-oriented non-decreasing returns to scale Data Envelopment Analysis (DEA) model with weight restrictions. In 2022, the regulator proposed a revised model that is currently under discussion. Besides five positive outputs, the model incorporates two negative outputs to account for the quality of provided services. To determine the weight restrictions' upper and lower bounds, a bootstrap technique is applied. The final efficiency scores are resulting from the average from 21 DEA models, each using different percentiles of upper and lower bounds for the weight restrictions. Our analysis of the respective methodology shows that the use of data-driven weight restrictions leads to several issues. One significant problem is that the direct identification of peers is no longer possible. Moreover, the methodology sets unreachable targets for the DSOs. Regarding the use of negative outputs, our analysis also reveals that the proposed methodology produces results contrary to what is intended by the regulator.

A. B. R. A. B. Barbosa · R. de Oliveira Gomes
CPFL Energia, Campinas, Brazil

A. Lopes-Ahn
Lopes-Ahn Consultoria e Pesquisa, Florianópolis, Brazil

B. Vilela (✉)
Universidade Federal do Espírito Santo (UFES), Vitória, Brazil
e-mail: bruno.vilela@ufes.br

M. Costa
Universidade Federal de Minas Gerais (UFMG), Belo Horizonte, Brazil

H. Ahn
Technische Universität Braunschweig, Braunschweig, Germany

A. Emrouznejad et al. (eds.), *Advances in the Theory and Applications of Performance Measurement and Management*, Lecture Notes in Operations Research,
https://doi.org/10.1007/978-3-031-61597-9_5

Keywords Regulation · Cost efficiency · Distribution service operators (DSOs) · Data envelopment analysis (DEA) · Bootstrap · Weight restrictions

1 Introduction

In regulation of natural monopolies, such as electricity distribution, determining fair tariffs while ensuring the economic sustainability of the agents involved is a great challenge. In accordance with the geographical boundaries of each monopoly area, Brazil is divided into electricity concessions areas. Each concession is managed by a single electricity distribution company, named distribution system operator (DSO). Furthermore, consumers cannot choose their energy supplier. In this context, regulation becomes crucial for coordinating, interconnecting, and standardizing rules, systems, and procedures to ensure efficient operation and service quality in the absence of market competition [1].

Brazilian electricity distribution companies have their prices regulated by ANEEL (*Agência Nacional de Energia Elétrica*), which employs mechanisms such as periodic tariff reviews and tariff readjustments to define energy tariffs. As a result, revenues are not strictly determined by the market. Instead, they are determined by a regulatory procedure to provide adequate coverage for the companies' efficient expenses/costs and to allow the firms to achieve a fair return, given the business risk.

ANEEL applies price cap regulation to incentivize cost reduction for DSOs. During tariff reviews, new benchmarks of productivity and efficiency are estimated by ANEEL. Furthermore, the allowed revenue of each DSO for upcoming years is determined based on efficient levels of operational costs.

In Brazil, the costs of regulated electricity distribution companies are divided into two components, Parcel A and Parcel B. Parcel A consists mainly of non-manageable costs related to the purchase and transmission of electricity. Parcel B, which is manageable, comprises administrative, operational, and maintenance costs, as well as the remuneration of investments and depreciation associated with the electrical assets of the concession.

At each tariff review and regulatory methodology change, the regulator initiates a public consultation to gather contributions from stakeholders and society on the proposed methodology. Thus, hearings and public consultations are instruments promoted by the Brazilian regulator to encourage the participation of the consumers, associations, and companies, facilitating the exchange of documents and contributions on themes under discussion.

The ANEEL Public Consultation No. 023 of 2014 (AP 23/2014) presented the current methodology for calculating the regulatory operating costs of the Brazilian DSOs, which are part of Parcel B. The methodology uses a Data Envelopment Analysis (DEA) model with non-decreasing returns to scale (NDRS), including quality variables as undesirable outputs and employing weight restrictions.

Starting in 2020, the Brazilian regulator began the process of reviewing the benchmarking model for determining efficient operating costs by opening the first phase of Public Consultation No. 62 (CP 62/2020) to discuss the proposed methodology and the presented database. Up to present moment, the final methodology has not yet been approved by the regulator.

The methodology proposal was published by the regulator in Technical Note No. 176 of 2022 (TN 176/2022) [2]. The previous DEA-NDRS model remained unchanged. Furthermore, the proposal maintained to draw on negative outputs to model the quality of electricity supply. As a new feature, a bootstrap technique and the estimation of 21 models based on different sets of percentiles were introduced to define lower and upper bounds of the weight restrictions.

The objective of our study is to address main shortcomings of the methodology presented by ANEEL to calculate the regulatory operational costs of DSOs. One main issue is that the direct identification of peers, which is one of the advantages of standard DEA, is no longer possible. Additionally, the methodology is affected by multiple solutions and sets unreachable targets for the DSOs. Regarding the use of negative outputs, our analysis also reveals that the proposed methodology produces results contrary to what is intended by the regulator: companies with poor quality indices in the supply of electricity show efficiency gains, while those with good quality indices face reductions in their efficiency scores.

This paper is organized as follows. Section 2 describes the characteristics of the Brazilian regulation framework and the proposed benchmarking model. Section 3 presents a critical analysis of the model's informational content, the proposed weight restrictions methodology using bootstrap, and the use of quality variables as negative outputs. Section 4 concludes the paper.

2 The Actual Regulatory DSO Benchmarking Proposal

2.1 *Basic Setting*

According to Bogetoft and Otto [3], benchmarking is the systematic comparison of performance among production units that perform the same function, i.e., they transform the same types of resources (inputs) into the same types of products or services (outputs). Benchmarking methods are used worldwide by regulators seeking to foster companies to pursue efficient costs, thereby preventing the misuse of natural monopoly control.

Since 2015, the Brazilian regulator has exclusively used DEA to determine efficient operational costs for 52 electricity distribution companies. The performance of each company is evaluated using a DEA-NDRS model in combination with weight restrictions that incorporate trade-off relationships between inputs and outputs taken into account in the model. The actual ANEEL proposal for revising the current methodology is described in Technical Note (TN) No. 176/2022. The database used in the proposed benchmarking analysis is available online under https://shre.ink/aBv8. It is compiled from periodic data submissions made by the distribution companies.

Operational expenditures ranging from R$ 2.2 million to R$ 2.4 billion, electrical assets, and quality indicator variables are quantified in the database.

For calculating the efficiency scores of operational costs, ANEEL proposes operational expenditures—OPEX—as input. OPEX includes costs related to personnel, third-party services, and labor-related fines, which are adjusted by the regional wage index. Included are also expenses for materials, insurance, taxes, communication services, contributions to the Electric Energy Trading Chamber, and asset depreciation.

Outputs are grouped into positive and negative ones. The former group comprises the total length of electricity distribution lines (in km), the number of consumer units, the weighted market share, the total power of distribution substation transformers (in MVA), and the number of distribution transformers. Regarding the negative outputs, distribution service interruptions (CHI) and non-technical losses (PNT) are proposed by the regulator. By incorporating those negative outputs in the proposed DEA modeling, ANEEL aims to include the quality of electricity supply in the calculation of efficiency scores.

2.2 *Definition of Weight Restrictions Using Bootstrap*

Trade-off weight restrictions [4] in the DEA benchmarking of operational costs are used since 2015. ANEEL justifies this with the arguments that the complete freedom to choose weight restrictions can lead to counterintuitive results (such as zero weights), reduces the discretionary power of the model and introduces an overestimation bias in the efficiencies [2]. Currently, engineering criteria from distribution companies and data from benchmarking companies are used to define the weight restrictions.

The new regulatory proposal also contains weight restrictions. Table 1 shows the respective output and input ratios suggested for the DEA model.

Table 1 Output and OPEX ratios proposed by ANEEL as weight restrictions

Output-input ratio	Restriction*	Output-input ratio	Restriction*
Network extension (km)/OPEX	v.network/u	Number of consumers/OPEX	v.cons/u
Total power (MVA)/OPEX	v.power/u	Non-technical losses (MWh)/OPEX	v.pnt/u
Number of transformers/OPEX	v.transf/u	distribution service interruptions/OPEX	v.chi/u
Weighted market share (MWh)/OPEX	v.market/u		

* v = output weight; u = input weight

Table 2 Different percentile combinations with respect to positive and negative outputs

		Upper and lower bounds for negative outputs		
		Percentiles 0.5 and 99.5	Percentiles 2.5 and 97.5	Percentiles 5 and 95
Upper and lower bounds for positive outputs	Percentiles 0.5 and 99.5	Model 1	Model 8	Model 15
	Percentiles 2.5 and 97.5	Model 2	Model 9	Model 16
	Percentiles 5 and 95	Model 3	Model 10	Model 17
	Percentiles 10 and 90	Model 4	Model 11	Model 18
	Percentiles 15 and 85	Model 5	Model 12	Model 19
	Percentiles 20 and 80	Model 6	Model 13	Model 20
	Percentiles 25 e 75	Model 7	Model 14	Model 21

The proposal presented by ANEEL in 2022 introduced a new approach by incorporating the bootstrap method to determine the lower and upper bounds of the weight restrictions. These bounds are calculated based on the non-zero weight ratios estimated using the unrestricted DEA model. The proposal applies 21 DEA models, drawing on different percentiles from the empirical distributions of the non-zero ratios, to determine the final efficiency scores. Table 2 shows the respective percentile combinations.

Table 2 shows that, e.g., model 1 uses percentiles 0.5 and 99.5 for positive outputs ratios and percentiles 0.5 and 99.5 for negative output ratios as lower and upper bounds, respectively. As a second example, Model 21 uses percentiles 25 and 75 for positive outputs ratios and percentiles 5 and 95 for negative output ratios. These variations of lower and upper bounds for the weight restrictions are obtained using the empirical distribution of non-zero weights, generated by applying 2,000 bootstrap simulations. The final efficiency for each DSO is the average efficiency from all 21 models.

2.3 *Treatment of Electricity Service Quality Using Negative Outputs*

In Technical Note 121/2020 [5], the regulator argues that there is a technical trade-off between OPEX and quality. Companies with better quality indices may have higher operational costs due to additional efforts for improving quality. Thus, these companies should not be penalized in the DEA benchmarking process. Correspondingly, the regulator states on page 16 of the Technical Note 121/2020:

"65. … Such an approach is, therefore, capable of internalizing in the proposed methodology the trade-off inherent in the premise that *companies which cut their costs to the extent of compromising the quality of the service provided may, artificially, appear more efficient than the other companies in the sample.*" (emphasis added).

The model proposed by ANEEL considers only one input and different outputs aiming to represent different characteristics associated with the distribution service. Regarding the quality variables, the proposal suggests introducing them as negative outputs, represented by distribution service interruptions (CHI) and non-technical losses (PNT). CHI is the average number of hours consumers suffer from energy interruption, PNT is the amount of electricity loss due to theft. The underlying idea of using these indicators is that higher operational costs would result in lower indicators of poor quality, thereby representing better quality of the distribution service.

2.4 Outline of the Proposed Steps for Computing the DSOs' Efficiency Scores

The primary justification provided by the regulator in Technical Note 176/2022 [3] for using the DEA model lies in its simplicity and flexibility, as well as its adoption worldwide in studies of a similar nature, i.e., of electricity regulation. The proposed steps for calculating the efficiency scores of distribution companies are:

Step 1 Bootstrap simulations of the DEA-NDRS model without weight restrictions to generate empirical distributions of the non-zero weight ratios.

Step 2 Determination of 21 configurations of the lower and upper bounds using the non-zero ratios between output and input weights estimated in Step 1.

Step 3 Calculation of 21 DEA-NDRS models using the 21 combinations of weight restrictions defined in Step 2.

Step 4 Definition of the final efficiency score of each company as the average of the 21 efficiency scores calculated in Step 3.

It is worth mentioning that the methodology proposed by the regulator generates 104,000 observations (52 companies $\times$ 2,000 bootstrap iterations) for each input–output ratio in order to create empirical distributions and calculate lower and upper bounds for weight restrictions in each of the 21 DEA models.

3 Analysis of the Regulatory DSO Benchmarking Proposal

3.1 *Loss of Peer Benchmarking*

By analyzing the step-by-step procedure for calculating the efficiency scores, the approach proposed by the Brazilian regulator lacks simplicity and reproducibility. Notably, the adoption of bootstrap for determining lower and upper bounds for weight restrictions and the use of weight restrictions as proposed by ANEEL stands as a unique case in the global regulatory context [6, 7] and requires further investigations.

Along with its complexity, the new approach loses a significant portion of the intuition and interpretability of the original DEA methodology. Especially, by modelling 21 different efficiency frontiers instead of a single one, the proposed approach no longer allows for the determination of straightforward benchmarks, i.e., peers, which is crucial for companies to evaluate best practices. For example, the most restrictive Model 21 leads to 8 benchmark companies, the intermediate Model 12 to 17 and the least restrictive Model 1 to 21 benchmark companies.

Hence, by including 21 different models and calculating the average efficiency as the final efficiency score, the regulatory proposal does not support to get insights into existing best practices. Thus, with an unclear determination of peers, one of the primary advantages of using the original DEA methodology is forfeited [8].

3.2 *Analysis of the Use and Estimation of Weight Restrictions*

Impact of the inclusion of weight restrictions into the model

Traditional approaches to determine weight restrictions are based on the concept of relative value (or price) of inputs and outputs, or on their relative importance [9, 10]. In order to preserve the interpretation of technological possibilities, Podinovski [4] suggests using trade-off relationships observed in the evaluated companies.

The Brazilian regulator proposal states that the approach of Podinovski [4] was used to estimate lower and upper limits for the weight restrictions in the DEA model [2]. However, a closer analysis of the proposal shows that different assumptions were used compared to those presented by Podinovski. In contrast to using observed and validated trade-off relationships, ANEEL incorporates data-driven model restrictions that may not necessarily represent observable trade-offs.

In order to estimate the number of companies affected by the use of weight restrictions, first the average value for lower and upper bounds for each weight restriction was calculated using the 21 models. Second, using the DEA model without weight restrictions, companies with output-input ratios below the lower bound or greater than the upper bound were identified. Table 3 presents the number of companies affected by the lower and upper bounds of the weight restrictions. It is evident that only the lower bound weight restrictions have an influence on the DSOs. Most of the companies are affected by the lower bounds associated with the number of transformers (31 DSOs) and the total power (26 DSOs). Furthermore, the proposed weight restrictions have successfully achieved the intended purpose of preventing zero weights.

The problem of multiple solutions in the DEA weight estimation

The existence of multiple solutions in linear programming models is a well-documented phenomenon [11, 12]. To assess the impact of multiple solutions on DEA results of the proposed model, Costa et al. [13] indicate reordering of the database prior to the DEA estimation. The authors conducted a sensitivity analysis of the methodology proposed in TN 176/2022 with respect to multiple solutions. They assessed the impact of changes in the order of the database rows on the calculation of weight restriction limits and, consequently, on the efficiency estimates. One hundred iterations of the entire procedure proposed by the regulator were performed, employing random switch of the database row.

Results show that the estimated efficiencies vary between −1.52% and +0.81% due to changes in the order of the database rows. This variation is a consequence of multiple solutions. Thus, the application of bootstrap to determine weight restrictions is subject to multiple solutions in the linear programming DEA model. Consequently, the calculation of DSOs' regulatory revenues is sensitive to the order in which the companies are presented in the database. This multiple solution sensitivity is still unexplored by the regulator.

The problem of unachievable efficiency targets

As previously discussed, the DEA methodology presented in TN 176/2022 involves the dynamic estimation of lower and upper bounds for weight restrictions, followed by the adjustment of 21 DEA models obtained using different combinations of percentiles to define the lower and upper bounds for the restrictions. In this context, the regulator's proposal was evaluated in terms of whether it allows companies to achieve efficiency, meaning whether companies can reach full efficiency by reducing their costs according to the efficiency scores estimated using the ANEEL's proposal.

To evaluate the achievability of the efficiency goals, the costs of each DSO were adjusted according to the efficiency score estimated by the proposed model. The cost adjustment was made to the input of a selected company while keeping the costs of the other companies unchanged. In sequence, the efficiency score was recalculated to evaluate if the selected DSO would become fully efficient, as postulated. The

Table 3 Number of companies affected by lower and upper bounds of the weight restrictions

	Weight restrictions						
Boundaries	v.network/u	v.power/u	v.transf/u	v.market/u	v.cons/u	v.pnt/u	v.chi/u
Lower	12	26	31	17	21	18	7
Upper	0	0	0	0	0	0	0

Table 4 Recalculated efficiencies using the adjusted cost and the regulatory methodology

Company	OPEX (R$ Mil)	Efficiency TN 176 (%)	Efficient OPEX (R$ Mil)	Recalculated efficiency score (%)	Difference to reach 100 (%)
DSO 1	52,430	91.13	47,780	95.73	4.27
DSO 2	5,179	78.86	4,084	96.47	3.53
DSO 3	342,514	62.92	215,526	96.90	3.10
DSO 4	328,439	96.08	315,556	96.94	3.06

described procedure was applied to each of the 52 electricity distribution companies. Table 4 shows the recalculated efficiency score for four selected DSOs.

For instance, the first company shown in Table 4 is estimated to have an efficiency score of 91.13% using the methodology proposed in NT 176/2022. Thus, it is required an 8.87% reduction in its OPEX to reach full efficiency of 100%. However, after applying the same methodology using the efficient OPEX (47,780 R$ Mil), the new efficiency score is estimated as 95.73%. Therefore, even after reducing its operational costs by 8.87%, the reapplied methodology indicates that the company still needs a 4.27% reduction of its OPEX to reach 100% efficiency.

With respect to estimating unachievable cost goals, the methodology proposed in TN 176/2022 resembles the Stochastic Frontier Analysis (SFA) method, in which efficiency estimation relies on conditional mean calculations. Consequently, SFA generates unachievable efficiency goals, as already reported in the SFA literature [3]. Similarly, results show that the methodology proposed in TN 176/2020 estimates unachievable cost targets for DSOs, leading to a diffuse regulatory signal that may have a limited impact on incentivizing companies to reduce their costs.

3.3 Regulatory Signal Using the Negative Outputs CHI and PNT

Regarding the quality indicators, some companies are aggressively reducing costs over the years while increasing—i.e., impairing—their System Average Interruption Duration Index (SAIDI). If quality indicators are not included in the efficiency model, these companies are expected to become more efficient at the expense of the quality of their services. As an example, Fig. 1 presents a DSO that reduced its OPEX significantly, but and at the same time showed a deterioration of its quality indicator. On the basis of these data, the DSO is becoming efficient under the regulator's proposal.

Figure 1 shows a cost reduction of 56.10% from 2018 to 2019 for the selected DSO. In the same period, the SAIDI indicator increased by 45.62%, indicating a deterioration in the quality of supplied electricity. The model proposed by the regulator, however, indicates an efficiency increase over the years at the expense of the

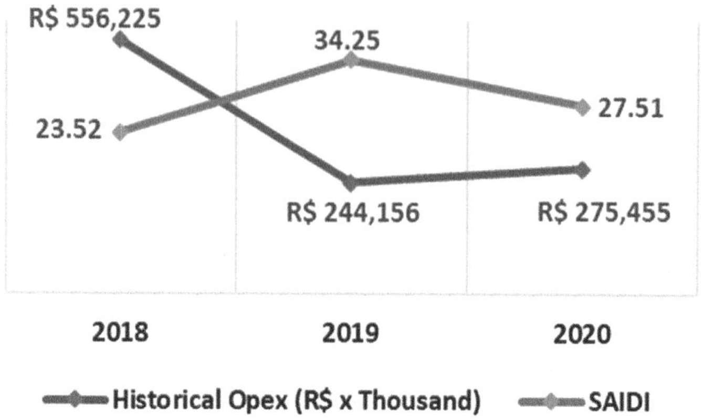

Fig. 1 Observed OPEX and SAIDI of a DSO becoming efficient under the regulator's proposal

decrease in quality of electricity supply. This situation can be observed for different DSOs.

To further investigate the effect of quality variables in the proposed DEA methodology, the impact on the efficiency scores was calculated using the DEA model with and without the proposed negative outputs. Using the DEA methodology without the negative outputs, only 7 DEA models are estimated since weight restrictions are applied to positive outputs only. Thus, the results presented in Fig. 2 compare the average efficiencies using the 21 DEA models (DEA_21) and the average of the 7 DEA models using only positive outputs (DEA_7).

On the left y-axis of Fig. 2, the scale of the differences between the average efficiencies (DEA_21–DEA_7) is presented. Results are arranged in ascending order of the differences, indicating that the more positive the difference, the greater the efficiency gain for a company when the PNT and CHI quality variables are included in the model. The standardized values for PNT and CHI quality variables on a 0–1 scale is shown on the right y-axis.

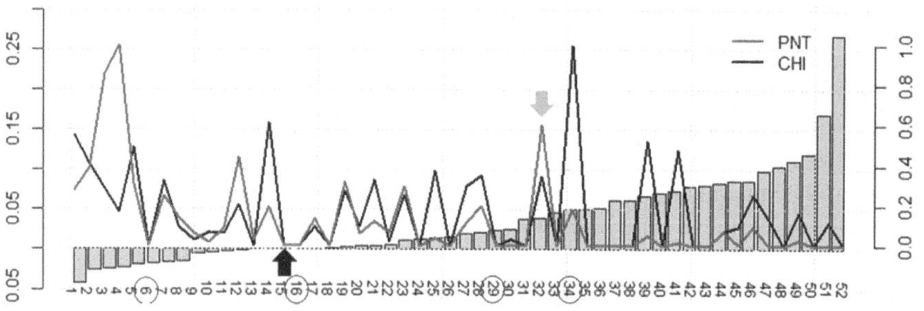

Fig. 2 Differences in efficiency scores between the average of the 21 models with the quality variables and the average of the 7 models without them

The PNT and CHI quality variables represent undesirable outputs, i.e., it is expected that the lower the quality values, the greater the efficiency score. However, this expected behavior is not observed. Some companies with counterintuitive results are indicated in Fig. 2. The dark arrow shows a company with zero quality values and no change in the efficiency score. On the contrary, this company with zero undesirable outputs was expected to have an increase in its efficiency score. The green arrow shows another problematic DSO—it has a greater undesirable PNT output and shows an increased efficiency when quality variables are included.

Figure 2 also shows that company number 3, which has the second greatest PNT value, has a significant gain in the efficiency score when quality variables are included in the DEA model. Similar inconsistencies are observed when including the CHI output. For instance, company number 34 has greater values for both CHI and PNT. However, it ranks 19th among companies with the greatest efficiency gains when the negative outputs are included. One would expect this company to have a smaller efficiency difference between the average of the 21 models and the average of the 7 models as compared to companies with lower values of CHI, such as those with indices 6, 16, and 29 in the figure.

Furthermore, the results of the proposed model (NT 176/2022) show that for companies with CHI = 0, variations ranging from a decrease in efficiency score of about 1.8% to an increase of 26.7% are observed. Thus, if two companies share the same CHI value and one significantly outperforms the other in PNT values, the one with poorer performance gains efficiency, while the other with a much better quality faces an unexpected reduction of −1.8% in its efficiency score.

Therefore, the results indicate that the proposed way to include quality variables in the DEA models leads to inconsistent changes in the efficiency scores, deviating from the expected impact intended by the regulator. Thus, it is recommended that the PNT and CHI quality variables should not be considered in the proposed methodology.

Finally, to evaluate whether the observed inconsistencies are related to the inclusion of quality variables and not related to the inclusion of weight restrictions, a study was conducted comparing the DEA model without weight restrictions and DEA model using the current methodology. Results indicate that indeed the inclusion of PNT and CHI quality variables in the model generates inconsistent and unexpected changes in the efficiency scores.

4 Conclusion

The regulation of electricity distributors in Brazil has fostered a democratic and participatory approach in the development of advanced studies and discussions over the years. Thus, it is important to acknowledge the initiative of the Brazilian regulatory agency in updating the methodology for calculating regulatory operational costs and the commitment to transparency.

In the present study, main revision points and inconsistences concerning the methodology proposed by ANEEL in 2022 were detected and evaluated. The results indicate that the regulator should carefully evaluate the need for using weight restrictions and check alternative approaches. Moreover, the proposed quality-related variables should be reassessed. Future research also should focus on the evaluation of the database's consistency and on sensitivity analyses concerning the selected input and output variables.

Acknowledgements The authors thank FAPES (Call FAPES no. 01/2023), CNPq/Brazil [Grant number: PQ-303119/2019-5] and CAPES/Brazil [PROBRAL Process number: 88881.370796/2019-01] for financial support. The project underlying this paper also has been supported with funds of the Federal Ministry of Education and Research of Germany. The financial support of Deutscher Akademischer Austauschdienst (DAAD), Germany for the respective Grant No. 57598225 is gratefully acknowledged. The responsibility for the content of this paper lies with the authors.

References

1. Pires, A., & Campos Filho, L. (2002). Investimentos em setores de infra-estrutura: A questão da regulação do monopólio natural e a defesa da concorrência. In A. C. Castro (Org.), *Desenvolvimento em debate* (vol. 2, pp. 281–303). Editora Mauad: Banco Nacional de Desenvolvimento Econômico e Social.
2. ANEEL: Agência Nacional de Energia Elétrica. (2022). NOTA TÉCNICA Nº 176/2022–SRM/ANEEL. Brasília, Brazil.
3. Bogetoft, P., & Otto, L. (2010). *Benchmarking with DEA, SFA, and R*. Springer.
4. Podinovski, V. V. (2004). Production trade-offs and weight restrictions in data envelopment analysis. *Journal of the Operational Research Society, 55*(12), 1311–1322.
5. ANEEL: Agência Nacional de Energia Elétrica. (2020). NOTA TÉCNICA Nº 121/2020–SRM/ANEEL. Brasília, Brazil.
6. Mesquita, R. D. B. (2017). *Regulação de tarifas de distribuição de energia elétrica: Uma análise comparativa entre reguladores europeus e latino-americanos (Tese de Doutorado)*. Universidade Federal de Minas Gerais.
7. Council of European Energy Regulators. (2022). *CEER: Report on regulatory frameworks for European energy networks 2021*. In Incentive regulation and benchmarking work stream.
8. Charnes, A., Cooper, W. W., & Rhodes, E. (1978). Measuring the efficiency of decision making units. *European Journal of Operational Research, 2*(6), 429–444.
9. Allen, R., Athanassopoulos, A., Dyson, R. G., Thanassoulis, E. (1997). Weights restrictions and value judgements in data envelopment analysis: Evolution, development and future directions. *Annals of Operations Research, 73*, 13–34.
10. Thanassoulis, E. (2001). *Introduction to the theory and application of data envelopment analysis*. Kluwer Academic Publishers.
11. Sueyoshi, T., & Sekitani, K. (2009). An occurrence of multiple projections in DEA-based measurement of technical efficiency: Theoretical comparison among DEA models from desirable properties. *European Journal of Operational Research, 196*(2), 764–794.
12. Fumero, F. (2004). Multiple solutions identification in data envelopment analysis. *Central European Journal of Operations Research, 12*(3), 307.
13. Costa, M. A., Lopes-Ahn, A. L., Kilger, A., & Micas, A. F. (2023). Limitations of weight restrictions in data envelopment analysis for benchmarking Brazilian electricity distribution system operators. *Utilities Policy, 82*, 101540.
14. Meeusen, W., & van Den Broeck, J. (1977). Efficiency estimation from Cobb-Douglas production functions with composed error. *International Economic Review, 18*(2), 435–444.

Empirical Estimation of the Production Frontier

Yu Zhao

Abstract Estimating production frontiers is essential for performance benchmarking and productivity analysis. Current approaches include Data Envelopment Analysis (DEA), Stochastic Nonparametric Envelopment of Data (StoNED), and Stochastic Frontier Analysis (SFA). This study briefly reviews these existing approaches and proposes two distinct nonparametric methods for estimating the production frontier, based on data-fitting techniques. The first method offers an asymptotically consistent estimator for a piece-wise linear frontier, assuming a random inefficiency term. The second method combines a modified ordinary least squares approach with a resampling technique to account for the impact of random noise and enhance estimation precision. To illustrate the benefits of these proposed methods, simulated and empirical examples are included.

Keywords Production frontier · Data-fitting approach · Nonparametric estimation · Stochatsic setting

1 Introduction

This study explores two distinct models for estimating the production frontier. The first model bears resemblance to data envelopment analysis (DEA, [4, 6]), with the exception that the inefficiency term is treated as a random variable. We propose a data-fitting approach to construct a piece-wise linear frontier. The proposed method offers more flexibility than DEA, as the shape of the estimated frontier can be controlled by a specific parameter. Furthermore, by incorporating the uncertainty of inefficiencies into the model, a variety of statistical methods (e.g., [3, 5]) can be employed to enhance the discussion of performance analysis. The second model additionally accounts for the impact of random noise. Earlier methods such as the stochastic frontier analysis (SFA, [1]) and the stochastic nonparametric envelopment of data (StoNED, [13]) have addressed this type of issue. However, unlike these methods,

Y. Zhao (✉)
Tokyo University of Science, 1-3 Kagurazaka, Shinjuku-ku, Tokyo 162-8601, Japan
e-mail: yu.zhao@rs.tus.ac.jp

A. Emrouznejad et al. (eds.), *Advances in the Theory and Applications of Performance Measurement and Management*, Lecture Notes in Operations Research,
https://doi.org/10.1007/978-3-031-61597-9_6

we integrate a resampling method into the model to improve measurement precision. Both simulated and empirical examples are provided to illustrate the advantages of the proposed methods.

Existing methods such as DEA, SFA, and StoNED have become important performance analysis tools for policymakers and regulators in various fields. This study introduces two alternative approaches to support the decision-making process from a methodological perspective. Furthermore, our proposed approach addresses several limitations of existing methods and offers decision-makers a general and flexible modeling approach in a stochastic context. We will discuss this in detail in Sect. 2.

It is worth noting that chance-constrained data envelopment analysis (CCDEA; see, e.g., [7, 14, 17]) is also known as a stochastic nonparametric approach. The CCDEA approach, similar to SFA and StoNED, recognizes that some data variations may be attributable to noise. However, the estimation problem in CCDEA requires strong assumptions about the noise terms, and it entails a greater computational burden [16]. In this study, we only compare the performance of our proposed approach with DEA, SFA, and StoNED. Once the frontier is estimated, existing metrics for efficiency or productivity from the benchmarking literature, such as distance functions (e.g., [8, 20]) or Malmquist-type indices (e.g., [15, 18, 22]), can be applied for further analysis.

The structure of the paper is as follows: Sections 2.1 and 2.2 detail two distinct models for empirically estimating the production frontier. Following the description of the proposed estimation methods, simulated and empirical examples are presented. The paper concludes with Sect. 3.

2 Methodology

2.1 *Estimating the Frontier with Consideration for Random Inefficiencies*

Description of Model (A) Let $(\boldsymbol{x}, y) \in \mathbb{R}_+^{m+1}$ represent an input-output vector and define the production possibility set as $P := \{(\boldsymbol{x}, y) \in \mathbb{R}_+^{m+1} \mid \boldsymbol{x} \text{ can produce } y\}$. The outer boundary of P is referred to as a production frontier, and can be mathematically defined as $f(\boldsymbol{x}) := \max\{y \in \mathbb{R}_+ | (\boldsymbol{x}, y) \in P\}$. Our focus lies on how to determine the level of maximum outputs for a given input vector. Let u be a negative random variable with $E[u] = \mu \leq 0$, $Var(u) = \sigma_u^2 < \infty$. We assume the following relationship

$$y = f(\boldsymbol{x}) + u. \tag{1}$$

In Eq. (1), u symbolizes the inefficiency term. In relation to $P = \{(\boldsymbol{x}, y) \in \mathbb{R}_+^{m+1} \mid y \leq f(\boldsymbol{x})\}$, we postulate the subsequent assumptions:

P.1 If $\boldsymbol{x}' \geq \boldsymbol{x}''$, then $f(\boldsymbol{x}') \geq f(\boldsymbol{x}'')$ for all $\boldsymbol{x}', \boldsymbol{x}'' \in \mathbb{R}_+^m$

P.2 If $0 \leq \theta \leq 1$, then $\theta f(\boldsymbol{x}') + (1-\theta) f(\boldsymbol{x}'') \leq f(\theta \boldsymbol{x}' + (1-\theta)\boldsymbol{x}'')$ for all $\boldsymbol{x}', \boldsymbol{x}'' \in \mathbb{R}^m_+$

P.3 For each observation $i = 1, \ldots, n$, $y_i \leq f(\boldsymbol{x}_i)$

This model has been explored in numerous previous studies, such as [2, 3, 10, 11]. It differs from the standard DEA models in that it does not consider the principle of minimal extrapolation.

Estimation method In the following, we describe the estimation approaches for a single input and output case for the sake of simplicity. The proposed approaches can be easily extended to a multi-input and single-output setting. We approximate the unknown function f by using a non-decreasing, concave, piecewise linear function $f^K(x)$ of the input level x. This function is given by

$$f^K(x) := a_k + b_k x, \quad (x \in I_k, k = 1, \ldots, K)$$

where $I_k := (w_1^k, w_2^k]$, $0 \leq w_1^1 \leq \cdots \leq w_1^K$, $0 \leq w_2^1 \leq \cdots \leq w_2^K$, and $(0 \leq) a_1 \leq \cdots \leq a_K, b_1 \geq \cdots \geq b_K \geq 0$. Let $u_i, i = 1, \ldots, n$ be a sequence of i.i.d. random variables with unknown density $g(u)$. The basic statistical model is given by

$$y_i = f^K(x_i) + u_i, \quad (i = 1, \ldots, n)$$

where x_i is a bounded deterministic variable and $w_1^1 \leq x_1 \leq \cdots \leq x_n \leq w_2^K$.

Taking into account that the unknown function $f^K(x)$ can be estimated only when $K \leq n$, we determine the intervals $I_k, k = 1, \ldots, K$ such that $\bigcup_{k=1}^K I_k = (w_1^1, w_2^K]$ and $\sum_{k=1}^K n_k \geq n$, where n_k is the number of data points in I_k.

For the k-th interval, we estimate the unknown parameters by

$$\hat{b}_k = \frac{\sum_{i=1}^{n_k} (y_i - \bar{y})(x_i - \bar{x})}{\sum_{i=1}^{n_k} (x_i - \bar{x})^2} \tag{2}$$

$$\hat{a}_k = \min_{i=1,\ldots,n_k} \{a | a + \hat{b}_k x_i \geq y_i\} \tag{3}$$

where $\bar{y} = \frac{1}{n_k} \sum_{i=1}^{n_k} y_i$ and $\bar{x} = \frac{1}{n_k} \sum_{i=1}^{n_k} x_i$.

Theorem 1 *Assume that u_i is a sequence of i.i.d. random variables with $E[u_i] = \mu$, $Var(u_i) = \sigma_u^2 < \infty$, the density $g(u)$ is bounded and smooth at $u = 0$, and x_i are bounded. Then, for each interval I_k,*

$$\begin{bmatrix} \hat{a}_k - a_k \\ \hat{b}_k - b_k \end{bmatrix} \xrightarrow{p} \mathbf{0}$$

as $n_k \to \infty$. Moreover, as $n_k \to \infty$,

$$\sqrt{n_k}(\hat{b}_k - b_k) \xrightarrow{w} N\left(0, \frac{\sigma_u^2}{M_x}\right)$$

where $M_x := \lim_{n_k \to \infty} \frac{1}{n_k} \sum_{i=1}^{n_k} (x_i - \bar{x})^2$ *is a positive. As* $n_k \to \infty$

$$n_k^t(\hat{a}_k - a_k) \xrightarrow{p} 0$$

for any $0 < t < \frac{1}{2}$.

The proof for Theorem 1 can be found in our discussion paper [12]. After we estimate the unknown parameters $\hat{a}_k, \hat{b}_k$ for all k, we can impose the assumptions of concavity and monotonicity by choosing $\hat{a}_k, \hat{b}_k$ such that $0 \leq a_1 \leq \cdots \leq a_K$ and $b_1 \geq \cdots \geq b_K \geq 0$. When multiple inputs $\boldsymbol{x}_i \in \mathbb{R}_+^{m \times 1}$ are available, the parameters are estimated by the linear regression equation

$$\hat{\boldsymbol{b}}_k = \left(\sum_{i=1}^{n_k} \boldsymbol{x}_i \boldsymbol{x}_i'\right)^{-1} \left(\sum_{i=1}^{n_k} \boldsymbol{x}_i y_i\right)$$

$$\hat{a}_k = \min_{i=1,\ldots,n_k} \{a | a + \hat{\boldsymbol{b}}_k' \boldsymbol{x}_i \geq y_i\}$$

and we can choose $\hat{a}_k, \hat{\boldsymbol{b}}_k$ by using the following inequalities

$$a_k + \boldsymbol{b}_k' \bar{\boldsymbol{x}}_k \leq a_h + \boldsymbol{b}_h' \bar{\boldsymbol{x}}_k, \quad (k, h = 1, \ldots, K, k \neq h)$$
$$\boldsymbol{0} \leq \boldsymbol{b}_k, \quad (k = 1, \ldots, K).$$

The Theorem 1 can be extended directly to the multi-input and single output case. That is, the order of the asymptotic distribution of b_{lk} and a_k, $(l = 1, \ldots, m, k = 1, \ldots, K)$ are $\sqrt{n}$ and n^t, $(t \geq \frac{1}{2})$, respectively.

In practice, the interval I_k can be determined by repeatedly dividing the input region using its mean value or median until a certain stopping rule is satisfied (e.g., repeat the process until $\hat{a}_k, \hat{b}_k$ no longer satisfy the conditions of concavity and monotonicity). One can also specify the number of intervals based on their prior knowledge or the desired shape of the production frontier. Figure 1 illustrates the proposed estimation method using 100 simulated data points in a single input and single output setting. The true technology is assumed to be $y = x^{0.5}$, and the non-positive inefficiency terms are extracted from a negative exponential distribution. As a result, we obtain four hyperplanes. By highlighting the common half-space of all four hyperplanes, we can empirically express the region of the production possibility set. In Fig. 1, the frontier of the empirically estimated production possibility set (which we refer to as the SDEA frontier) is represented by the red lines.

Empirical example We further applied both the BCC model and model (A) to 40 Japanese life insurance companies to elucidate the differences between the production possibility set of model (A) and the traditional DEA. Capital (total shareholders' equity) was used as an input and ordinary income as an output to estimate the efficient frontier in the fiscal year 2021. Figure 2 illustrates the two frontiers derived from the DEA and the proposed model (A). We observe that the production possibility set of

Fig. 1 Simulated example

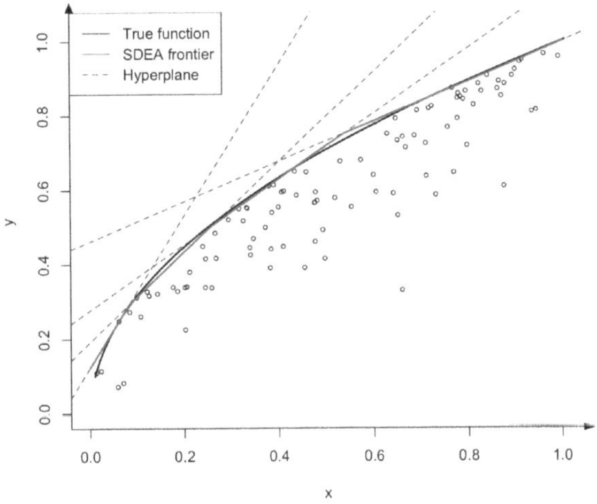

Fig. 2 Results of BCC and model (A)

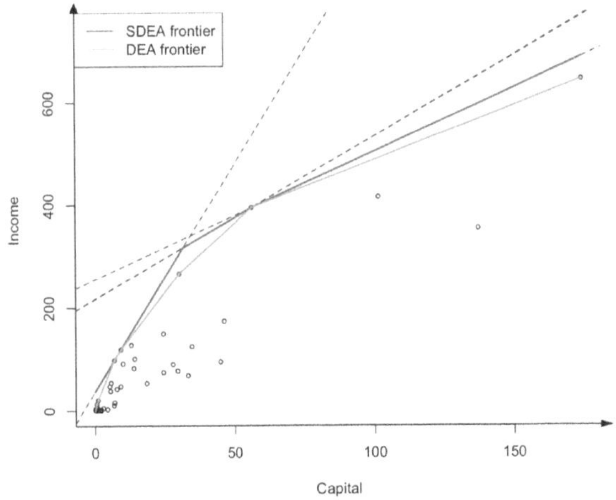

the DEA is smaller than that of model (A). Indeed, due to the DEA's enforcement of the principle of minimal extrapolation, we have $P^{\text{DEA}} \subseteq P^{\text{SDEA}}$, leading to $u_i^{\text{SDEA}} \geq u_i^{\text{DEA}}$. This relationship is graphically demonstrated in the left plot of Fig. 3. The right plot in Fig. 3 displays the empirical distribution of inefficiency, and we find that for every level of inefficiency, the proportion of Japanese life insurance companies at that level or lower is greater for P^{SDEA} than for P^{DEA}.

To further investigate whether the principle of minimal extrapolation is essential for constructing the production possibility set, we tested the hypothesis $H_0 : g^{\text{SDEA}} = g^{\text{DEA}}$ against $H_1 : g^{\text{SDEA}} \neq g^{\text{DEA}}$, where g^{SDEA} and g^{DEA} are the distributions of inefficiencies under P^{SDEA} and P^{DEA}, respectively. Neither the two-sample Kolmogorov-Smirnov test (with a p-value of 1) nor the Kruskal-Wallis

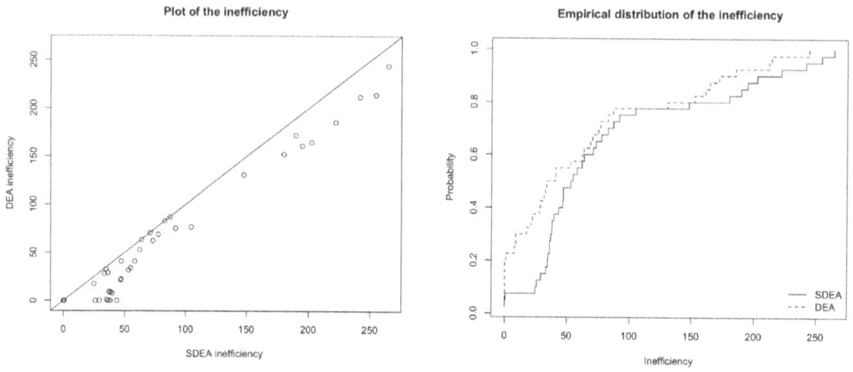

Fig. 3 Results of inefficiencies

rank sum test (with a p-value of 0.06724) reject the null hypothesis. This implies that no significant difference in the distribution of inefficiency could be identified, whether the principle of minimal extrapolation was assumed or not. Therefore, the principle of minimal extrapolation is not necessary, at least in this empirical example. As a result, model (A) provides a more appropriate maximum level of outputs, as P^{SDEA} is larger than P^{DEA}.

2.2 *Estimating the Frontier with Consideration for Random Noise*

Description of Model (B) Sometimes the noisy data or modeling errors cannot be ignored. Let v be a random variable with $E[v] = 0$ and $Var(v) = \sigma_v^2 < \infty$. Another model for determining the maximum outputs is given by

$$y = f(\boldsymbol{x}) + u + v. \tag{4}$$

This model relies solely on assumptions (P.1) and (P.2). In other words, any deviation from the unknown production frontier consists of both inefficiency (i.e., u) and noise (i.e., v), and it can be estimated based on the maximum-likelihood principle using methods such as the SFA [1] and StonNED [13].

Estimation method Unlike SFA or StoNED, the model (B) adopts a more flexible estimation of the frontier f with consideration for the prediction accuracy. For any observed (x_i, y_i), the tangent line passing through $(x_i, f(x_i))$ represents the upper bound of the data point (x_i, y_i) and $y_i \leq f(x_i)$ holds. To estimate the upper bound for any given (x_i, y_i), we consider the following minimization problem [19]:

$$\min \sum_{i=1}^{n} (\alpha + \beta x_i - y_i)^2 \tag{5}$$

$$\text{s.t. } \alpha + \beta x_i - y_i \geq 0, \quad i = 1, \ldots, n \tag{6}$$

$$\beta \geq 0. \tag{7}$$

Let $e_i := \alpha + \beta x_i - y_i$ for all $i = 1, \ldots, n$ and let $(\hat{\boldsymbol{e}}, \hat{\alpha}, \hat{\beta})$ be the optimal solutions to the problem (5)–(7). Then, $y_i + \hat{e}_i (= \hat{\alpha} + \hat{\beta} x_i)$ is the estimated upper-bound value of (x_i, y_i).

To approximate the unknown function f, we consider the problem of estimating the tangent line locally at any given (x_i, y_i) such as

$$\min \sum_{j \in G_i} \left(\alpha + \beta x_j - y_j\right)^2 \tag{8}$$

$$\text{s.t. } \alpha + \beta x_j - y_j \geq 0, \quad j \in G_i \tag{9}$$

$$\beta \geq L, \tag{10}$$

where $L \in \mathbb{R}_+$ is the lower bound of the parameter β and G_i satisfies $i \in G_i \subseteq \{1, \ldots, n\}$. The lower bound L can be set to zero to ensure the non-decreasing assumption or other positive numbers to avoid the existence of output slacks in (x_i, y_i). The set G_i can be determined based on a priori information about the production function or the similarity of the observed data point. Here, we use a naive method for determining G_i for the purpose of clarification. For some pre-assigned number $1 \leq k \leq n$, define

$$G_i := \left[\max(i - k, 1), \min(i + k, n)\right] \quad i \in \{1, \ldots, n\}.$$

If $k = n$, the problem (8)–(10) coincides with the problem (5)–(7).

The problem (8)–(10) estimates an upper bound for any observed point (x_i, y_i) at a given range G_i. Solving the problem (8)–(10) for n observed points leads to n pairs of estimated coefficients. Let $(\tilde{\alpha}_i, \tilde{\beta}_i)$, $i = 1, \ldots, n$ be the optimal solutions of the problem (8)–(10) for all observed points. Define

$$(\bar{\alpha}_i, \bar{\beta}_i) := \begin{cases} (\tilde{\alpha}_i, \tilde{\beta}_i) & \tilde{\alpha}_i + \tilde{\beta}_i x_i \leq \tilde{\alpha}_h + \tilde{\beta}_h x_i \\ (\tilde{\alpha}_h, \tilde{\beta}_h) & \tilde{\alpha}_i + \tilde{\beta}_i x_i > \tilde{\alpha}_h + \tilde{\beta}_h x_i \end{cases}$$

for $i, h = 1, \ldots, n$ and $i \neq h$. The set of $(\bar{\alpha}_i, \bar{\beta}_i)$ can be used to construct a concave function f. Using $(\bar{\alpha}_i, \bar{\beta}_i)$, $i = 1, \ldots, n$, the estimated outputs are computed by $\bar{y}_i = \bar{\alpha}_i + \bar{\beta}_i x_i$, $\forall i$. We then construct the production frontier with the lower bound of the estimated outputs:

$$\bar{f}(x) := \min\{\alpha + \beta x \mid \alpha + \beta x_i \geq \bar{y}_i, \ \forall i\}.$$

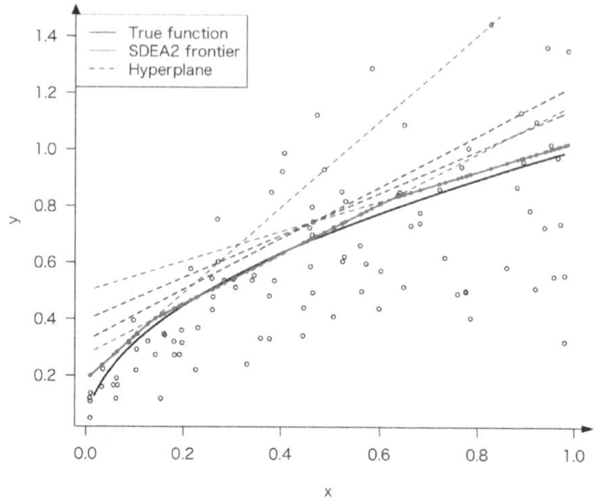

Fig. 4 Simulated example

In Fig. 4, we randomly generated 100 data points using $y = x^{0.5}$ (i.e., the solid black line) and the data are perturbed by $u \sim Exp[\mu = 1/6]$ and $v \sim N(0, 1/6)$ in a multiplicative fashion. By using a pre-assigned number $k = 4$, we solve the problem (8)–(10) for each data point. By further imposing the concavity assumption, the upper bounds for determining the frontier are reduced to five lines (i.e., the dashed lines). Consequently, the frontier is estimated as the lower bound of those five dashed lines (i.e., the red line, which we refer to as the SDEA2 frontier), which is actually a piece-wise linear frontier.

As for the determination of k, we introduce two methods. If the decision-makers have knowledge on the level of random noise, k can be appropriately determined. Let J be the number of points satisfying $y_i \leq \bar{f}(x_i), \forall i$. Then, $\rho := 100(J/n)$ is the percentage of the observations enveloped by $\bar{f}$. Suppose the interest is in estimating a frontier that envelops 80% observations. In that case, one can start by selecting $k = 1$ and repeat the estimation procedure by changing the value of k until $\rho \geq 80\%$ is achieved. Another method is to use the cross-validation approach to determine k automatically. In the case of using K-fold cross-validation, we can alter the value of k from 1 to $n(K-1)/K$ to find the k that achieves the smallest MSE at each iteration.

For a detailed comparison of the proposed estimation method for model (B) with DEA, SFA, and StoNED, refer to the simulated example in [21].

Empirical example The proposed method was applied to evaluate the research and development investment efficiency in the automotive industry, focusing on major companies in the automotive and auto parts sectors in Japan and the United States (29 companies in Japan and 21 companies in the US). The conventional evaluation method uses the R&D intensity, calculated as "R&D Expenses"/"Net Sales"×100%. For this reason, an analysis was conducted using R&D expenses (R&D 2020) as input and net sales as output. This data can be obtained from The 2021 EU Industrial R&D Investment Scoreboard.We applied K-fold cross-validation to determine

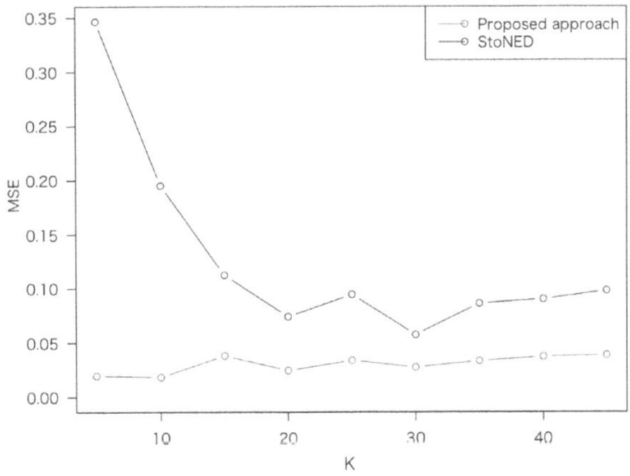

Fig. 5 *K*-fold cross-validation

k automatically. Figure 5 illustrates the relationship between K and the Mean Squared Error (MSE). We found that 10-fold cross-validation results in the smallest MSE, with the corresponding k being 4. This suggests that about nine data points located around each observation will be used to estimate the hyperplane, which will be used to construct the final frontier. Further, our results also indicate that the proposed approach for model (B) yields a smaller and more stable MSE than the StoNED approach.

To further understand the disparity between the proposed inefficiency measurement and the commonly used R&D intensity index, we first calculate the point estimates of inefficiencies for our proposed approach, StoNED, SFA, and Ordinary Least Squares (OLS) respectively, following the method outlined in [9]. Subsequently, we transform the R&D intensity into a form that represents inefficiency, using the formula:

$$\text{Ranking(R\&D intensity)} = \frac{\max(\text{R\&D intensity}) - \text{R\&D intensity}}{\max(\text{R\&D intensity})}.$$

We then compute both Spearman's rank correlation coefficients and Kendall's rank correlation coefficients. The results of these computations are summarized in Tables 1 and 2.

Table 1 Spearman's rank correlation coefficients

	Model(B)	SFA	StoNED	OLS	Ranking(R&D intensity)
Model(B)	1.0000				
SFA	0.5406	1.0000			
StoNED	0.7315	0.8581	1.0000		
OLS	0.6006	0.9060	0.9104	1.0000	
Ranking(R&D intensity)	0.8273	0.8256	0.8599	0.8099	1.0000

Table 2 Kendall's rank correlation coefficients

	Model(B)	SFA	StoNED	OLS	Ranking(R&D intensity)
Model(B)	1.0000				
SFA	0.3989	1.0000			
StoNED	0.6654	0.6817	1.0000		
OLS	0.5014	0.7819	0.7922	1.0000	
Ranking(R&D intensity)	0.7451	0.6925	0.7417	0.6846	1.0000

Both rank correlation coefficients indicate that the proposed method has the strongest correlation with R&D intensity, followed by StoNED, SFA, and OLS. This suggests that the inefficiency measurement can serve as a valuable alternative for quantifying R&D performance.

3 Conclusion

In this paper, we investigate two models for estimating the production frontier. We utilize a data-fitting approach that offers the advantages of flexibility in controlling the shape of the frontier. In conclusion, we highlight some significant features of the proposed model (A): (i) the inefficiency term is estimated as a random variable; (ii) the coefficients of the facets of the frontier are asymptotically consistent; (iii) the principle of minimal extrapolation is disregarded. These features enable the performance of statistical tests for the inefficiency term and allow for an alternative estimation method in DEA. It should be noted that the proposed estimation approach in model (A) is best suited for small datasets. When dealing with a large amount of data, this approach can be expanded to incorporate a statistical extreme value method (see details in the discussion paper [12]). Subsequently, the proposed model (B) has the following features: (i) it considers both random inefficiency and random noise; (ii) the coefficients of the facets of the frontier are estimated locally based on a modified OLS problem, which is easy to implement; (iii) resampling methods such as K-fold cross-validation can be easily incorporated to improve estimation precision. Both simulated and empirical examples show better performance compared to existing methods.

The models introduced in this paper are primarily described in a single-input and single-output setting and can be readily extended to a multi-input and single-output setting. Further investigation is required for multiple outputs.

References

1. Aigner, D., Lovell, C. K., & Schmidt, P. (1977). Formulation and estimation of stochastic frontier production function models. *Journal of Econometrics, 6*(1), 21–37.
2. Banker, R. D. (1993). Maximum likelihood, consistency and data envelopment analysis: A statistical foundation. *Management Science, 39*(10), 1265–1273.
3. Banker, R. D. (1996). Hypothesis tests using data envelopment analysis. *Journal of Productivity Analysis, 7*, 139–159.
4. Banker, R. D., Charnes, A., & Cooper, W. W. (1984). Some models for estimating technical and scale inefficiencies in data envelopment analysis. *Management Science, 30*(9), 1078–1092.
5. Bogetoft, P., & Otto, L. (2010). *Benchmarking with DEA, SFA, and R* (vol. 157). Springer
6. Charnes, A., Cooper, W. W., & Rhodes, E. (1978). Measuring the efficiency of decision making units. *European Journal of Operational Research, 2*(6), 429–444.
7. Cooper, W. W., Huang, Z., Lelas, V., Li, S. X., & Olesen, O. B. (1998). Chance constrained programming formulations for stochastic characterizations of efficiency and dominance in DEA. *Journal of Productivity Analysis, 9*(1), 53–79.
8. Färe, R., & Grosskopf, S. (2000). Theory and application of directional distance functions. *Journal of Productivity Analysis, 13*(2), 93–103.
9. Jondrow, J., Lovell, C. K., Materov, I. S., & Schmidt, P. (1982). On the estimation of technical inefficiency in the stochastic frontier production function model. *Journal of Econometrics, 19*(2–3), 233–238.
10. Kneip, A., Park, B. U., & Simar, L. (1998). A note on the convergence of nonparametric DEA estimators for production efficiency scores. *Econometric Theory, 14*(6), 783–793.
11. Korostelëv, A. P., Simar, L., & Tsybakov, A. B. (1995). Efficient estimation of monotone boundaries. *The Annals of Statistics*, 476–489
12. Kunitomo Naoto, Y. Z. (2022). A statistical data envelopment analysis, Discussion paper, The Institute of Statistical Mathematics (ISM), Tokyo, Japan (pp. 1–26).
13. Kuosmanen, T., & Johnson, A. L. (2010). Data envelopment analysis as nonparametric least-squares regression. *Operations Research, 58*(1), 149–160.
14. Land, K. C., Lovell, C. K., & Thore, S. (1993). Chance-constrained data envelopment analysis. *Managerial and Decision Economics, 14*(6), 541–554.
15. Malmquist, S. (1953). Index numbers and indifference surfaces. *Trabajos de Estadistica y de Investigacion Operativa, 4*(2), 209–242.
16. Olesen, O. B., & Petersen, N. C. (2016). Stochastic data envelopment analysis-a review. *European Journal of Operational Research, 251*(1), 2–21.
17. Olesen, O. B., & Petersen, N. (1995). Chance constrained efficiency evaluation. *Management Science, 41*(3), 442–457.
18. Pastor, J. T., & Lovell, C. K. (2005). A global Malmquist productivity index. *Economics Letters, 88*(2), 266–271.
19. Sekitani, K., & Zhao, Y. (2021). Performance benchmarking of achievements in the Olympics: An application of data envelopment analysis with restricted multipliers. *European Journal of Operational Research, 294*(3), 1202–1212.
20. Shephard, R. W. (1970). *Theory of cost and production functions*. Princeton, New Jersey: Princeton University Press.
21. Zhao, Y. (2022). Nonparametric estimation of the production frontier using a data-fitting technique. In *Modern Management based on Big Data III* (pp. 9–20). IOS Press.
22. Zhao, Y., Morita, H., & Maruyama, Y. (2019). The measurement of productive performance with consideration for allocative efficiency. *Omega, 89*, 21–39.

Towards Gaining Robustness in Inverse Data Envelopment Analysis Models

Adel Hatami-Marbini, Aliasghar Arabmaldar, and Matthias Klumpp

Abstract The inverse data envelopment analysis (IDEA) approach aims to estimate input and/or output levels when preserving efficiency scores. The input-oriented IDEA version seeks the input level for producing expected output and the output-oriented one estimates the output level under a given input level while efficiency remains unchanged. However, in many real-world applications, full and precise information may not be available to guarantee successful IDEA implementation. This study presents a novel approach to combat inherent uncertainty, resultantly, enabling a move towards robustness of IDEA models. We particularly focus on two cases in this research. The first case occurs where the amount of extra input is not certain and almost impossible to be precisely determined due to restricted budget, market volatility, political issues and other external factors. The second case is observed in situations where input and/or output data encompasses uncertainty that might be resulting from errors in data measurement, data clearing, vagueness in variables (e.g., customer satisfaction or quality) or other internal factors from organizations.

Keyword Inverse data envelopment analysis · Uncertainty · Robustness

A. Hatami-Marbini
Department of Logistics, Marketing, Hospitality and Analytics, Huddersfield Business School, University of Huddersfield, Huddersfield, UK
e-mail: a.hatamimarbini@hud.ac.uk

A. Arabmaldar (✉)
Hertfordshire Business School, University of Hertfordshire, Hatfield AL10 9AB, UK
e-mail: a.arabmaldar@herts.ac.uk

M. Klumpp
School of Management, Politecnico di Milano, Milan, Italy
e-mail: matthias.klumpp@polimi.it

A. Emrouznejad et al. (eds.), *Advances in the Theory and Applications of Performance Measurement and Management*, Lecture Notes in Operations Research,
https://doi.org/10.1007/978-3-031-61597-9_7

1 Introduction

Data envelopment analysis (DEA) is a common non-parametric technique to measure the relative efficiency of a set of homogenous decision-making units (DMUs) in a multiple-input–output production technology. There are two types of DMUs in any DEA study; (i) the efficient or benchmark DMUs with an objective function value of one that are located on the efficient frontier, and (ii) inefficient DMUs with an objective function value of less than one. Following [1] who originally developed the DEA concept, many extensions of DEA models have been presented in the literature. Therefore, other researchers have improved the method using different techniques or applications.

Inverse DEA (IDEA) is a post-DEA sensitivity analysis approach developed initially for solving resource allocation problems [2]. The main objective of IDEA is to determine the optimal quantity of inputs and/or outputs for each DMU under input and/or output perturbation(s) that would allow them to reach a given efficiency target [2, 3].

It is notable that the aim of conventional IDEA models is to estimate the input and output levels as deterministic types, which does not take random errors of data into account in the production analysis process. Due to the inherent complexity within the real world, IDEA problems often involve some random or ordinal variables. However, existing IDEA studies have not paid enough attention to imprecise information. Therefore, it is suitable to incorporate different types of uncertain data and novel uncertainty programming techniques into IDEA.

This study presents a theoretical and practical framework of IDEA incorporating polyhedral uncertainty data and focuses on the key thematic areas of resource allocation management, handling for example investment analysis problems. In other words, IDEA models are proposed in this study to manage resource allocation and investment analysis issues, considering the real-world situation of imprecise data at some point or another. Our proposed models are constructed by employing the robust optimisation approach as a powerful and ever-increasing technique to deal with uncertainty in order to estimate inputs and output changes to keep the focal decision-making unit's efficiency score and those of other DMUs unchanged.

1.1 Deterministic IDEA Models

In contrast with classical DEA models, the IDEA investigates the efficiency concept from another viewpoint. As it is clear from the name of this method, IDEA is a reversed optimisation approach of classic DEA. While classic DEA models focus on a post-hoc assessment of organisational performance, IDEA assists managers by providing useful information on the values of the resources (or outcomes) which should be invested or produced to achieve a certain level of efficiency. The initial idea of the IDEA appeared in [4] used the multi-objective linear programming (MOLP) technique to estimate the output levels by changing inputs and preserving the efficiency score. In addition, it can estimate the input level by changing outputs and preserving the efficiency scores. In other words, IDEA models aim to investigate the scenario of changing the input or output data of one DMU to a maximum/minimum possible value to provide a constant relative efficiency.

IDEA has been widely investigated in the previous literature. As an example, [2] applied IDEA for resource allocation and production estimation. Jahanshahloo et al. [5] studied IDEA for sensitivity investigation of efficient and inefficient DMUs. This study determined the important policies over inputs, outputs and DMUs using preference cones. In another study, Hadi-Vencheh & Foroughi (2006) proposed a MOLP to estimate input/output levels, regardless of the efficiency or inefficiency of DMUs. Hadi-Vencheh et al. [6] introduced sufficient conditions for input estimation when the output is increased, and an IDEA model was converted to a MOLP and solved accordingly.

Lin [7] investigated the efficiency measurement and revenue setting problems. A home improvement company with 22 chain stores in Taiwan was considered in this research. Lertworasirikul et al. [8] applied an IDEA for variable returns to scale technology. Gattoufi et al. [9] suggested an innovative application of IDEA in strategic decision making and a set of 42 banking units are applied to show the usefulness of the proposed approach. Lim [10] developed an IDEA model for new product target setting. The proposed model was discussed by using a case study of vehicle engine development. Ghiyasi [11] proposed a new IDEA model in the presence of technical efficiency, cost efficiency and revenue efficiency. Amin et al. [12] applied IDEA to anticipate the major or minor consolidation of a merger. Ghiyasi [13] showed the drawbacks of the method suggested by [8] and modified the application of MOLP in a proposed inverse BCC model. Jahanshahloo et al. [14] considered the IDEA under inter-temporal dependence assumption considering both input and output estimation problem.

Emrouznejad et al. [15] proposed a new IDEA model to achieve our goal to allocate CO2 emission quota under several assumptions. Amin et al. [16] suggested a novel method to deal with target setting in mergers using goal programming (GP) and IDEA. Zhang and Cui [17] proposed an integrated framework of IDEA called inverse non-radial DEA since it is based on non-radial DEA by MOLP, which covers existing IDEA models. Amin and Ibn Boamah [18] introduced a new IDEA based on a cost-efficiency model for estimating potential gains from mergers. Lim [19] proposed an inverse optimisation model for operational planning by taking into account frontier changes in conjunction with environmental factors. Chen et al. [20] established a new measurement method by employing an IDEA with frontier changes to evaluate China's CO2 emissions reduction. Recently, [21] developed a network IDEA model to evaluate sustainable supply chain performance considering the nature of network systems and then, a two-phase method is introduced to solve the proposed model to estimate the inputs and outputs while efficiency improves or remains unchanged. Amin and Ibn Boamah [3] defined various types of strategic alliance collaborations and partnerships between different DMUs and develops IDEA models to guide partners on how redistributing their inputs and outputs could enhance their performance.

Emrouznejad et al. [22] provided a comprehensive survey of the latest theoretical and practical advancements in IDEA, while also highlighting potential areas for future research and development in this field.

1.2 Imprecise DEA and IDEA Models

The traditional DEA measure of efficiency assume that input and output data are precise and are measured with full certainty. However, in real-life situations, the general production relationships are frequently uncertain, and the concept of efficiency is inextricably related to how firms deal with such uncertainty. Therefore, when data uncertainties exist, efficiency evaluation in a deterministic DEA setting becomes sensitive to such uncertainties. To account for uncertain variations in inputs and outputs, there are several approaches considered in the DEA setting. They are imprecise DEA [23], fuzzy DEA [24], stochastic DEA [25], bootstrap DEA [26] and robust DEA [27–31].

Similar to classic DEA models, IDEA models assume the perfect information of input and output values. However, this assumption is not accurate in many situations. Hadi-Vencheh et al. [32] extended the IDEA model to interval data for evaluating the relative efficiency of DMUs. The proposed model determines the inputs' lower and upper bounds of a given DMU when its interval outputs are modified. Ghomi et al. [33] addressed the IDEA in the presence of stochastic data. Under maintaining the efficiency score, the inputs/outputs-estimation problem is investigated when some or all of its outputs/inputs increase. A novel optimality concept for MOLP is introduced to derive necessary and sufficient conditions for input/output estimation. Younesi et al. [34] dealt with an IDEA based on the non-radial slacks-based model in the presence of uncertainty employing both integer and continuous interval data. Except the above-mentioned studies in this subsection, all aforementioned studies, in the IDEA literature, assume perfect information about inputs and outputs, which would not be valid for all real-life situations.

1.3 Research Aims and Contributions

According to previous literature, many real-world data are available in an uncertain form. Therefore, the observed values of the data can be defined in different imprecise forms instead of crisp numbers. There is a research gap in the current literature on investigating the IDEA models for imprecise input and output data. Trying to fill this specified literature gap, this paper aims to determine the required input modification of a DMU to maintain its relative efficiency if the output modifications occur in an imprecise environment. The models are developed through the lens of robust optimisation. In addition, this study deals with the effects of perturbation on the efficiency level of other DMUs. In summary, it is shown that proposed models preserve efficiency level for considered DMU and also other DMUs.

The rest of the paper is organised as follows. Section 2 provides preliminary DEA and IDEA models, as well as existing robust DEA models. The proposed robust DEA and IDEA models of this study are discussed in Sect. 3, and, finally, the paper concludes in Sect. 4.

2 Basic Models and Motivation

In this section, we review basic models including basic DEA and IDEA models that are required throughout the paper.

2.1 Basic DEA Models

Suppose that there exists n decision making units (DMUs), $\{\text{DMU}_j : j = 1,2, \ldots, n\}$, in which DMU_j consumes multiple positive inputs $x_{ij}(i = 1, \ldots, m)$ to produce multiple positive outputs $y_{rj}(r = 1, \ldots, s)$. Let input and output of DMU_j be denoted by $X_j = (x_{1j}, x_{2j}, \ldots, x_{mj})^t$ and $Y_j = (y_{1j}, y_{2j}, \ldots, y_{sj})^t$, respectively. We consider the following input-oriented DEA model, in which DMU_o, $o \in \{1,2, \ldots, n\}$, is the unit under assessment:

$$
\begin{aligned}
&\theta_o = \min \theta \\
&\text{s.t.} \\
&\sum_{j=1}^{n} \lambda_j x_{ij} \leq \theta x_{io}, i = 1, \ldots, m, \\
&\sum_{j=1}^{n} \lambda_j y_{rj} \geq y_o, r = 1, \ldots, s, \\
&\lambda_j \geq 0, j = 1, \ldots, n.
\end{aligned}
\tag{1}
$$

This model is the first basic DEA model, provided by Charnes, Cooper, and Rhodes (CCR model), and measures the efficiency under the constant returns to scale (CRS) assumption.

2.2 Basic IDEA Models

The first study in IDEA has been done by [4]. They dealt with the following important question: if the efficiency index o remains unchanged, but the outputs increase, how much should the inputs of DMU_o increase?

To solve this problem, suppose the outputs of DMU_o are increased from Y_o to $\alpha_o = Y_o + \Delta Y_o$, where $\Delta Y_o \geq 0$ and $\Delta Y_o \neq 0$. We need to estimate the input vector α_o^* provided that the efficiency index of DMU_o is still θ_o. In fact, $\alpha_o^* = (\alpha_{1o}^*, \alpha_{2o}^*, \ldots, \alpha_{mo}^*)^t = X_o + \Delta X_o$, $\Delta X_o \geq 0$. Suppose DMU_{n+1} represents DMU_o after changing the inputs and outputs. The following model measures the efficiency score of DMU_{n+1}:

$$
\begin{aligned}
&\min \theta, \\
&s.t. \\
&\textstyle\sum_{j=1}^{n} \lambda_j x_{ij} + \lambda_{n+1}\alpha_o^* \leq \theta\alpha_o^*, i = 1, \ldots, m, \\
&\textstyle\sum_{j=1}^{n} \lambda_j y_{rj} + \lambda_{n+1}\beta_o \geq \beta_o, r = 1, \ldots, s, \\
&\lambda_j \geq 0, j = 1,2, \ldots, n+1.
\end{aligned} \tag{2}
$$

The variables of the above model are θ, $\lambda_1, \lambda_2, \ldots, \lambda_{n+1}$. If the optimal values of models (1) and (2) are equal, the efficiency is unchanged, i.e., $\theta\left(\alpha_o^*, \beta_o\right) = \theta(X_o, Y_o)$.

Hadi-Vencheh et al. [6] considered the following MOLP problem to respond to the aforesaid question:

$$
\begin{aligned}
&\min\ (\alpha_{1o}, \alpha_{2o}, \ldots, \alpha_{mo}), \\
&\text{s.t.} \\
&\textstyle\sum_{j=1}^{n} \lambda_j x_{ij} \leq \theta_o\alpha_o, i = 1, \ldots, m, \\
&\textstyle\sum_{j=1}^{n} \lambda_j y_{rj} \geq \beta_o, r = 1, \ldots, s, \\
&\alpha_{io} \geq x_{io}, i = 1, \ldots, m, \\
&\lambda_j \geq 0, j = 1,2, \ldots, n.
\end{aligned} \tag{3}
$$

The variables vector in MOLP (3) is (α_o, λ), while θ_o is fixed.

In the classic DEA, input and output data are expressed in a deterministic form. In practice, uncertain information which is expressed in the form of bounded data, ordinal data or ratio-bounded data is common with regard to uncertainty issues. This type of uncertain information is unavoidable and should be considered as imprecise data. Different techniques have been introduced to deal with such uncertainty issues in the DEA literature. In the next subsection, we overview existing robust DEA models as an ever-increasing stream in the context of imprecise DEA.

2.3 Existing Robust DEA Models

This subsection first reviews the robust counterpart of the deterministic CCR model (1) that was introduced in the robust DEA literature and developed employing the budgeted uncertainty sets proposed by [35]. Then, we develop new robust DEA model by considering the variable budgeted uncertainty (VBU) set.

The robust counterpart of the CCR model (1), based on [35]'s approach, has been extended in the robust DEA literature as follows (for more details see [28, 36]):

$$
\begin{aligned}
\theta_o^R &= \min \theta, \\
&\text{s.t.} \\
\sum_{j=1}^{n} \lambda_j x_{ij} - \theta x_{io} + \sum_{j=1}^{n} p_{ij}^x + \Gamma_i^x q_i^x &\leq 0, && \forall i, \\
\sum_{j=1}^{n} \lambda_j y_{rj} - y_{ro} - \sum_{j=1}^{n} p_{rj}^y - \Gamma_r^y q_r^y &\geq 0, && \forall r, \\
p_{ij}^x + q_i^x &\geq \hat{x}_{ij} \lambda_j, && \forall i; \forall j, j \neq o, \\
p_{io}^x + q_i^x &\geq \hat{x}_{io}(\theta - \lambda_o), && \forall i, \\
p_{rj}^y + q_r^y &\geq \hat{y}_{rj} \lambda_j, && \forall r; \forall j, j \neq o, \\
p_{ro}^y + q_r^y &\geq \hat{y}_{ro}(1 - \lambda_o), && \forall r, \\
p_{ij}^x, q_i^x, p_{rj}^y, q_r^y, \lambda_j &\geq 0, && \forall i; \forall r; \forall j.
\end{aligned}
\tag{4}
$$

Model (4) is a linear model and the variables q_i^x, q_r^y, p_{ij}^x, and p_{rj}^y measure the robustness of model (4) when the level of uncertainty budgeting is varied by an infinitesimally small amount. The terms $\sum_{j=1}^{n} p_{ij}^x + \Gamma_i^x q_i^x$, $\forall i$ and $\sum_{j=1}^{n} p_{rj}^y - \Gamma_r^y q_r^y$, $\forall r$ demonstrate the worst-case deviations of the uncertain inputs and outputs from their nominal values for the budgeted uncertainty and are added to the robust model (4) to immunise the model against violation of the ith and rth constraints, respectively. Furthermore, the pre-defined robust parameters $\Gamma_i^x \left(\Gamma_r^y\right)$ reflect the maximum number of uncertain inputs (outputs) that are protected against perturbations. In addition, to protect constraints against data perturbations, the objective function value of the robust counterpart model is penalised by loss of optimality compared to the deterministic model which is called the price of robustness [35, 39].

3 Proposed Robust DEA and Robust IDEA Models

This section presents the robust counterpart of model (1) based on the variable budgeted uncertainty proposed by [37, 38] rather than the budgeted uncertainty introduced by [35]. To do so, assume that the true values of uncertain input and output data are formulated as $\tilde{x}_{ij} = x_{ij} + z_{ij}^{x}\hat{x}_{ij}(\forall i; \forall j)$, and $\tilde{y}_{rj} = y_{rj} + z_{rj}^{y}\hat{y}_{rj}(\forall r; \forall j)$, respectively. We also assume that $\sum_{j=1}^{n} z_{ij}^{x} \leq \gamma^{x}(\lambda)$, and $\sum_{j=1}^{n} z_{rj}^{y} \leq \gamma^{y}(\lambda)$, where $\gamma^{x}(\lambda)$ and $\gamma^{y}(\lambda)$ are given non-negative functions on the feasibility region λ that put a limitation on the amount of uncertainty. By adopting the variable budgeted uncertainty proposed by [37, 38] as a generalization of the budgeted uncertainty defined by Bertsimas and Sim [35], the following variable budgeted uncertainty sets can be defined for the uncertain input and output data:

$$\mathcal{U}^{\gamma^{x}}(\lambda) = \left\{ \left(\widetilde{x_i}\right) | \tilde{x}_{ij} = x_{ij} + z_{ij}^{x}\hat{x}_{ij}, 0 \leq z_{ij}^{x} \leq 1, \sum_{j=1}^{n_1} z_{ij}^{x} \leq \gamma^{x}(\lambda), \forall i, \forall j \right\},$$

$$\mathcal{U}^{\gamma^{y}}(\lambda) = \left\{ \left(\widetilde{y_r}\right) | \tilde{y}_{rj} = y_{rj} + z_{rj}^{y}\hat{y}_{rj}, 0 \leq z_{rj}^{y} \leq 1, \sum_{j=1}^{n_1} z_{rj}^{y} \leq \gamma^{y}(\lambda), \forall r, \forall j \right\}. \quad (5)$$

In (5), the uncertainty sets $\mathcal{U}^{\gamma^{x}}(\lambda)$ and $\mathcal{U}^{\gamma^{y}}(\lambda)$ are multi-functions of the decision variable λ and are defined as an alternative for the uncertainty sets $\mathcal{U}_{\Gamma_i^x}$ and $\mathcal{U}_{\Gamma_r^y}$ bounded by the static number Γ_i^x (Γ_r^y). In other words, for the decision variable λ, uncertainty sets $\mathcal{U}^{\gamma^{x}}(\lambda)$ and $\mathcal{U}^{\gamma^{y}}(\lambda)$ contain all feasible values for the uncertain input and output parameters, respectively. Furthermore, if $\gamma^{x}(\lambda)$ ($\gamma^{y}(\lambda)$) is set constantly equal to $\Gamma_i^x(\Gamma_r^y)$, then it is trivial that $\mathcal{U}^{\gamma^{x}}(\lambda)$ and $\mathcal{U}^{\gamma^{y}}(\lambda)$ coincide with $\mathcal{U}_{\Gamma_r^x}$ and $\mathcal{U}_{\Gamma_r^y}$, respectively, for any λ. In general, the use of $\mathcal{U}^{\gamma^{x}}(\lambda)$ and $\mathcal{U}^{\gamma^{y}}(\lambda)$ enables us to avoid the over-conservatism issue when there are decision-variable vectors λ with few components in each constraint, providing a new framework that is less conservative than $\mathcal{U}_{\Gamma_r^x}$ and $\mathcal{U}_{\Gamma_r^y}$. The following theorem provides the robust formulation of model (1) based on the uncertain input and output data in (5).

As discussed by [37, 38], it is required that functions $\gamma^{x}(\lambda)$ and $\gamma^{y}(\lambda)$ involved in the definition of $\mathcal{U}^{\gamma^{x}}(\lambda)$ and $\mathcal{U}^{\gamma^{y}}(\lambda)$ be an affine function of λ so as to satisfy the probability bound proposed by Bertsimas and Sim [35]. By following [37, 38], herein we consider the case where $\gamma^{x}(\lambda)$ and $\gamma^{y}(\lambda)$ are affine functions of λ as follows: $\gamma^{x}(\lambda) = \gamma_0^{x} + \sum_{j=1}^{n_1} \gamma_j^{x}\lambda_j$ and $\gamma^{y}(\lambda) = \gamma_0^{y} + \sum_{j=1}^{n_1} \gamma_j^{y}\lambda_j$.

The robust counterpart of model (1) is equivalent to the following robust (hereafter, robust VBU) model:

$$
\begin{array}{lr}
\theta_o^{VBU*} = \min \ \theta_o^{VBU} & \\
s.t. & \\
\sum_{j=1}^{n} \lambda_j x_{ij} + p_i^{\prime x} \gamma_0^x + \sum_{j=1}^{n} \gamma_j^x w_{ij}^x + \sum_{j=1}^{n} q_{ij}^{\prime x} \le \theta_o^{VBU} x_{io}, & \forall i, \\
\sum_{j=1}^{n} \lambda_j y_{rj} - p_r^{\prime y} \gamma_0^y - \sum_{j=1}^{n} \gamma_j^y w_{rj}^y - \sum_{j=1}^{n} q_{rj}^{\prime y} \ge y_{ro}, & \forall r, \\
p_i^{\prime x} + q_{ij}^{\prime x} \ge \lambda_j \hat{x}_{ij}, & \forall i;\ \forall j, \\
p_i^{\prime x} + q_{io}^{\prime x} \ge \left(\theta_o^{VBU} - \lambda_o\right) \hat{x}_{io}, & \forall i;\ j = o, \\
w_{ij}^x - p_i^{\prime x} \ge -M_x\left(1 - \lambda_j\right), & \forall i;\ \forall j, \\
p_r^{\prime y} + q_{rj}^{\prime y} \ge \lambda_j \hat{y}_{rj}, & \forall r;\ \forall j, \\
p_r^{\prime y} + q_{ro}^{\prime y} \ge (1 - \lambda_o) \hat{y}_{ro}, & \forall r;\ j = o, \\
w_{rj}^y - p_r^{\prime y} \ge -M_y\left(1 - \lambda_j\right), & \forall r;\ \forall j, \\
\lambda_j \le M_x^{\prime} h_j^x, & \forall j, \\
\lambda_j \le M_y^{\prime} h_j^y, & \forall j, \\
p_i^{\prime x},\ q_{ij}^{\prime x},\ p_r^{\prime y},\ q_{rj}^{\prime y},\ w_{ij}^x,\ w_{rj}^y \ge 0, & \forall i;\ \forall r;\ \forall j, \\
\lambda_j,\ h_j^y, h_j^x \in \{0, 1\}^n, \quad \ge 0, & \forall j,
\end{array}
\tag{6}
$$

Model (6) is always feasible and its optimal objective function is greater than or equal to model (1).[1]

3.1 Proposed Robust IDEA

This sub-section thinks of question in sub-Sect. 2.2 through a robust framework. Considering model (2) under uncertain situation, therefore we arrive at the following model:

$$
\begin{array}{l}
\min \theta, \\
s.t. \\
\sum_{j=1}^{n} \lambda_j \tilde{x}_{ij} + (\lambda_{n+1} - \theta) \tilde{\alpha}_o^{*} \le 0, i = 1, \ldots, m, \\
\sum_{j=1}^{n} \lambda_j \tilde{y}_{rj} + (\lambda_{n+1} - 1) \tilde{\beta}_o \ge 0, r = 1, \ldots, s, \\
\lambda_j \ge 0, j = 1,2, \ldots, n + 1.
\end{array}
\tag{7}
$$

Considering the uncertainty sets for the input and output data introduced in (5) and following similar methods according to subsect. 3.1, the robust IDEA model can be formulated as follows:

[1] The proofs will be presented in the full paper.

$$
\begin{aligned}
&\theta_o^{IR*} = \min \ \theta_o^{IR} \\
&s.t. \\
&\sum_{j=1}^{n} \lambda_j x_{ij} + p_i'^{x} \gamma_0^{x} + \sum_{j=1}^{n} \gamma_j^{x} w_{ij}^{x} + \sum_{j=1}^{n} q_{ij}'^{x} \leq \theta_o^{IR} x_{io}, && \forall i, \\
&\sum_{j=1}^{n} \lambda_j y_{rj} - p_r'^{y} \gamma_0^{y} - \sum_{j=1}^{n} \gamma_j^{y} w_{rj}^{y} - \sum_{j=1}^{n} q_{rj}'^{y} \geq y_{ro}, && \forall r, \\
&p_i'^{x} + q_{ij}'^{x} \geq \lambda_j \hat{x}_{ij}, && \forall i;\ \forall j, \\
&p_i'^{x} + q_{io}'^{x} \geq \left(\theta_o^{IR} - \lambda_{n+1}\right) \hat{\alpha}_{io}, && \forall i;\ j = n+1, \\
&w_{ij}^{x} - p_i'^{x} \geq -M_x \left(1 - \lambda_j\right), && \forall i;\ \forall j, \\
&p_r'^{y} + q_{rj}'^{y} \geq \lambda_j \hat{y}_{rj}, && \forall r;\ \forall j, \\
&p_r'^{y} + q_{ro}'^{y} \geq (1 - \lambda_o) \hat{\beta}_{ro}, && \forall r;\ j = n+1, \\
&w_{rj}^{y} - p_r'^{y} \geq -M_y \left(1 - \lambda_j\right), && \forall r;\ \forall j, \\
&\lambda_j \leq M_x' h_j^{x}, && \forall j, \\
&\lambda_j \leq M_y' h_j^{y}, && \forall j, \\
&p_i'^{x},\ q_{ij}'^{x},\ p_r'^{y},\ q_{rj}'^{y},\ w_{ij}^{x},\ w_{rj}^{y} \geq 0, && \forall i;\ \forall r;\ \forall j, \\
&\lambda_j \geq 0, && \forall j, \\
&h_j^{y}, h_j^{x} \in \{0, 1\}^n && \forall j,
\end{aligned}
\tag{8}
$$

If the optimal values of the problems (6) and (8) are the same regarding the level of uncertainty Γ, it is said that the efficiency index remains unchanged.

4 Conclusion

In the current study, IDEA problems are developed for input estimation in the presence of polyhedral uncertain data. We employ a robust optimization approach to combat to data uncertainty. An extension of existing robust DEA models and novel robust IDEA model is proposed. The suggested models can be utilized in real-world management applications, such as resource allocation issues, sensitivity analysis, preserving (or improving) efficiency values, determining revenue targets, merging the banks, and firm restructuring. Identification of the type of distribution of the efficiency random variable and also obtaining similar models in dynamic and network DEA frameworks can be considered as promising future research aspects.

References

1. Charnes, A., Cooper, W. W., & Rhodes, E. (1978). Measuring the efficiency of decision-making units. *European Journal of Operational Research, 2*(6), 429–444. https://doi.org/10.1016/0377-2217(78)90138-8
2. Yan, H., Wei, Q., & Hao, G. (2002). Continuous optimization DEA models for resource reallocation and production input/output estimation. *European Journal of Operational Research, 136*(1), 19–31. www.elsevier.com/locate/dsw
3. Amin, G. R., & Ibn Boamah, M. (2023). Modeling business partnerships: A data envelopment analysis approach. *European Journal of Operational Research, 305*(1), 329–337. https://doi.org/10.1016/j.ejor.2022.05.036
4. Wei, Q., Zhang, J., & Zhang, X. (2000). Theory and methodology an inverse DEA model for inputs/outputs estimate. *European Journal of Operational Research, 121*, 151–163. www.elsevier.com/locate/orms
5. Jahanshahloo, G. R., Lotfi, F. H., Shoja, N., Tohidi, G., & Razavyan, S. (2005). Sensitivity of efficiency classifications in the inverse DEA models. *Applied Mathematics and Computation, 169*(2), 905–916. https://doi.org/10.1016/j.amc.2004.09.093
6. Hadi-Vencheh, A., Foroughi, A. A., & Soleimani-damaneh, M. (2008). A DEA model for resource allocation. *Economic Modelling, 25*(5), 983–993. https://doi.org/10.1016/j.econmod.2008.01.003
7. Lin, H. T. (2010). An efficiency-driven approach for setting revenue target. *Decision Support Systems, 49*(3), 311–317. https://doi.org/10.1016/j.dss.2010.03.006
8. Lertworasirikul, S., Charnsethikul, P., & Fang, S. C. (2011). Inverse data envelopment analysis model to preserve relative efficiency values: The case of variable returns to scale. *Computers and Industrial Engineering, 61*(4), 1017–1023. https://doi.org/10.1016/j.cie.2011.06.014
9. Gattoufi, S., Amin, G. R., & Emrouznejad, A. (2014). A new inverse DEA method for merging banks. *IMA Journal of Management Mathematics, 25*(1), 73–87. https://doi.org/10.1093/imaman/dps027
10. Lim, D. J. (2016). Inverse DEA with frontier changes for new product target setting. *European Journal of Operational Research, 254*(2), 510–516. https://doi.org/10.1016/j.ejor.2016.03.059
11. Ghiyasi, M. (2017). Inverse DEA based on cost and revenue efficiency. *Computers and Industrial Engineering, 114*, 258–263. https://doi.org/10.1016/j.cie.2017.10.024
12. Amin, G. R., Emrouznejad, A., & Gattoufi, S. (2017). Minor and major consolidations in inverse DEA: Definition and determination. *Computers and Industrial Engineering, 103*, 193–200. https://doi.org/10.1016/j.cie.2016.11.029
13. Ghiyasi, M. (2015). On inverse DEA model: The case of variable returns to scale. *Computers and Industrial Engineering, 87*, 407–409. https://doi.org/10.1016/j.cie.2015.05.018
14. Jahanshahloo, G. R., Soleimani-Damaneh, M., & Ghobadi, S. (2015). Inverse DEA under intertemporal dependence using multiple-objective programming. *European Journal of Operational Research, 240*(2), 447–456. https://doi.org/10.1016/j.ejor.2014.07.002
15. Emrouznejad, A., Yang, G. L, & Amin, G. R. (2019). A novel inverse DEA model with application to allocate the CO2 emissions quota to different regions in Chinese manufacturing industries. *Journal of the Operational Research Society, 70*(7), 1079–1090. https://doi.org/10.1080/01605682.2018.1489344
16. Amin, G. R., Al-Muharrami, S., & Toloo, M. (2019). A combined goal programming and inverse DEA method for target setting in mergers. *Expert Systems with Applications, 115*, 412–417. https://doi.org/10.1016/j.eswa.2018.08.018
17. Zhang, G. J., Cui, J. C. (2020). A general inverse DEA model for non-radial DEA. *Computers and Industrial Engineering, 142*. https://doi.org/10.1016/j.cie.2020.106368
18. Amin, G. R., & Ibn Boamah, M. (2020). A new inverse DEA cost efficiency model for estimating potential merger gains: A case of Canadian banks. *Annals of Operations Research, 295*(1), 21–36. https://doi.org/10.1007/s10479-020-03667-9

19. Lim, D. J. (2020). Inverse data envelopment analysis for operational planning: The impact of oil price shocks on the production frontier. *Expert Systems with Applications, 161*. https://doi.org/10.1016/j.eswa.2020.113726
20. Chen, Y., Chen, M., Li, T. (2021). China's CO2 emissions reduction potential: A novel inverse DEA model with frontier changes and comparable value. *Energy Strategy Reviews, 38*. https://doi.org/10.1016/j.esr.2021.100762
21. Moghaddas, Z., Tosarkani, B. M., & Yousefi, S. (2022). Resource reallocation for improving sustainable supply chain performance: An inverse data envelopment analysis. *International Journal of Production Economics, 252*, 108560. https://doi.org/10.1016/j.ijpe.2022.108560
22. Emrouznejad, A., Amin, G. R., Ghiyasi, M., & Michali, M. (2023). A review of inverse data envelopment analysis: Origins, development and future directions. *IMA Journal of Management Mathematics, 34*(3), 421–440. https://doi.org/10.1093/imaman/dpad006
23. Toloo, M., Keshavarz, E., & Hatami-Marbini, A. (2021). An interval efficiency analysis with dual-role factors. *OR Spectrum, 43*, 255–287. https://doi.org/10.1007/s00291-020-00606-9
24. Hatami-Marbini, A., Emrouznejad, A., & Tavana, M. (2011). A taxonomy and review of the fuzzy data envelopment analysis literature: Two decades in the making. *European Journal of Operational Research, 214*(3), 457–472. https://doi.org/10.1016/j.ejor.2011.02.001
25. Olesen, O. B., & Petersen, N. C. (2016). Stochastic data envelopment analysis–A review. *European Journal of Operational Research, 251*(1), 2–21. https://doi.org/10.1016/j.ejor.2015.07.058
26. Daouia, A., Florens, J. P., & Simar, L. (2020). Robust frontier estimation from noisy data: A Tikhonov regularization approach. *Econometrics and Statistics, 14*, 1–23. https://doi.org/10.1016/j.ecosta.2018.07.003
27. Arabmaldar, A., Sahoo, B. K., & Ghiyasi, M. (2023). A generalized robust data envelopment analysis model based on directional distance function. *European Journal of Operational Research, 311*(2), 617–632. https://doi.org/10.1016/j.ejor.2023.05.005
28. Hatami-Marbini, A., & Arabmaldar, A. (2021). Robustness of Farrell cost efficiency measurement under data perturbations: Evidence from a US manufacturing application. *European Journal of Operational Research, 295*(2), 604–620. https://doi.org/10.1016/j.ejor.2021.03.019
29. Toloo, M., Mensah, E. K., & Salahi, M. (2022). Robust optimization and its duality in data envelopment analysis. *Omega, 108*, 102583. https://doi.org/10.1016/j.omega.2021.102583
30. Hatami-Marbini, A., Arabmaldar, A., & Asu, J. O. (2022). Robust productivity growth and efficiency measurement with undesirable outputs: Evidence from the oil industry. *OR Spectrum, 44*(4), 1213–1254. https://doi.org/10.1007/s00291-022-00683-y
31. Arabmaldar, A., Hatami-Marbini, A., Loske, D., Hammerschmidt, M., & Klumpp, M. (2024). Robust data envelopment analysis with variable budgeted uncertainty. *European Journal of Operational Research, 315*(2), 626–641. https://doi.org/10.1016/j.ejor.2023.11.043
32. Hadi-Vencheh, A., Hatami-Marbini, A., Ghelej Beigi, Z., & Gholami, K. (2015). An inverse optimization model for imprecise data envelopment analysis. *Optimization, 64*(11), 2441–2454. https://doi.org/10.1080/02331934.2014.974599
33. Ghomi, A., Ghobadi, S., Behzadi, M. H., & Rostamy-Malkhalifeh, M. (2021). Inverse data envelopment analysis with stochastic data. *RAIRO - Operations Research, 55*(5), 2739–2762. https://doi.org/10.1051/ro/2021135
34. Younesi, A., Lotfi, F. H., & Arana-Jiménez, M. (2023). Using slacks-based model to solve inverse DEA with integer intervals for input estimation. *Fuzzy Optimization and Decision Making*. https://doi.org/10.1007/s10700-022-09403-1
35. Bertsimas, D., & Sim, M. (2004). The price of robustness. *Operations Research, 52*(1), 35–53. https://doi.org/10.1287/opre.1030.0065
36. Salahi, M., Toloo, M., & Torabi, N. (2020). A new robust optimization approach to common weights formulation in DEA A new robust optimization approach to common weights formulation. *Journal of the Operational Research Society, 72*(6), 1390–1402. https://doi.org/10.1080/01605682.2020.1718016

37. Poss, M. (2013). Robust combinatorial optimization with variable budgeted uncertainty. *4OR, 11*(1), 75–92. https://doi.org/10.1007/s10288-012-0217-9
38. Poss, M. (2014). Robust combinatorial optimization with variable cost uncertainty. *European Journal of Operational Research, 237*(3), 836–845. https://doi.org/10.1016/j.ejor.2014.02.060

A Selecting Data Envelopment Analysis Approach Under an Onion Framework to Select Quality Service Criteria and Evaluate the Perception of Airport Concessionaires' Users About Brazilian Airports

Paulo Nocera Alves Junior, Isotilia Costa Melo, Tatiana Kimura Kodama, Jessica Suarez Campoli, and Marcelo Scido Nagano

Abstract Traditional methods of cost–benefit analysis in business administration usually measure the performance of activities exclusively in monetary terms, though the airport users' perception of quality service is relevant to airlines' operations. This paper aims to present a novel Onion framework for Selecting Data Envelopment Analysis (Selecting DEA) model to select multiple quality service criteria, compare, and evaluate the perception of airport concessionaires' users about Brazilian airports through layers of analysis. This Onion framework consists of running the DEA again, in loop, where each loop is considered an onion layer, and each loop only with the criteria not selected for any of the Decision-Making Units (DMUs) in the previous loop, and to repeat it until all criteria is selected for at least one DMU and there are no more criteria for another layer. The first step was the compilation of the answers of 31 criteria, collected monthly from the Brazilian National Civil Aviation Agency (ANAC) for the main airports, deemed as Decision-Making Units (DMUs) in this study, from 2015 to 2017 for the empirical application (there are more measures than DMUs). The results showed 14 layers of the Selecting DEA, the last layer had most of the important quality service criteria. It can support decision-makers to select criteria to evaluate DMUs and support management best practices.

P. Nocera Alves Junior (✉)
Universidad Católica del Norte (UCN), Escuela de Ingeniería de Coquimbo (EIC), Coquimbo, Chile
e-mail: paulo.alves@ucn.cl

I. Costa Melo
ICN Business School, CEREFIGE—Université de Lorraine, Paris, France

T. Kimura Kodama · J. Suarez Campoli · M. Seido Nagano
São Carlos School of Engineering (EESC), University of São Paulo (USP), São Carlos, Brazil

A. Emrouznejad et al. (eds.), *Advances in the Theory and Applications of Performance Measurement and Management*, Lecture Notes in Operations Research,
https://doi.org/10.1007/978-3-031-61597-9_8

Keywords Selecting data envelopment analysis (DEA) · Onion framework · Quality service criteria of Brazilian airports

1 Introduction

According to the Brazilian National Civil Aviation Agency (ANAC), the airports' concessions in Brazil began in 2011. It aimed to invest in airport infrastructure to improve it and meet this increased demand for air transportation and the growth of the sector in the country [1]. Also, private airports tend to be more efficient than public administration [10].

To ensure that the concessioned airports also invest in quality for the services and products offered, it was established, in the concession contracts, the incidence of periodic assessment of the level of passenger satisfaction and efficiency of airport operations. These quality levels were based on international standards [1].

Besides, today's travelers have significant options among airports and there is an increasing urgency among them to differentiate themselves in seeking to meet the customers' needs concerning their competitors. While passengers' perception of service quality is only one of many variables (routes, times, locations, and prices) that contribute to the general attractiveness of the airport, it is nevertheless a vital variable in the growing consumer orientation as a competitive advantage in this industry [11].

Based on the monthly quality service surveys of [2], for six main airports in Brazil (deemed as Decision-Making Units—DMUs), from 2015 to 2017, this paper aims to propose a methodology for selecting important criteria and evaluating the efficiency of the DMUs through multi-criteria service quality indicators (SQI) constructed with Selecting Data Envelopment Analysis (SelectDEA) under Variable Returns to Scale (VRS), i.e. the Selecting BCC (SelectBCC) model, evaluating the quality of the retailing facilities and airport services jointly.

2 Literature Review

2.1 Selecting DEA Model

Charnes et al. [5] proposed Data Envelopment Analysis (DEA), considered a non-parametric approach that has been widely used as an effective decision support system to measure the relative efficiency of homogeneous decision-making units (DMUs).

DMUs uses multiple inputs to produce various outputs. Although the application of DEA identifies efficient and inefficient DMUs, the analysis needs to discriminate between efficient DMUs, which is the primary concern in problems based on DEA. One of the concerns when applying DEA is the issue of selecting variables and number of efficient DMUs, which requires the maximum discrimination power in DEA [17].

In conventional DEA models, the data (inputs and outputs or simply performance factors) play a decisive role in the performance evaluation. Therefore, selecting performance factors is crucial for successful DEA applications and adding or removing variables. In general, if the number of performance factors is considerably more significant than the number of DMUs, then a large part of the DMUs will be identified as efficient, which is questionable [12].

To solve this problem, we must select among the performance factors in a way that has the least impact on the efficiency scores while providing the DMU tie-breaking. For this, [16] suggested selecting DEA models in multiplier (optimistic) and envelopment (pessimistic) forms. Toloo and Allahyar [18] developed the idea of [16] to correct the problems of infeasibility of the models. They presented models under Variable Returns to Scale (VRS). Therefore, this study fits into the application of the Selecting DEA models, related to the BCC model [3] with Variable Returns to Scale (VRS) and is based on the model proposed by [18].

2.2 *Brazilian Airports and DEA*

Given our focus on Brazilian airports, we present a brief review of the literature on the application of the DEA to measure the efficiency of Brazilian airports.

Fernandes and Pacheco [7] applied Data Envelope Analysis (DEA) to analyze 35 Brazilian airports and determine which ones were efficient in terms of the number of passengers processed, and the result was a determination of the periods in which the capacity expansion requirements to maintain services in the currently detected standard passengers for each airport.

Pacheco and Fernandes [13] studied the same 35 Brazilian airports from [7] to identify improvements in two dimensions: (a) the first one is improved management, which shows the airport's capacity for financial return, (b) the second is a physical dimension, which shows the level of use of the airport infrastructure. The results are shown as positions of Brazilian airports and the ways to improve their performance.

Pacheco [14] analyzed the changes that occurred between 1998 and 2001 in the airport system managed by the Brazilian Airport Infrastructure Management Agency (INFRAERO—Empresa Brasileira de Infraestrutura Aeroportuária) because, although INFRAERO has not been privatized, it has undergone changes in the way airport business in Brazil by employees of the government and private entrepreneurs in two aspects: inputs and revenues. The results showed that the financial performance improved, despite the decline in operational performance in the analyzed period.

Wanke [19] listed different approaches to measure the efficiency of 65 Brazilian airports and Wanke [20] pointed out the lack of infrastructure due to a capacity deficit in Brazilian airports, where the short-term potential for the recovery of passengers/cargo by landing/take-off is practically nonexistent.

Wanke [21] indicates that contextual variables or efficiency factors, such as hub operations and airport location, affect the efficiency levels of physical infrastructure and flight consolidation differently.

Périco et al. [15] analyzed the efficiency of 16 Brazilian airports. The results showed that the size of an airport was not decisive for attributing efficiency, although it is a relevant criterion to drive improvements in its performance.

Fernandes and Pacheco [8] examined the performance of 60 airports under the state-owned INFRAERO's management in 2009 and 2015, three years before and three years after the concessions began to operate. The results show that this set of airports lost market share, and significantly decreased performance. In the period studied, the companies' overall performance index fell from 71.48 to 62.73%, as it did not find ways to improve the scenario of airport concessions for private administration.

Fernandes et al. [9] analyzed the geographical changes in the Brazilian air transport network through an analysis of the domestic airport network of 2007 and 2016, correlating issues of the country's regional economies with changes in spatial patterns of infrastructure and air flows in 2007 and 2016. They showed that, although the airports in the Southeast region are more important in terms of airflow and number of connections, there is a slight decentralization of air flows and connection with the Midwest and Northeast regions of the country, evidenced by the growing network capacity of the airports and regional economic changes, leading to a change in the hierarchical position of the airports in these regions and a structural change in the aerial network.

The DEA studies and models applied and their forms of analysis, as well as the number of airports studied, and the period of data analyzed are summarized in Table 1.

It is noted that, among the analyzed papers, no paper applied a Selecting DEA model, and no paper addressed the efficiency of service quality indicators in Brazilian airports. Thus, this article can contribute significantly to this issue.

3 Methodology and Procedures

The main airports (DMUs) in Brazil that have regulated metering for SQI, from 2015 to 2017, are Brasília (BSB or SBBR), Confins (CFN or SBCF), Galeão (GIG or SBGL), Guarulhos (GRU or SBGR), São Gonçalo do Amarante (SGA or SBSG) and Viracopos (VCP or SBKP).

There are 31 question items in common for the six airports. The airports SBGR, SBBR, and SBKP use the same codification of quality questions, so, we used their codification in Table 2. The other airports use a different codification for the same questions, so, we standardized them as the other three. For example, ITEM 12_1 ("How do you evaluate the quality of signage?") in Table 2 is the ITEM 12_1 for the airports SBGR, SBBR, and SBKP, but ITEM 14_1 for the airports SBCF and

Table 1 Summary of studies on the application of DEA at Brazilian airports

Author (s)	DEA method applied	Number of airports analyzed	Analyzed period
Fernandes and Pacheco [7]	BCC model, output-oriented approach using a variable return to scale (VRS)	35	1998
Pacheco and Fernandes [13]	BCC model, input-oriented approach using a VRS	35	1998
Pacheco et al. [14]	BCC model, input-oriented approach using a VRS	58	1998 and 2001
Wanke [19]	BCC and CCR model approach using constant return to scale (CRS) and VRS	65	2009
Wanke [20]	BCC model, output-oriented approach using a VRS	63	2009
Wanke [21]	Two-stage network DEA model	63	2009
Périco et al. [15]	BCC model, output-oriented approach using a VRS and bootstrap	16	2010–2012
Fernandes and Pacheco [8]	BCC model, output-oriented approach using a VRS and Malmquist analysis	60	2009 and 2015
Fernandes et al. [9]	BCC model, output-oriented approach and Malmquist analysis	132 airports in 2007 and 112 airports in 2016	2007–2016

SBGL and ITEM 11_1 for the airport SBSG. The answers are numeric from 1 to 5 (assuming very bad to very good).

We collected the data (SQI) from the monthly quality survey of the six airports from January 2015 until December 2017 from [2]. We manually compiled the answers to 31 items of these airports (in an Excel spreadsheet), translating to the codification adopted by SBGR, SBBR, and SBKP. Table 2 shows the code and description of the quality service criteria.

Usually, when applying DEA models, a rule of thumb is usually required for increasing the discrimination power of the DEA result, so, the number of DMUs is at least three times higher than the number of variables, or higher than the number of

Table 2 Description of the quality service criteria

Code	Description
ITEM_12_1	"How do you evaluate the quality of signage?"
ITEM_12_2	"How do you evaluate the quality of flight information?"
ITEM_12_3	"How do you evaluate the quality of the passenger announcement system?"
ITEM_13_1	"How do you evaluate the availability of restrooms?"
ITEM_13_2	"How do you evaluate restroom cleanliness?"
ITEM_14_1	"How do you evaluate the availability of seats in the boarding lounge?"
ITEM_14_2	"How do you evaluate the comfort of seats in the boarding lounge?"
ITEM_14_3*	"How do you evaluate the availability of seats in other public areas?"
ITEM_14_4*	"How do you evaluate the comfort of seats in other public areas?"
ITEM_15	"How do you evaluate the overall cleanliness of the airport?"
ITEM_16	"How do you evaluate the friendliness of airport staff?"
ITEM_17_1	"How do you evaluate the variety of stores?"
ITEM_17_2	"How do you evaluate the quality of stores?"
ITEM_18_1	"How do you evaluate the cost–benefit of the stores?"
ITEM_18_2	"How do you evaluate the cost–benefit of restaurants and bars?"
ITEM_19*	"How do you evaluate the availability of spaces in the parking lot?"
ITEM_20*	"How do you evaluate the availability of baggage carts?"
ITEM_21	"How do you evaluate overall satisfaction with the airport?"
ITEM_22_1	"How do you evaluate thermal comfort?"
ITEM_22_2	"How do you evaluate acoustic comfort?"
ITEM_23	"How do you evaluate the perception of security?"
ITEM_24_1*	"How do you evaluate the parking lot options?"
ITEM_24_2*	"How do you evaluate the cost–benefit of the parking lot?"
ITEM_25_1*	"How do you evaluate the availability of curb space for boarding?"
ITEM_25_2*	"How do you evaluate the availability of curb space for disembarking?"
ITEM_26	"How do you evaluate the terminal transportation equipment?"
ITEM_27*	"How do you evaluate the organization of the security inspection queue?"
ITEM_28_1	"How do you evaluate the availability and convenience of banking services?"
ITEM_28_2	"How do you evaluate the location of banking services?"
ITEM_29_1	"How do you evaluate the availability of wireless network and other internet connections?"
ITEM_30*	"How do you evaluate the availability of hotel services?"

inputs multiplied to the number of outputs despite some real cases may not follow it [6, 18, 17] and it is part of the formulation of the Selecting DEA models. We adopted each airport each year as one DMU, so we had 18 DMUs and 31 variables. The VRS assumption is justified because of the relationship among the perception of service quality. We applied the SelectBCC model to reduce the number of variables as follows in Eqs. (1)–(7), and the formulation is from [18], also presented in [17].

$$Minimize\ \theta \tag{1}$$

Subject to:

$$\sum_{k=1}^{z} \lambda_k x_{ik} + S_i^- = \theta x_{i0} + \mathrm{M}(1 - p_i) \quad i = 1, 2, \ldots, m \tag{2}$$

$$\sum_{k=1}^{z} \lambda_k y_{rk} - S_r^+ = y_{r0} - \mathrm{M}(1 - q_r) \quad r = 1, 2, \ldots, s \tag{3}$$

$$\sum_{k=1}^{z} \lambda_k = 1 \tag{4}$$

$$\sum_{i=1}^{m} p_i + \sum_{r=1}^{s} q_r \leq \min\left\{\left[\frac{n}{3}\right], 2\sqrt{n}\right\} \tag{5}$$

$$\lambda_k \geq 0, S_i^- \geq 0, S_r^+ \geq 0 \tag{6}$$

$$p_i, q_r \epsilon \{0, 1\} \tag{7}$$

where θ is the efficiency, S_i^- is the slack of the ith input, S_r^+ is the slack of the rth output, λ_k is the contribution of the kth DMU to the analyzed DMU, x_{i0} is the ith input of the DMU under analysis, y_{r0} is the rth output of the DMU under analysis, x_{ik} is the ith input of the kth DMU, y_{rk} is the rth output of the kth DMU, p_i is a binary variable to indicate if the ith input should be selected or not, q_r is a binary variable to indicate if the rth output should be selected or not, M is a large positive number, m is the number of inputs, s is the number of outputs, z is the number of DMUs.

Applying this model, a situation happened. Apparently not all the performance measures were the best ones to evaluate the airport only with a few indicators, because, for example, there are indicators about the retailing facilities in the airports and, at the same time, more important indicators about the comfort and security of the airports. And the model selected only variables related to the retailing facilities of the airports.

With this in mind, we applied an Onion Framework based on [4] to investigate the layers of the selected performance measures. In this Onion framework for SelectBCC (OF-SelectBCC), each layer consists of applying the SelectBCC again, but without the previously selected variables.

4 Results

Table 3 shows the variables from the first and last layers of the OF-SelectBCC application. Applying the SelectBCC resulted in the selection of the following variables in layer 0 (Table 3): ITEM_18_1, ITEM_18_2, ITEM_24_2*, ITEM_28_2, and ITEM_30*. These variables were related to the retailing facilities of the airports, like stores, restaurants, bars, parking lot, hotel, and banking services.

Depending on the aim of the study, not all the performance measures were considered the best ones to evaluate the airports. For example, these are not good indicators to evaluate the comfort and security of the airports could be good indicators, if the study aims to evaluate the retailing facilities of the airports, but. One possible cause for selecting only indicators related to the retailing facilities of the airports was because all airports try at first to focus on the more important measures (related to comfort and security), so the retailing facilities indicators would discriminate more the airports.

This was the reason for applying the OF-SelectBCC model with the Onion framework based on [4] to investigate the layers of the selected performance measures. After obtaining all layers, apparently the last layer was the one mixing more important indicators.

Applying the OF-SelectBCC resulted in the selection of the following variables in the last layer (14): ITEM_12_3, ITEM_14_2, ITEM_21, ITEM_22_2, and ITEM_26 (Table 3). These variables were related to the airports' services and its retailing facilities, like the quality of the passenger announcement system, the comfort of seats in the boarding lounge, satisfaction with the airport, acoustic comfort, and the terminal transportation equipment.

Table 3 Variables from the first and last layers

Layer	Code	Description
0	ITEM_18_1	"How do you evaluate the cost–benefit of the stores?"
0	ITEM_18_2	"How do you evaluate the cost–benefit of restaurants and bars?"
0	ITEM_24_2*	"How do you evaluate the cost–benefit of the parking lot?"
0	ITEM_28_2	"How do you evaluate the location of banking services?"
0	ITEM_30*	"How do you evaluate the availability of hotel services?"
14	ITEM_12_3	"How do you evaluate the quality of the passenger announcement system?"
14	ITEM_14_2	"How do you evaluate the comfort of seats in the boarding lounge?"
14	ITEM_21	"How do you evaluate overall satisfaction with the airport?"
14	ITEM_22_2	"How do you evaluate acoustic comfort?"
14	ITEM_26	"How do you evaluate the terminal transportation equipment?"

5 Final Remarks

In summary, the results pointed out that it was observed that the last layer of the OF-SelectBCC had more kinds of important variables than the first one, because the layer 0 did not have all kinds of important variables to evaluate airports. For further studies, after COVID-19, we recommend investigating if there are changes in customers' behavior, mainly regarding their predisposition to consume in the airport, that would affect the evaluation and selection of variables too. Also, we recommend verifying the results of applying a Selecting SBM or other methods of reducing the number of variables, as DEA-PCA.

References

1. ANAC. (2019). Concessões. Retrieved March 13, 2020, from https://www.anac.gov.br/assuntos/paginas-tematicas/concessoes
2. ANAC. (2020). Dados Dos Indicadores de Qualidade de Serviço. Retrieved March 12, 2020, from https://www.anac.gov.br/assuntos/paginas-tematicas/concessoes/aeroportos-concedidos/guarulhos/arquivos/03indicadores-de-qualidade-de-servico/dados-dos-indicadores-de-qualidade-de-servico-iqs
3. Banker, R. D., Charnes, A., & Cooper, W. W. (1984). Some models for estimating technical and scale inefficiencies in data envelopment analysis. *Management Science, 30*(9), 1078–1092.
4. Barr, R. S., Durchholz, M. L., & Seiford, L. (2000). Peeling the DEA onion: Layering and rank-ordering DMUs using tiered DEA. *Southern Methodist University Technical Report*, 5(1–24).
5. Charnes, A., Cooper, W. W., & Rhodes, E. (1978). Measuring the efficiency of decision making units. *European Journal of Operational Research, 2*, 429–444.
6. Cook, W. D., Tone, K., & Zhu, J. (2014). Data envelopment analysis: Prior to choosing a model. *Omega (United Kingdom), 44*, 1–4.
7. Fernandes, E., & Pacheco, R. R. (2002). Efficient use of airport capacity. *Transportation Research Part A: Policy and Practice, 36*(3), 225–38. https://linkinghub.elsevier.com/retrieve/pii/S096585640000046X
8. Fernandes, E., & Pacheco, R. R. (2018). Managerial performance of airports in Brazil before and after concessions. *Transportation Research Part A: Policy and Practice, 118*, 245–57. https://linkinghub.elsevier.com/retrieve/pii/S0965856417315732
9. Fernandes, V. A., Pacheco, R. R., Fernandes, E., & da Silva, W. R. (2019). Regional change in the hierarchy of Brazilian airports 2007–2016. *Journal of Transport Geography, 79*, 102467. https://linkinghub.elsevier.com/retrieve/pii/S096669231930242X
10. Fiaco, R. M. D., Campos, V. B. G., & Silva, P. A. L. (2016). Aplicação Do Método Estatístico ANOVA Quanto à Percepção Dos Usuários Na Fila Do Check-in Em Aeroportos Públicos e Concessionados. In: Anais Do XIV Rio de Transportes, pp 1–14
11. Fodness, D., & Murray, B. (2007). Passengers' Expectations of airport service quality. *Journal of Services Marketing, 21*(7), 492–506.
12. Golany, B., & Roll, Y. (1989). An application procedure for DEA. *Omega, 1*(3), 237–250.https://doi.org/10.1016/0305-0483(89)90029-7
13. Pacheco, R. R., & Fernandes, E. (2003). Managerial efficiency of Brazilian airports. *Transportation Research Part A: Policy and Practice, 37*(8), 667–80. https://linkinghub.elsevier.com/retrieve/pii/S0965856403000132
14. Pacheco, R. R., Fernandes, E., & de Santos, M. P. S. (2006). Management style and airport performance in Brazil. *Journal of Air Transport Management, 12*(6), 324–30. https://linkinghub.elsevier.com/retrieve/pii/S0969699706000603

15. Périco, A. E., Santana, N. B., & do Rebelatto, D. A. N. (2017). Eficiência Dos Aeroportos Internacionais Brasileiros: Uma Análise Envoltória de Dados Com Bootstrap. *Gestão & Produção, 24*(2), 370–81. http://www.scielo.br/scielo.php?script=sci_arttext&pid=S0104-530X2017000200370&lng=pt&tlng=pt
20. Toloo, M., & Tichý, T. (2015). Two alternative approaches for selecting performance measures in data envelopment analysis. *Measurement, 65*, 29–40. https://doi.org/10.1016/j.measurement.2014.12.043
17. Toloo, M., Tone, K., & Izadikhah, M. (2023). *Selecting slacks-based data envelopment analysis models., 308*(3), 1302–1318. https://doi.org/10.1016/j.ejor.2022.12.032
18. Toloo, M., & Allahyar, M. (2018). A simplification generalized returns to scale approach for selecting performance measures in data envelopment analysis. *Measurement: Journal of the International Measurement Confederation, 121*, 327–334. https://doi.org/10.1016/j.measurement.2018.02.056
19. Wanke, P. F. (2012a). Capacity Shortfall and Efficiency Determinants in Brazilian Airports: Evidence from Bootstrapped DEA Estimates. *Socio-Economic Planning Sciences 46*(3), 216–29. https://linkinghub.elsevier.com/retrieve/pii/S0038012112000043
20. Wanke, P. F. (2012b). Efficiency of Brazil's Airports: Evidences from Bootstrapped DEA and FDH Estimates. *Journal of Air Transport Management, 23*: 47–53. https://linkinghub.elsevier.com/retrieve/pii/S0969699712000154
21. Wanke, P. F. (2013). Physical infrastructure and flight consolidation efficiency drivers in Brazilian airports: A two-stage network-DEA approach. *Journal of Air Transport Management, 31*, 1–5. https://linkinghub.elsevier.com/retrieve/pii/S0969699712001196

Applying Path-Based Models to Negative Data with a Focus on Super-Efficiency

Mária Trnovská and Margaréta Halická

Abstract Negative data are encountered in various applications, such as insurance, accounting, or finance. Although Data Envelopment Analysis (DEA) models have been originally designed for non-negative data, the path-based models (such as the input or output radial models, the directional distance function models or the hyperbolic distance function model) can be modified to handle negative data. We analyse the extension of these models for the super-efficiency measurement and demonstrate the results on a numerical example.

Keywords Path-based models · Negative data · Directional distance function · Hyperbolic distance function · Super-efficiency

1 Introduction

Various approaches have been developed in Data Envelopment Analysis (DEA) to address negative data, which have been discussed in numerous publications (see e.g. [8–15, 17, 18]). This article is motivated by two of them.

The first motivation comes from the recently published article [8], in which the so-called path-based models are analysed from the point of view of ten properties. The models are formulated through a general scheme that includes the input or output radial models, the directional distance function (DDF) models, and the hyperbolic distance function (HDF) model. When a model is rewritten in terms of the general scheme, it can be simply modified to handle arbitrary (even negative) data: the key step is to change the direction vector prescription so that it remains non-negative, e.g. by taking the absolute value. The results of [8] also include a sufficient condition for the well-definedness of the super-efficiency models, but the properties of

M. Trnovská (✉) · M. Halická
Faculty of Mathematics, Physics and Informatics, Comenius University in Bratislava, Bratislava, Slovakia
e-mail: trnovska@fmph.uniba.sk

M. Halická
e-mail: halicka@fmph.uniba.sk

A. Emrouznejad et al. (eds.), *Advances in the Theory and Applications of Performance Measurement and Management*, Lecture Notes in Operations Research,
https://doi.org/10.1007/978-3-031-61597-9_9

the super-efficiencies are not examined. The second motivation comes from [10, 11], where the authors propose DDF super-efficiency models on arbitrary data and introduce possible direction vectors so that the corresponding models have desirable properties. However, the proposed direction vectors are unnecessarily complex and depend on several technical parameters that must be predetermined, see [10, Sect. 3], [11, Sect. 4].

In this paper, we build on the general scheme proposed in [8], but with a focus on the characteristics of super-efficiency. We provide a necessary and sufficient condition for the feasibility of the super-efficiency models that fit our general scheme, as well as a formula for estimating the upper bound of the super-efficiency score. We also analyse the properties of the models in six different directions, five of which are the usual choices, and the sixth is designed to ensure that the super-efficiency score is always well defined, bounded, translation, and unit invariant, and fulfils the property of monotonicity.

2 Preliminaries

Consider a set of n decision making units DMU$_j$ ($j \in J = \{1, \dots, n\}$) with observed input–output vectors $(\boldsymbol{x}_j, \boldsymbol{y}_j) \in \mathbb{R}^m \times \mathbb{R}^s$, where $\mathbf{x}_j := (x_{1j}, \dots, x_{mj})^\top \in \mathbb{R}^m$ and $\mathbf{y}_j := (y_{1j}, \dots, y_{sj})^\top \in \mathbb{R}^s$. The input–output data of $(\text{DMU}_j)_{j=1}^n$ is arranged into the input/output matrices $\mathbf{X} := (\mathbf{x}_1, \dots \mathbf{x}_n) \in \mathbb{R}^{m\times n}$ and $\mathbf{Y} := (\mathbf{y}_1, \dots \mathbf{y}_n) \in \mathbb{R}^{s\times n}$. We admit negative values for input and output data and assume (AS): for any input or output, the values are not the same for all j.

Based on the given data set we define the technology set

$$\mathcal{T} = \left\{(\boldsymbol{x}, \boldsymbol{y}) \in \mathbb{R}^m \times \mathbb{R}^s \mid \boldsymbol{X\lambda} \le \boldsymbol{x},\ \boldsymbol{Y\lambda} \ge \boldsymbol{y},\ \boldsymbol{\lambda} \ge \mathbf{0},\ \boldsymbol{e}^T\boldsymbol{\lambda} = 1\right\}, \tag{1}$$

which corresponds to variable returns to scale (VRS). Note that the common non-negativity of $(\boldsymbol{x}, \boldsymbol{y})$ is not imposed here. The input/output vectors of $\mathcal{T}$ will be called units. By $(\mathbf{x}_o, \mathbf{y}_o)$ we denote the unit under evaluation.

A unit $(\mathbf{x}, \mathbf{y})$ is called dominating the unit $(\mathbf{x}_o, \mathbf{y}_o)$ if and only if $\mathbf{x} \le \mathbf{x}_o$ and $\mathbf{y} \ge \mathbf{y}_o$. Moreover, a unit $(\mathbf{x}_o, \mathbf{y}_o) \in \mathcal{T}$ is called efficient if and only if there is no other unit in $\mathcal{T}$ dominating $(\mathbf{x}_o, \mathbf{y}_o)$, i.e. if $(\mathbf{x}, \mathbf{y}) \in \mathcal{T}$ is dominating $(\mathbf{x}_o, \mathbf{y}_o)$, then $(\mathbf{x}, \mathbf{y}) = (\mathbf{x}_o, \mathbf{y}_o)$. Obviously, an efficient unit lies on the boundary of $\mathcal{T}$. The set of all efficient units is called the efficient boundary of $\mathcal{T}$. The remaining part of the boundary of $\mathcal{T}$ consists of the so-called weakly efficient units.

By the general scheme (GS) model applied to $(\boldsymbol{x}_o, \boldsymbol{y}_o) \in \mathcal{T}$ with directions $\boldsymbol{g}_o = (\boldsymbol{g}_o^x, \boldsymbol{g}_o^y) \gneqq 0$ that may depend on $(\boldsymbol{x}_o, \boldsymbol{y}_o)$, we understand

$$(\text{GS})_o \qquad \min \theta \tag{2a}$$

$$\boldsymbol{X\lambda} \le \boldsymbol{x}_o + (\psi^x(\theta) - 1)\boldsymbol{g}_o^x, \tag{2b}$$

$$\boldsymbol{Y\lambda} \ge \boldsymbol{y}_o + (\psi^y(\theta) - 1)\boldsymbol{g}_o^y, \tag{2c}$$

$$e^T\boldsymbol{\lambda} = 1, \quad \boldsymbol{\lambda} \geq \mathbf{0}. \tag{2d}$$

Here, the real functions ψ, their domains (dom), and their images (im) satisfy the following assumptions: 1. $\mathrm{dom}(\psi^x)$, $\mathrm{dom}(\psi^y)$ and $\mathrm{im}(\psi^x)$, $\mathrm{im}(\psi^y)$ are either $(0, \infty)$ or $(-\infty, \infty)$; 2. ψ^x is smooth, concave, increasing and ψ^y is smooth, convex, decreasing; 3. $\psi^x(1) = \psi^y(1) = 1$; The right-hand sides of (2b) and (2c) denoted as

$$\boldsymbol{\phi}_o^x(\theta) := \boldsymbol{x}_o + (\psi^x(\theta) - 1)\boldsymbol{g}_o^x \quad \text{and} \quad \boldsymbol{\phi}_o^y(\theta) := \boldsymbol{y}_o + (\psi^y(\theta) - 1)\boldsymbol{g}_o^y \tag{3}$$

define a continuous path in the input-output space $\mathbb{R}^m \times \mathbb{R}^s$ parametrized by θ, $\theta \in \mathcal{D} = \mathrm{dom}(\psi^x) \cap \mathrm{dom}(\psi^y)$: $\boldsymbol{\phi}_o : \theta \mapsto (\boldsymbol{\phi}_o^x(\theta), \boldsymbol{\phi}_o^y(\theta))$.

Assumptions 1.-3. imposed on the functions ψ^x and ψ^y secure the well-definiteness of the program $(GS)_o$ as well as other useful properties of the path $\boldsymbol{\phi}_o$: the path $\boldsymbol{\phi}_o$ passes through the point $(\boldsymbol{x}_o, \boldsymbol{y}_o) \in \mathcal{T}$ at $\theta = 1$ (i.e. $\boldsymbol{\phi}_o(1) = (\boldsymbol{x}_o, \boldsymbol{y}_o)$), and for decreasing values of θ it moves towards the boundary of T gradually passing through points that dominate one another. Finally, the path leaves $\mathcal{T}$ at some $\boldsymbol{\phi}_o(\theta_o^*) \in \partial\mathcal{T}$, where $\theta_o^* \leq 1$, and θ_o^* is the optimal value in $(GS)_o$. The optimal value θ_o^* is called the *efficiency score*, or alternatively, the *value of the efficiency measure* for $(\boldsymbol{x}_o, \boldsymbol{y}_o)$. The point $(\boldsymbol{\phi}_o^x(\theta_o^*), \boldsymbol{\phi}_o^y(\theta_o^*))$ on the path $\boldsymbol{\phi}_o$ is called the *projection* of $(\boldsymbol{x}_o, \boldsymbol{y}_o)$ in the GS model.

It is easy to see that the BCC input and output models [3], the hyperbolic distance function model (HDF) [7] as well as the general directional distance measure model (DDF) [4] can be equivalently rewritten in the form of the general scheme (2). The scheme also includes the so-called generalised distance function model (GDF) introduced by [6]. The corresponding parameterizations are shown in Table 1.

Table 1 Parameterization of the standard path-based models

Model	$\boldsymbol{\phi}_o^x(\theta)$	$\boldsymbol{\phi}_o^y(\theta)$
BCC-I [3]	$\boldsymbol{x}_o + (\theta - 1)\boldsymbol{x}_o$	$\boldsymbol{y}_o$
BCC-O [3]	$\boldsymbol{x}_o$	$\boldsymbol{y}_o + (\frac{1}{\theta} - 1)\boldsymbol{y}_o$
DDF-g [4]	$\boldsymbol{x}_o + (\theta - 1)\boldsymbol{g}_o^y$	$\boldsymbol{y}_o + (2 - \theta - 1)\boldsymbol{g}_o^y$
HDF-g [7, 8]	$\boldsymbol{x}_o + (\theta - 1)\boldsymbol{g}_o^x$	$\boldsymbol{y}_o + (\frac{1}{\theta} - 1)\boldsymbol{g}_o^y$
GDF [6] $p \in [0, 1]$	$\boldsymbol{x}_o + (\theta^{1-p} - 1)\boldsymbol{x}_o$	$\boldsymbol{y}_o + (\theta^{-p} - 1)\boldsymbol{y}_o$

3 Super-Efficiency for the Path-Based Models

For the super-efficiency assessment, we now apply the $(GS)_o$ model on $(\boldsymbol{x}_o, \boldsymbol{y}_o) \notin \mathcal{T}$. Unlike the case $(\boldsymbol{x}_o, \boldsymbol{y}_o) \in \mathcal{T}$, where the $(GS)_o$ model was well defined, in the case

$(\boldsymbol{x}_o, \boldsymbol{y}_o) \notin \mathcal{T}$, the $(GS)_o$ model may be infeasible. The following theorem provides the necessary and sufficient conditions for feasibility in this case.

Theorem 1 *Consider $(GS)_o$, the general model (2) applied to $(\boldsymbol{x}_o, \boldsymbol{y}_o) \notin \mathcal{T}$, with $\boldsymbol{g}_o = (\boldsymbol{g}_o^x, \boldsymbol{g}_o^y) \gneq 0$ and denote $I_o = \{i \mid \boldsymbol{g}_{io}^x = 0\}$, $R_o = \{r \mid \boldsymbol{g}_{ro}^y = 0\}$. Then $(GS)_o$ is feasible if and only if there exists $(\boldsymbol{x}^*, \boldsymbol{y}^*) \in \mathcal{T}$ such that*

$$x_i^* \leq x_{io}\ \forall i \in I_o,\ y_r^* \geq y_{ro}\ \forall r \in R_o, \ \textit{and} \tag{4}$$

$$y_r^* > y_{ro} - g_{ro}^y,\ \forall r \notin R_o,\ \textit{if}\ \operatorname{im}(\psi^y) = (0, \infty). \tag{5}$$

Proof ($\Rightarrow$): Assume that $(GS)_o$ is feasible, that is, there exists $(\theta^o, \boldsymbol{\lambda}^o)$ such that $\boldsymbol{\lambda}^o \geq 0$, $\mathbf{e}^T\boldsymbol{\lambda}^o = 1$, $\boldsymbol{X}\boldsymbol{\lambda}^o \leq \boldsymbol{x}_o + (\psi^x(\theta^o) - 1)\boldsymbol{g}_o^x$ and $\boldsymbol{Y}\boldsymbol{\lambda}^o \geq \boldsymbol{y}_o + (\psi^y(\theta^o) - 1)\boldsymbol{g}_o^y$. Take $\boldsymbol{x}^* = \boldsymbol{X}\boldsymbol{\lambda}^o$, $\boldsymbol{y}^* = \boldsymbol{Y}\boldsymbol{\lambda}^o$. Then for $i \in I_0$ and $r \in R_0$ we clearly have

$$x_i^* \leq x_{io} + (\psi^x(\theta^o) - 1)g_{io}^x = x_{io}, \quad y_r^* \geq y_{ro} + (\psi^y(\theta^o) - 1)g_{ro}^y = y_{ro}$$

and hence $(\boldsymbol{x}^*, \boldsymbol{y}^*)$ satisfies (4). For the case $\operatorname{im}(\psi^y) = (0, \infty)$ we assume by contradiction that there exists $r \notin R_0$ such that $y_r^* = (\boldsymbol{Y}\boldsymbol{\lambda}^o)_r \leq y_{ro} - g_{ro}^y$. Then from the feasibility, we have $y_{ro} + (\psi^y(\theta^o) - 1)g_{ro}^y \leq (\boldsymbol{Y}\boldsymbol{\lambda}^o)_r \leq y_{ro} - g_{ro}^y$, implying that $\psi^y(\theta^o)\boldsymbol{g}_{ro}^y \leq 0$, which is a contradiction.

($\Leftarrow$): Assume that there exists $(\boldsymbol{x}^*, \boldsymbol{y}^*) \in \mathcal{T}$ such that (4) and (5) hold. Since $(\boldsymbol{x}^*, \boldsymbol{y}^*) \in \mathcal{T}$, we have that there exists $\boldsymbol{\lambda}^* \geq 0$ satisfying $\mathbf{1}^T\boldsymbol{\lambda}^* = 1$ such that

$$\boldsymbol{X}\boldsymbol{\lambda}^* \leq \boldsymbol{x}^*, \quad \boldsymbol{Y}\boldsymbol{\lambda}^* \geq \boldsymbol{y}^*. \tag{6}$$

For all $i \in I_o$ and $r \in R_o$, we have $g_{io}^x = 0$ and $g_{ro}^y = 0$ and hence (4) and (6) imply that the inequalities

$$[\boldsymbol{X}\boldsymbol{\lambda}^*]_i \leq x_i^* \leq x_{io} = x_{io} + (\psi^x(\theta) - 1)g_{io}^x, \tag{7}$$

$$[\boldsymbol{Y}\boldsymbol{\lambda}^*]_r \geq y_r^* \geq y_{ro} = y_{ro} + (\psi^y(\theta) - 1)g_{ro}^y, \tag{8}$$

hold for all θ. On the other hand, for all $i \notin I_o$ and $r \notin R_o$, we have $g_{io}^x > 0$ and $g_{ro}^y > 0$, and hence the properties of ψ^x and ψ^y guarantee that

$$\lim_{\theta\to\infty} \psi^x(\theta)g_{io}^x = +\infty, \tag{9}$$

$$\lim_{\theta\to\infty} \psi^y(\theta)g_{ro}^y = \begin{cases} -\infty & \text{if } \operatorname{im}(\psi^y) = (-\infty, \infty), \\ 0 & \text{if } \operatorname{im}(\psi^y) = (0, \infty). \end{cases} \tag{10}$$

Now (9) and (10) imply that the inequalities

$$[\boldsymbol{X}\boldsymbol{\lambda}^*]_i \leq x_{io} - g_{io}^x + \psi^x(\theta)g_{io}^x, \tag{11}$$

$$[\boldsymbol{Y}\boldsymbol{\lambda}^*]_r \geq \begin{cases} y_{ro} - g_{ro}^y + \psi^y(\theta)g_{ro}^y, & \text{if } \operatorname{im}(\psi^y) = (-\infty, \infty) \\ y_r^* > y_{ro} + (\psi^y(\theta) - 1)g_{ro}^y & \text{if } \operatorname{im}(\psi^y) = (0, \infty). \end{cases} \tag{12}$$

hold for θ large enough. Therefore, there exists θ^* large enough so that $(\theta^*, \boldsymbol{\lambda}^*)$ is a feasible solution for $(GS)_o$. □

Note that as soon as the model $(GS)_o$, where $(\boldsymbol{x}_o, \boldsymbol{y}_o) \notin \mathcal{T}$, is feasible, then it is well defined, i.e. the optimal value θ_o^* (called *the super-efficiency score*) is attained and $\theta_o^* \geq 1$. The proof of this claim is analogous to the proof for case $(\boldsymbol{x}_o, \boldsymbol{y}_o) \in \mathcal{T}$ provided in [8, Theorem 1]. Moreover, the theoretical results of [8] that allow to verify the validity of the unit invariance [8, Theorem 7], the translation invariance [8, Theorem 8] and the monotonicity [8, Theorem 9] of the efficiency score (the case $(\boldsymbol{x}_o, \boldsymbol{y}_o) \in \mathcal{T}$) can easily be adopted for the super-efficiency (the case $(\boldsymbol{x}_o, \boldsymbol{y}_o) \notin \mathcal{T}$).

Regarding monotonicity, it is generally true for the path-based models that if the directions are independent of the evaluated unit, then the monotonicity applies.[1] If the directions are dependent, then the monotonicity of the path (as a function of the evaluated unit for the fixed θ) must be examined, since the result may depend on the range for the super-efficiency score.

We now establish a formula for the upper bound of the DDF-g super-efficiency and a sufficient condition for the upper bound equal to 2. The latter result is in accordance with [11, Theorem 4]. By a similar technique, the upper bound for the HDF-g super-efficiency can be obtained.

Theorem 2 *Assume that the DDF-g super-efficiency model for $(\boldsymbol{x}_o, \boldsymbol{y}_o) \notin \mathcal{T}$ is feasible and denote by θ_o^* the corresponding super-efficiency score. Then*

$$\theta_o^* \leq \max\left\{\max_{i \notin I_0} \frac{x_i^{max} - x_{io} + g_{io}^x}{g_{io}^x}, \max_{r \notin R_0} \frac{y_{ro} - y_r^{min} + g_{ro}^y}{g_{io}^x}\right\} =: \theta^o. \tag{13}$$

Especially, if $x_i^{max} - x_{io} \leq g_{io}^x$ for $i \notin I_0$ and $y_{ro} - y_r^{min} \leq g_{ro}^y$ for $r \notin R_0$, then $\theta_o^* \leq 2$.

Proof From the definition of θ^0 introduced in (13) it follows that

$$\theta^o \geq \frac{x_i^{max} - x_{io} + g_{io}^x}{g_{io}^x} \; \forall i \notin I_0. \quad \theta^o \geq \frac{y_{ro} - y_r^{min} + g_{ro}^y}{g_{io}^x} \forall r \notin R_0,$$

which implies that for all $\boldsymbol{\lambda} \geq 0$, $\mathbf{e}^T\boldsymbol{\lambda} = 1$ and for all $i \notin I_0, r \notin R_0$ it holds:

$$[\boldsymbol{X\lambda}]_i \leq x_i^{max} \leq x_{io} + (\theta^o - 1)g_{io}^x, \quad [\boldsymbol{Y\lambda}]_r \geq y_r^{min} \geq y_{ro} + (1 - \theta^o)g_{ro}^y.$$

Also, since we assumed feasibility, we see that there exists $\boldsymbol{\lambda}^o \geq 0$ satisfying $\mathbf{e}^T\boldsymbol{\lambda}^o = 1$ such that

$$[\boldsymbol{X\lambda}^o]_i \leq x_{io} \; \forall i \in I_0, \quad [\boldsymbol{Y\lambda}^o]_r \geq y_{ro} \; \forall r \in R_0.$$

Therefore $(\boldsymbol{\lambda}^o, \theta^o)$ is feasible and hence $\theta_o^* \leq \theta^o$. □

---

[1] By *monotonicity* we mean weak monotonicity. Strict monotonicity fails in the class of path-based models (see [8]).

4 Analysing the Super-Efficiency for Models with Different Direction Vectors

Although the data in X and Y may contain negative values, the directions must be non-negative and non-zero even though they may depend on the assessed unit $(\boldsymbol{x}_o, \boldsymbol{y}_o)$, that is, $\boldsymbol{g}_o = (\boldsymbol{g}_o^x, \boldsymbol{g}_o^y) \gneqq 0$. The directions considered in this study are listed in Table 2. For a description of the input part of the direction, we have used virtual input vectors $\boldsymbol{x}^{\min}$ and $\boldsymbol{x}^{\max}$, defined for $i = 1, \dots m$ as $x_i^{\min} = \min_j x_{ij}$ and $x_i^{\max} = \max_j x_{ij}$. The notation $\boldsymbol{y}^{\min}$ and $\boldsymbol{y}^{\max}$ is introduced analogously for the output vectors.

Next, we analyse properties of super-efficiency scores for the path-based models with directions (G1)–(G6). We will focus on the well-definedness, unit invariance, translation invariance, boundedness, and monotonicity of the super-efficiency score.

Models with $(G1)$ direction. To analyse the feasibility of $(GS)_o$ for $(\boldsymbol{x}_o, \boldsymbol{y}_o) \notin \mathcal{T}$, we observe that the conditions (4) and (5) can be rewritten as

$$x_i^* \leq 0 \; \forall i \in I_o, \; y_r^* \geq 0 \; \forall r \in R_o, \; \text{ and} \tag{14}$$

$$y_r^* > y_{ro} - |y_{ro}| = \begin{cases} 0 & y_{ro} > 0, \\ 2y_{ro} & y_{ro} < 0 \end{cases}, \; \forall r \notin R_o, \; \text{if } \operatorname{im}(\psi^y) = (0, \infty). \tag{15}$$

Now, if $(\boldsymbol{x}_o, \boldsymbol{y}_o)$ has all inputs and outputs non-zero, then the condition (14) is trivially fulfilled and so the super-efficiency models with $\operatorname{im}(\psi^y) = (-\infty, \infty)$ (e.g. DDF-(G1)) are feasible. However, if $\operatorname{im}(\psi^y) = (0, \infty)$, then condition (15) is satisfied, provided that there exists $(\boldsymbol{x}^*, \boldsymbol{y}^*) \in \mathcal{T}$, such that $\boldsymbol{y}^* > \boldsymbol{0}$. On the other hand, if some components of $(\boldsymbol{x}_o, \boldsymbol{y}_o)$ are zero, then condition (14) must be verified. There are cases where (14) is satisfied for all $(\boldsymbol{x}_o, \boldsymbol{y}_o) \notin \mathcal{T}$ or fails for some $(\boldsymbol{x}_o, \boldsymbol{y}_o)$ (see Fig. 1).

It is easy to see that the super-efficiencies of the path-based models with direction (G1) are unit but not translation invariant. The super-efficiency scores in the DDF-(G1) and HDF-(G1) models may attain values greater than 2. The model is monotone if each component of the path is an increasing function of the input-output component at each possible value of super-efficiency. A simple sufficient condition

Table 2 The directions (G1)–(G5) are commonly encountered choices of directions g_o. The direction (G6) is a new one. Here $\beta = 1$ if $\operatorname{im}(\psi^y) = (-\infty, \infty)$ (DDF model) and $\beta \geq 1$ if $\operatorname{im}(\psi^y) = (0, \infty)$ (HDF model). The references correspond to directions for non-negative data

Notation	$g_{io}^x, i \in \{1, \dots, m\}$	$g_{ro}^y, r \in \{1, \dots, s\}$	Reference
G1	$\lvert x_{io}\rvert$	$\lvert y_{ro}\rvert$	[5, 9]
G2	$\lvert x_{io} - x_i^{\min}\rvert$	$\lvert \boldsymbol{y}_r^{\max} - \boldsymbol{y}_{ro}\rvert$	[12]
G3	$x_i^{\max} - x_i^{\min}$	$y_r^{\max} - y_r^{\min}$	[12]
G4	$\frac{1}{n}\sum_j \lvert x_{ij}\rvert$	$\frac{1}{n}\sum_j \lvert x_{ij}\rvert$	[2]
G5	1	1	[5]
G6	$\max\{\boldsymbol{x}_i^{\max} - \boldsymbol{x}_{io}, 0\}$	$\max\{\beta(\boldsymbol{y}_{ro} - \boldsymbol{y}_r^{\min}), 0\}, \beta \geq 1$	

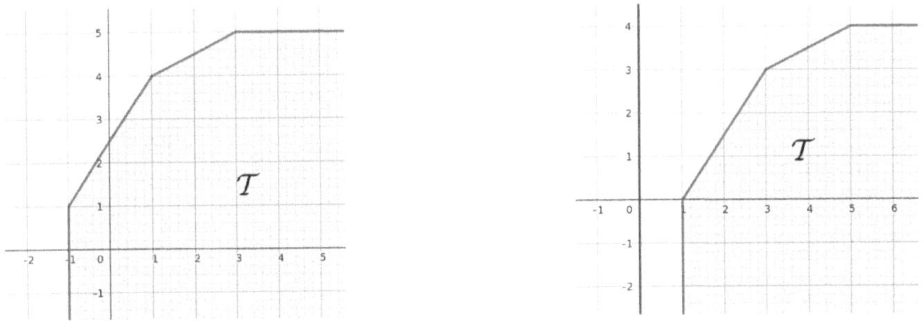

Fig. 1 Feasibility of the path based super-efficiency models with $(G1)$ directions. Left: technology set for which the super-efficiency models are feasible for all units; Right: technology set for which the models are not feasible for the units on the vertical axis

for monotonicity is if for all $\theta \leq \theta^o$ it holds

$$\psi^x(\theta) < 2 \text{ and } \psi^y(\theta) \in (0, 2). \tag{16}$$

Since the upper bound equal to 2 is not insured by the (G1) direction case, it is seen that also the fulfilment of (16) is not ensured and hence the monotonicity may fail.

Models with (*G*2) direction. We first observe that if the evaluated unit $(\boldsymbol{x}_o, \boldsymbol{y}_o) \notin \mathcal{T}$ has all inputs and outputs different from minimal and maximal (within the technologies set $\mathcal{T}$), respectively, the condition (4) is empty. Therefore, in this case, the DDF-(G2) model (and any other model with $\text{im}(\psi^y) = (-\infty, \infty)$) is well defined. If $I_0 \neq \emptyset$ and/or $R_0 \neq \emptyset$, then the condition (4) to be verified can be written as

$$x_i^* = x_i^{min} = x_{io} \; \forall i \in I_0, \; y_r^* = y_r^{max} = y_{ro} \; \forall r \in R_0.$$

In the case of HDF-(G2) model (and any other model with $\text{im}(\psi^y) = (0, \infty)$), the condition (5) for $r \notin R_0$ is equivalent to

$$y_r^* > y_{ro} - |y_r^{max} - y_{ro}| = \begin{cases} 2y_{ro} - y_r^{max} & y_{ro} < y_r^{max}, \\ y_r^{max} & y_{ro} > y_r^{max}. \end{cases} \tag{17}$$

It is seen from (17) that if $(\boldsymbol{x}_o, \boldsymbol{y}_o)$ is such that $y_{ro} > y_r^{max}$ for some r, then there does not exists $(\boldsymbol{x}^*, \boldsymbol{y}^*) \in \mathcal{T}$ such that $y_r^* > y_r^{max}$ and hence for this $(\boldsymbol{x}_o, \boldsymbol{y}_o)$ the HDF-(G2) model (and any other model with $\text{im}(\psi^y) = (0, \infty)$) is infeasible (see Fig. 2).

Evidently, the super-efficiencies of the path-based models with direction (G2) are both unit and translation invariant. The super-efficiency scores may attain values greater than 2. Similarly as in the (G1) case, the monotonicity is satisfied under condition (16).

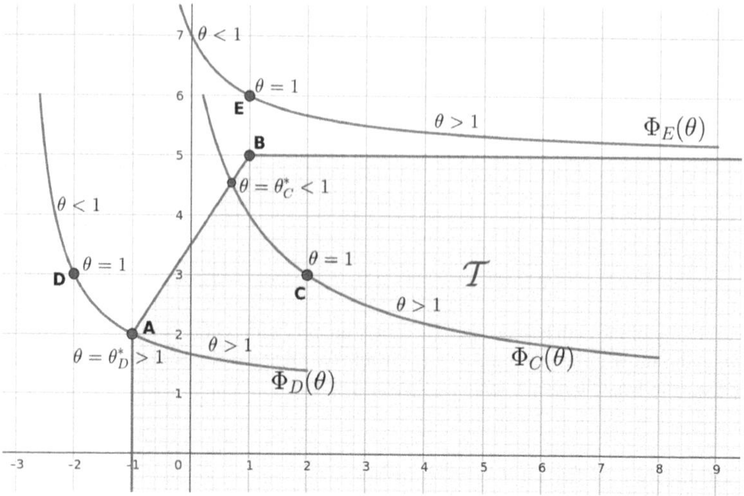

Fig. 2 Hyperbolic paths defined using (G2) directions for three units C, D, E. Unit C lies inside the technology set $\mathcal{T}$ (generated by units A, B) and the path $\Phi_C(\theta)$ determines the efficiency score θ_C^*. Units D, E lie outside the technology set $\mathcal{T}$; in this case we are interested in super-efficiency. While the path $\Phi_D(\theta)$ enters $\mathcal{T}$ for values $\theta > 1$ large enough and therefore the super-efficiency score θ_D^* is well defined, the path $\Phi_E(\theta)$ approaches its asymptote $y = 5$ and the super-efficiency score is not attained. This is because condition (17) is violated

Models with (*G*3) direction. The direction vector $(G3)$ does not depend on the evaluated unit and due to assumption (AS) the direction vector is positive. This implies that the DDF-(G3) super-efficiency model (and any other model with $\text{im}(\psi^y) = (-\infty, \infty)$) is well defined for all units. The HDF-(G3) model (and any other model with $\text{im}(\psi^y) = (0, \infty)$), is well defined, provided that there exists $(\boldsymbol{x}^*, \boldsymbol{y}^*) \in \mathcal{T}$ such that $y_r^* > y_{ro} - y_r^{max} + y_r^{min}$, for all r.

Obviously, the super-efficiencies of the path-based models with direction (G3) are both unit and translation invariant. The values of the super-efficiency score may attain values greater than 2. Since the direction vector (G3) does not depend on the evaluated unit, the super-efficiency models meet the monotonicity.

Models with (*G*4) direction. The direction vector $(G4)$ does not depend on the evaluated unit and due to assumption (AS) it is positive. Therefore, the super-efficiency model DDF-(G4) (and any other model with $\text{im}(\psi^y) = (-\infty, \infty)$) is well defined for all units. The condition (5) for the HDF-(G4) model (and any other model with $\text{im}(\psi^y) = (0, \infty)$) can also be easily verified.

It is easy to see that the super-efficiencies of the path-based models with direction (G3) are both unit and translation invariant. Even in this case, the values of the super-efficiency score may reach values greater than 2. Since the direction vector (G3) does not depend on the evaluated unit, the super-efficiency models are monotone.

Models with (*G5*) direction. The direction vector $(G5) = \mathbf{e}$ is positive, which means that the DDF-(G5) super-efficiency model (and any other model with $\mathrm{im}(\psi^y) = (-\infty, \infty)$) is well defined for all units. The HDF-(G5) model (and any other model with $\mathrm{im}(\psi^y) = (0, \infty)$) is well defined if for some $(\boldsymbol{x}^*, \boldsymbol{y}^*) \in \mathcal{T}$, it holds $y_r^* > y_{ro} - 1$, for all r.

Clearly, the super-efficiencies of the path-based models with direction (G5) are neither unit nor translation invariant. Even in this case, the values of the super-efficiency score may attain values greater than 2. Since the direction vector (G5) does not depend on the evaluated unit, the super-efficiency models are monotone.

Models with (*G6*) direction. Note that the direction (G6) is determined by the worst unit among the units generating a given technology and the unit in evaluation. It can be easily seen by setting $(\boldsymbol{x}^*, \boldsymbol{y}^*) = (\boldsymbol{x}^{\max}, \boldsymbol{y}^{\min}) \subset \mathcal{T}$ that the super-efficiency models are always feasible. Note that if $\beta > 1$, then for $r \notin R_0$ it holds

$$y_r^* = y_r^{\min} > y_{ro} - \beta(y_{ro} - y_r^{\min}),$$

and therefore (5) is satisfied.

Apparently, the super-efficiencies of the path-based models with direction (G6) are both unit and translation invariant. Contrary to the other choices of directions, the directions (G6) are designed so that the respective super-efficiencies attain values smaller than 2. This fact ensures monotonicity despite the fact that the direction vector (G6) depends on the evaluated unit.

Table 3 summarises properties of super-efficiencies discussed above for the six choices of directions with the focus on the DDF-g and HDF-g models. In empirical applications, super-efficiencies are calculated only for a finite number of units outside a specific technology set, and hence these units and the technology set may admit the fulfilment of the property even if in general the property is not satisfied. For example, the DDF-(G1) super-efficiency score is well defined, provided $(\boldsymbol{x}_o, \boldsymbol{y}_o)$ has non-zero elements and the HDF-(G1) superefficiency-score has the property if, in addition, $\mathcal{T}$

Table 3 Super-efficiency DDF and HDF models with directions (G1)–(G6) from the point of view of fulfilling properties: well definiteness (WD), unit invariance (UI), translation invariance (TI), boundedness from above by 2 (UB), (weak) monotonicity (WM)

Model property	DDF					HDF				
	WD	UI	TI	UB	WM	WD	UI	TI	UB	WM
(G1)	×*	✓	×	×*	×*	×*	✓	×	×*	×*
(G2)	×*	✓	✓	×*	×*	×*	✓	✓	×*	×*
(G3)	✓	✓	✓	×*	✓	×*	✓	✓	×*	✓
(G4)	✓	✓	✓	×*	✓	×*	✓	✓	×*	✓
(G5)	✓	×	×	×*	✓	×*	×	×	×*	✓
(G6)	✓	✓	✓	✓	✓	✓	✓	✓	✓	✓

✓—the property is satisfied for each $(\boldsymbol{x}_o, \boldsymbol{y}_o) \notin \mathcal{T}$ and any configuration of $\mathcal{T}$
×—the property does not hold in general
×*—the property need not be satisfied for each $(\boldsymbol{x}_o, \boldsymbol{y}_o) \notin \mathcal{T}$

includes points with positive output values. The DDF-(G2) super-efficiency score is well defined provided the elements of $(\boldsymbol{x}_o, \boldsymbol{y}_o)$ differ from those of $(\boldsymbol{x}^{\min}, \boldsymbol{y}^{\max})$. Also, the DDF-(G3) super-efficiency score is bounded above by 2 if $(\boldsymbol{x}_o, \boldsymbol{y}_o)$ dominates $(\boldsymbol{x}^{\min}, \boldsymbol{y}^{\max})$. These conditions are mild and may be satisfied for empirical data-sets. On the other hand, as can be seen in Table 3, the super-efficiency models with (G6) direction have all desirable properties that hold unconditionally.

5 Illustrative Numerical Examples

To illustrate our theoretical results, we have applied the DDF-g ($\psi^x(\theta) = \theta, \psi^y(\theta) = 2 - \theta$) and the HDF-g ($\psi^x(\theta) = \theta, \psi^y(\theta) = \frac{1}{\theta}$) efficiency and super-efficiency models with directions (G1)–(G6) to two data-sets.

The first one is a small data-set from [1] containing 11 DMUs, where negative data is present in both input and output variables; see the table below.

DMU	1	2	3	4	5	6	7	8	9	10	11
x	−7	−6	−5	−2	1	3	−2	−3	3	1	2
y	−6	−3	−1	2	4	5	1	−4	−2	3	−4

We have found that the first six units are efficient, which confirms the results in [1, Table 2]. For these units, we have computed the super-efficiency scores; the results are included in Table 4.

In the second example we have used the data-set of 30 DMUs containing negative data from [17]. The ranges of the particular inputs and outputs are shown in the next table.

	Inputs		Outputs	
Variable	x_1	x_2	y_1	y_2
min	418,242	1,319	−561,965	−0.2583
max	16,655,569	634,436	3,092,358	3.53714

Table 4 Super-efficiency scores for the efficient units evaluated for a small data-set form [1] using the DDF-g and HDF-g models with directions (G1)–(G6). All models, except for HDF-g with (G2) directions corresponding to DMUs with minimal input and maximal output values, are feasible

	DDF-g						HDF-g					
DMU_o	(G1)	(G2)	(G3)	(G4)	(G5)	(G6)	(G1)	(G2)	(G3)	(G4)	(G5)	(G6)
1	1.286	3.000	1.222	1.629	1.222	1.222	2.155	NaN	1.836	3.282	3.402	1.100
2	1.111	1.146	1.052	1.169	1.052	1.052	2.147	5.657	1.736	2.904	2.991	1.020
3	1.167	1.125	1.048	1.157	1.048	1.048	2.153	3.166	1.659	2.588	2.639	1.051
4	1.214	1.111	1.041	1.135	1.041	1.041	2.415	1.786	1.581	2.014	2.017	1.042
5	1.250	2.000	1.091	1.314	1.091	1.091	2.406	1.498	1.772	1.947	2.010	1.018
6	1.200	2.000	1.100	1.314	1.100	1.100	2.783	NaN	2.007	3.193	2.783	1.100

Table 5 Super-efficiency scores for the efficient units evaluated for dataset from [16] using the DDF-g and HDF-g models with directions (G1)–(G6). The models corresponding to DMUs with maximal output values are infeasible for HDF-g models with directions (G2), (G4) and (G5)

	DMU_o		(G1)	(G2)	(G3)	(G4)	(G5)	(G6)
DDF-g	2	$y_{1o} = y_1^{max}$	1.57	2	1.932	5.544	1762900.012	1.585
	4		1.011	1.007	1.003	1.017	15732.892	1.006
	5	$x_{2o} = x_2^{min}$	3.403	2	1.012	1.07	6317.092	1.012
	12		1.341	1.09	1.004	1.025	2877.955	1.004
	14	$y_{2o} = y_2^{max}$	1.739	3.129	1.374	4.509	4.063	1.293
	19		1.283	1.121	1.021	1.113	186639.681	1.037
	22		1.075	1.047	1.02	1.137	31030.253	1.043
	23		1.322	1.161	1.044	1.3	48996.725	1.075
	26	$x_{1o} = x_1^{min}$	1.477	2	1.012	1.057	141814.325	1.013
HDF-g	2	$y_{1o} = y_1^{max}$	2.326	NaN	14.716	NaN	NaN	1.932
	4		1.011	1.007	1.003	1.017	109428.4	1.005
	5	$x_{2o} = x_2^{min}$	3.951	2	1.012	1.07	6317.092	1.012
	12		1.358	1.097	1.004	1.025	2904.732	1.004
	14	$y_{2o} = y_2^{max}$	2.653	NaN	1.598	NaN	NaN	1.374
	19		1.326	1.136	1.021	1.117	649661.173	1.032
	22		1.079	1.049	1.02	1.146	52536.014	1.036
	23		1.455	1.185	1.045	1.339	58337.288	1.066
	26	$x_{1o} = x_1^{min}$	1.484	2	1.012	1.057	174729.93	1.013

It was shown (see also the results in [8, Table B.6]) that the efficient units are those with numbers 2, 4, 5, 12, 14, 19, 22, 23, 26. For these nine efficient units, we provide the super-efficiency scores in Table 5.

The numerical results are in accordance with our theoretical findings. The super-efficiency scores are well defined and lie in the interval $(1, 2)$ only for models with direction (G6) and DDF-(G3). While the boundedness in DDF-(G3) is just a consequence of the specific data configuration, the good properties of models with (G6) directions follow from the reasonable construction of the direction vector (G6). The high super-efficiency values in Table 5 in the direction vector (G5) are caused by the inappropriate size of the direction vector equal to 1 in all components compared to the size of the individual inputs and outputs.

6 Conclusion

The findings of this contribution demonstrate that the path-based models and their generalisations to arbitrary data studied in [8] can be used directly for the assessment of super-efficiency. Five desirable properties of path-based super-efficiency models, the validity of which depends on the selection of the direction vector, have been investigated. The study provides the necessary and sufficient conditions for the well-definedness of the models and allows for identification of the problematic cases. It was shown that for the traditional directions (G1)–(G5), the super-efficiency score may not be well defined, bounded above (by 2), or monotone for each unit. Additionally, a new direction (G6) has been suggested that has all desirable properties and hence is suitable for the super-efficiency measurement in the class of path-based models over arbitrary data.

We hope that the outcomes presented here will be of value to researchers and practitioners interested in the super-efficiency measurement to navigate among different path-based models over arbitrary data.

Acknowledgements We thank two anonymous reviewers for their comments. The research was supported by APVV-20-0311 and VEGA 1/0611/21 grants.

References

1. Allahyar, M., & Rostamy-Malkhalifeh, M. (2015). Negative data in data envelopment analysis: Efficiency analysis and estimating returns to scale. *Computers & Industrial Engineering, 82*, 78–81.
2. Aparicio, J., Pastor, J. T., & Ray, S. C. (2013). An overall measure of technical inefficiency at the firm and at the industry level: The 'lost profit on outlay'. *European Journal of Operational Research, 226*(1), 154–162.
3. Banker, R. D., Charnes, A., & Cooper, W. W. (1984). Some models for estimating technical and scale inefficiencies in data envelopment analysis. *Management Science, 30*(9), 1078–1092.
4. Chambers, R. G., Chung, Y., & Färe, R. (1996). Benefit and distance functions. *Journal of Economic Theory, 70*(2), 407–419.
5. Chambers, R. G., Färe, R., & Grosskopf, S. (1996). Productivity growth in APEC countries. *Pacific Economic Review, 1*(3), 181–190.
6. Chavas, J.-P., & Cox, T. L. (1999). A generalized distance function and the analysis of production efficiency. *Southern Economic Journal, 66*(2), 294–318.
7. Färe, R., Grosskopf, S., & Lovell, C. A. K. (1985). *The Measurement of Efficiency of Production*. Netherlands: Springer.
8. Halická, M., Trnovská, M., & Černý, A. (2023). A unified approach to radial, hyperbolic, and directional efficiency measurement in data envelopment analysis. *European Journal of Operational Research*.
9. Kerstens, K., & Van de Woestyne, I. (2011). Negative data in DEA: A simple proportional distance function approach. *Journal of the Operational Research Society, 62*(7), 1413–1419.
10. Lin, R., & Chen, Z. (2017). A directional distance based super-efficiency DEA model handling negative data. *Journal of the operational Research Society, 68*, 1312–1322.
11. Lin, R., & Liu, Y. (2019). Super-efficiency based on the directional distance function in the presence of negative data. *Omega, 85*, 26–34.
12. Portela, M. C. A. S., Thanassoulis, E., & Simpson, G. (2004). Negative data in DEA: A directional distance approach applied to bank branches. *Journal of the Operational Research Society, 55*(10), 1111–1121. Oct.
13. Tavana, M., Izadikhah, M., Toloo, M., & Roostaee, R. (2021). A new non-radial directional distance model for data envelopment analysis problems with negative and flexible measures. *Omega, 102*, 102355.

14. Toloo, M., & Mensah, E. K. (2019). Robust optimization with nonnegative decision variables: A DEA approach. *Computers & Industrial Engineering, 127*, 313–325.
15. Toloo, M., Zandi, A., & Emrouznejad, A. (2015). Evaluation efficiency of large-scale data set with negative data: An artificial neural network approach. *The Journal of Supercomputing, 71*, 2397–2411.
16. Tone, K. (2001). A slacks-based measure of efficiency in data envelopment analysis. *European Journal of Operational Research, 130*(3), 498–509.
17. Tone, K., Chang, T.-S., & Wu, C.-H. (2020). Handling negative data in slacks-based measure data envelopment analysis models. *European Journal of Operational Research, 282*(3), 926–935.
18. Tone, K., Toloo, M., & Izadikhah, M. (2020). A modified slacks-based measure of efficiency in data envelopment analysis. *European Journal of Operational Research, 287*(2), 560–571.

Fuzzy Utility Mining on DEA

Neda Sadeghi, Mohsen Rostamy-Malkhalifeh, and Mohammad Hassan Behzadi

Abstract Data Envelopment Analysis is used as an efficient tool in determining the productivity of organizations as a non-parametric method. In today's world, where competition in business and trade has become a fundamental implication, and any organization or brand that does not act intelligently in this competition is forced to accept failure, the concept of <productivity> has taken a different position. In the studies carried out in the evaluation of organizational productivity, usually the parts of the organization are investigated, which are located in the departments of the organizational queue, and less attention has been paid to the headquarters units of the organization in the evaluation of their productivity. Also, in most of the studies conducted in data envelopment analysis using fuzzy data, only the membership function has been used for quantification. But in some studies, it is necessary to use internal and external utility in addition to the membership function. In this article, the output indicators of the model are used by considering the internal utility as the utility that DMU has in the organization and the external utility as the value that DMU has for the organization. Then, by valuing the membership function, the output values for each headquarters have been obtained, and finally, by using the classic analysis of the BCC model, the productivity has been evaluated and the DMUs have been compared with each other.

Keywords Data envelopment analysis · Internal utility · External utility

N. Sadeghi (✉) · M. Rostamy-Malkhalifeh · M. H. Behzadi
Department of Statistics, Science and Research Branch, Islamic Azad University, Tehran, Iran
e-mail: neda.sadeghi@srbiau.ac.ir

M. Rostamy-Malkhalifeh
e-mail: rostamy@srbiau.ac.ir

M. H. Behzadi
e-mail: behzadi@srbiau.ac.ir

A. Emrouznejad et al. (eds.), *Advances in the Theory and Applications of Performance Measurement and Management*, Lecture Notes in Operations Research,
https://doi.org/10.1007/978-3-031-61597-9_10

1 Introduction

1.1 Importance of Study

Data envelopment analysis is a method to measure the relative productivity of DMUs and evaluate their performance. DMUs can be banks, companies, universities, hospitals, departments, or any other type of institution or DMU that operates in a process of producing or providing services.

The DEA purpose is to examine the DMUs based on their inputs and outputs and compare their productivity. This comparison is based on the concept of envelopment efficiency, where DMUs are categorized as "efficient" or "inefficient".

Using DEA, we can answer important questions, such as:

- Which DMUs are productivity best and which DMUs can be improved?
- How much inputs and outputs should be changed for the DMUs to achieve higher productivity?
- Which DMU is the best in productivity and can be considered as an example for other DMUs?

The concept of envelopment efficiency in DEA is calculated based on linear transformations from inputs to outputs, and the results of this analysis show how much of each DMU's production potential has been produced as output. This method can be used in many areas and industries including management, economy, health, production and services.

Using fuzzy data in data envelopment analysis is one of the important approaches in this field. In conventional DEA, evaluation of DMU efficiency is done based on accurate data and accurate values are considered for inputs and outputs. But by using fuzzy data, this approach is improved and can better cope with the uncertainties and ambiguities in the data.

Fuzzy data in DEA, instead of having exact values for inputs and outputs, have fuzzy values that show the membership of each input or output in the period of time. In other words, for each input or output, the fuzzy value indicates that input or output may occur with different membership values.

The use of fuzzy data in DEA causes:

Improving the accuracy of the model: by considering the uncertainties and ambiguities in the data, the DEA model becomes more accurate and provides better results.

More flexibility: the use of fuzzy data allows us to check the performance of DMUs with different fuzzy values for inputs and outputs and to evaluate the DMU in several different modes.

Analysis of non-linear behavior: If the behavior of the DMUs is non-linear, the use of fuzzy data in DEA allows us to better deal with non-linear changes in the data.

Through the use of fuzzy data in DEA, we can achieve a more accurate and reliable analysis of the DMUs' performance and a better understanding of their efficiency, considering the uncertainties and ambiguities in the data. This approach is very useful in cases where the data is incomplete or associated with many uncertainties.

1.2 The History of Using Fuzzy Data in DEA

An overview of the use of fuzzy data in data envelopment analysis according to the Table 1.

Table 1 The history of using fuzzy data in DEA

Title	Year	Explanations
The evaluation of productive efficiency using a fuzzy mathematical programming approach: the case of the newspaper preprint insertion process	1999	Using multiple mathematical programming formulas to overcome excessive measurement inaccuracies
Fuzzy DEA: a perceptual evaluation method	2001	Using fuzzy numbers for fluctuating data and reflecting some general feeling or experience of experts
A multiple criteria decision-making model based on fuzzy multiple objective DEA	2006	Using fuzzy multi-objective data envelopment analysis model and determining a set of joint weights so that the efficiency of all DMUs is maximized simultaneously by maximizing the fuzzy degree of all returns
Fuzzy assessment of performance of a decision-making unit using DEA	2010	Development of new model under uncertainty using fuzzy DEA for evaluation of DMUs
The analysis of bank business performance and market risk—applying fuzzy DEA	2013	Using the expanding fuzzy Slack-Based Measurement (fuzzy SBM) to manage risk and its uncertainty
Dynamic inverse DEA in the presence of fuzzy data	2014	The use of dynamic DEA, which is a technique for calculating the relative efficiency of a set of decision-making units (DMUs) in the presence of the time factor and the existence of intertemporal dependence
A cross-efficiency fuzzy data envelopment analysis technique for performance evaluation of decision-making units under uncertainty	2015	Using a new mutual efficiency fuzzy data envelopment analysis (DEA) technique under uncertainty that uses triangular fuzzy numbers

(continued)

Table 1 (continued)

Title	Year	Explanations
A fuzzy data envelopment analysis model for evaluating the efficiency of socially responsible and conventional mutual funds	2016	Using a fuzzy DEA model that allows evaluating the relative performance of mutual funds in a more realistic and flexible way and focuses on non-financial criteria
Development of intuitionistic fuzzy super-efficiency slack based measure with an application to health sector	2018	Using Slack based measure (SBM) model performance efficiency that deals with input surplus and output shortage and are consistent with real-world applications and in the form of fuzzy/ intuitive number (IF)
Estimating most productive scale size in DEA under uncertainty	2019	Using the Most Productive Scale Size Estimation (MPSS) by presenting a new approach in the presence of uncertain data whose upper and lower bounds are known only
A ranking system based on inverse data envelopment analysis	2020	Introducing a new ranking system using inverse (DEA) to estimate the expected levels of input/ output changes when the efficiency score is unchanged
Undesirable factors in stochastic DEA cross-efficiency evaluation: an application to thermal power plant energy efficiency	2021	Using input-oriented data envelopment analysis (DEA) model with undesirable outputs and introducing a new stochastic model called the expected ranking criterion, for which statistical techniques are taken in males

2 Article Technique

In many of the articles on data development analysis and the use of fuzzy data, only the values from the membership function are used, but in some analyzes, it is necessary to look for internal and external utility as a useful element in addition to membership. This means that if in obtaining the output values obtained in the organization, the internal utility is used as the amount of utility that the DMU involved in the organization has and also the external utility as the value that the DMU involved has for the organization, and then on Based on the obtained values and including it in the fuzzy membership function, the obtained fuzzy values can be used to measure the effectiveness and efficiency of all aspects, and finally, the performance evaluation can be done with high accuracy. The reason for using fuzzy logic is that the total utilities obtained from the organization's point of view in different numerical ranges can have different values, so the membership function has been used to show this issue.

3 Practical Example

In data envelopment analysis studies, the DMUs involved in the departments of the organization are usually studied, which are located in the queue department of the organization. The evaluation of the performance of the headquarters of an organization is always associated with complications, which in this article will be done using the mentioned technique.

As a practical example, active staffs on headquarters have been studied in a bank, which is in accordance with the following Table 2.

Table 2 Headquarters DMUs

DMUs	Headquarters DMUs
DMU1	Credits
DMU2	The international
DMU3	Development of systems and technology
DMU4	Branches and support
DMU5	Connections
DMU6	Information security
DMU7	Inspection matters
DMU8	Branch affairs
DMU9	Financial department
DMU10	Digital banking
DMU11	Investment banking
DMU12	Logistics and support
DMU13	International support
DMU14	Funding
DMU15	Economic research
DMU16	Analysis
DMU17	Business development
DMU18	Audit and internal controls
DMU19	Legal
DMU20	Electronic services and development of banking products
DMU21	Modern services
DMU22	Risk
DMU23	Organization and methods
DMU24	Human capital
DMU25	Peripheral systems
DMU26	The operation

(continued)

Table 2 (continued)

DMUs	Headquarters DMUs
DMU27	Information technology
DMU28	Reports, currency and stock accounting
DMU29	Anti-money laundering and law enforcement
DMU30	Currency transactions
DMU31	Business intelligence
DMU32	Customer contact center
DMU33	Central secretariat

To evaluate the performance of the active headquarters of a bank, five projects referred to each headquarters are considered.

The evaluated indicators for the input variables are as follows:

Workforce: the number of personnel in each headquarters
Fixed costs: Fixed costs in the time period studied
Technology and equipment: which is being used by every headquarters
Variety of services: available knowledge and services that each headquarters can provide
Headquarters capital: the current capital of each headquarters

Table 3 shows the input simulated data.

Also, the following items have been considered as effective indicators in the output:

- The duration of the project.
- The total cost of each project for the headquarters.
- Labor productivity.
- Penalty/correction rate.
- The role of the project in upgrading the headquarters.

Based on the technique of the article that was explained first, the work process is that for each output index, internal and external utility values are assigned to the desired headquarters based on organizational data. Then, based on the overall utility obtained for each index, fuzzy values have been calculated according to the membership function. Then, the general fuzzy utility for each project and in total for all five projects are considered as the output of the headquarters. The total fuzzy utility number of all five files is included as an output in data envelopment analysis.

The internal utility of each item is based on the internal desirability of the organization as follows (Table 4).

The external utility of each item is based on the organizational utility as follows (Table 5).

And the membership function used is based on the study and organizational utility according to the following function (Fig. 1).

Based on the mentioned cases, the overall fuzzy utility table of the output data based on the five investigated projects is given in Table 6.

Table 3 The input data

DMUs	Headquarters DMUs	Workforce	Fixed costs$	Technology and equipment$	Variety of services	Headquarters capital$
DMU1	Credits	9	10,679	5,760	4	60,266
DMU2	The international	16	14,351	14,028	8	65,884
DMU3	Development of systems and technology	6	5,347	15,763	4	41,932
DMU4	Branches and support	20	10,361	17,952	10	49,825
DMU5	Connections	5	6,092	19,038	2	51,389
DMU6	Information security	9	18,011	11,291	10	26,328
DMU7	Inspection matters	5	7,779	11,929	9	42,934
DMU8	Branch affairs	10	7,339	13,048	9	57,960
DMU9	Financial department	15	7,677	6,958	2	27,065
DMU10	Digital banking	17	15,188	12,229	10	28,211
DMU11	Investment banking	13	11,967	6,798	10	54,647
DMU12	Logistics and support	10	5,830	14,479	7	64,932
DMU13	International support	6	12,627	14,445	3	28,659
DMU14	Funding	5	7,789	9,632	2	54,507
DMU15	Economic research	10	7,383	11,949	1	66,155
DMU16	Analysis	12	17,429	16,162	5	27,320
DMU17	Business development	7	13,145	12,373	3	30,329
DMU18	Audit and internal controls	4	17,025	13,356	1	27,315
DMU19	Legal	5	12,399	11,570	6	27,382
DMU20	Electronic services and development of banking products	8	11,159	18,804	8	53,157

(continued)

Table 3 (continued)

DMUs	Headquarters DMUs	Workforce	Fixed costs$	Technology and equipment$	Variety of services	Headquarters capital$
DMU21	Modern services	11	11,270	15,124	8	61,650
DMU22	Risk	9	17,447	11,725	8	51,230
DMU23	Organization and methods	9	7,832	6,025	4	44,080
DMU24	Human capital	11	11,889	7,140	10	61,921
DMU25	Peripheral systems	5	19,296	12,208	8	55,743
DMU26	The operation	17	10,123	7,128	3	28,634
DMU27	Information technology	5	17,712	18,202	3	32,749
DMU28	Reports, currency and stock accounting	12	5,019	7,057	3	50,295
DMU29	Anti-money laundering and law enforcement	6	11,971	13,783	9	67,459
DMU30	Currency transactions	5	10,850	16,326	2	68,270
DMU31	Business intelligence	4	16,935	14,050	2	44,540
DMU32	Customer contact center	15	17,564	8,582	4	40,345
DMU33	Central secretariat	8	6,741	14,994	9	69,720

In the following, classic DEA methods, BCC model and R software have been used to evaluate the productivity of active employees in the bank, the results of which are shown in the diagram below (Fig. 2).

Table 4 The internal utility

Number assigned for internal utility	The duration of the project	The total cost of each project for the headquarters	Labor productivity (%)	Penalty/ correction rate	The role of the project in upgrading the headquarters (%)
10	In the estimated time for the project	In the estimated cost for the project	100–90	Without penalty	100–90
9	With 5 h delaying	$100 difference	90–80	Less than 1%	90–80
8	With 10 h delaying	$200 difference	80–70	1–2%	80–70
7	With 15 h delaying	$300 difference	70–60	2–3%	70–60
6	With 20 h delaying	$400 difference	60–50	3–4%	60–50
5	With 25 h delaying	$500 difference	50–40	4–5%	50–40
4	With 30 h delaying	$600 difference	40–30	5–6%	40–30
3	With 35 h delaying	$700 difference	30–20	6–7%	30–20
2	With 40 h delaying	$800 difference	20–10	7–8%	20–10
1	Over 45 h delaying	Over $800 difference	10–0	More than 8%	10–0

Table 5 The external utility

Items	The duration of the project	The total cost of each project for the headquarters	Labor productivity	Penalty/ correction rate	The role of the project in upgrading the headquarters
External utility	4$	7$	5$	2$	3$

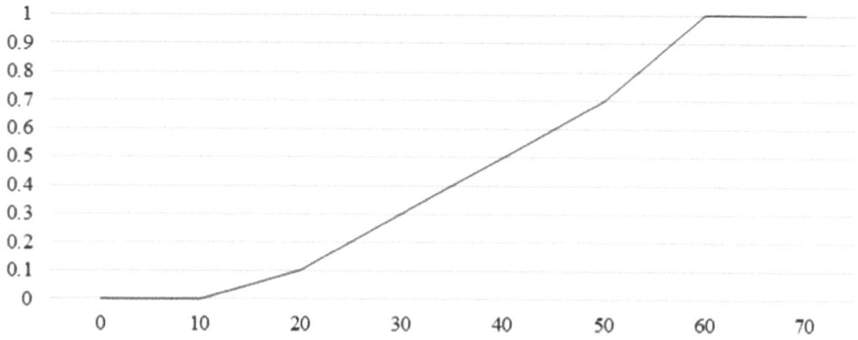

Fig. 1 The membership function

4 Discussion

By using the article technique and paying attention to the internal and external advantages of the organization, in addition to increasing the accuracy, we were able to include the important indicators and items from the point of view of the organization in the data and evaluate the performance for each organization based on the effective indicators of the same organization. Get unique for the same organization. In the following and for further studies, it is possible to achieve more achievements in this field by using this technique in big data and applying the relevant algorithms.

Table 6 The total utility

DMUs	Headquarters units	Total utility
DMU1	Credits	5.63
DMU2	The international	2.94
DMU3	Development of systems and technology	2.85
DMU4	Branches and support	3.89
DMU5	Connections	4.09
DMU6	Information security	3.21
DMU7	Inspection matters	4.77
DMU8	Branch affairs	3.38
DMU9	Financial department	4.19
DMU10	Digital banking	3.84
DMU11	Investment banking	5.17
DMU12	Logistics and support	5
DMU13	International support	3.85
DMU14	Funding	3.21
DMU15	Economic research	3.25
DMU16	Analysis	4.36
DMU17	Business development	2.74
DMU18	Audit and internal controls	6.2
DMU19	Legal	4.65
DMU20	Electronic services and development of banking products	4.09
DMU21	Modern services	2.93
DMU22	Risk	2.65
DMU23	Organization and methods	4.08
DMU24	Human capital	3.38
DMU25	Peripheral systems	6.12
DMU26	The operation	4.56
DMU27	Information Technology	3.07
DMU28	Reports, currency and stock accounting	5.19
DMU29	Anti-money laundering and law enforcement	3.34
DMU30	Currency transactions	5.45
DMU31	Business intelligence	4.57
DMU32	Customer contact center	3.9
DMU33	Central secretariat	2.88

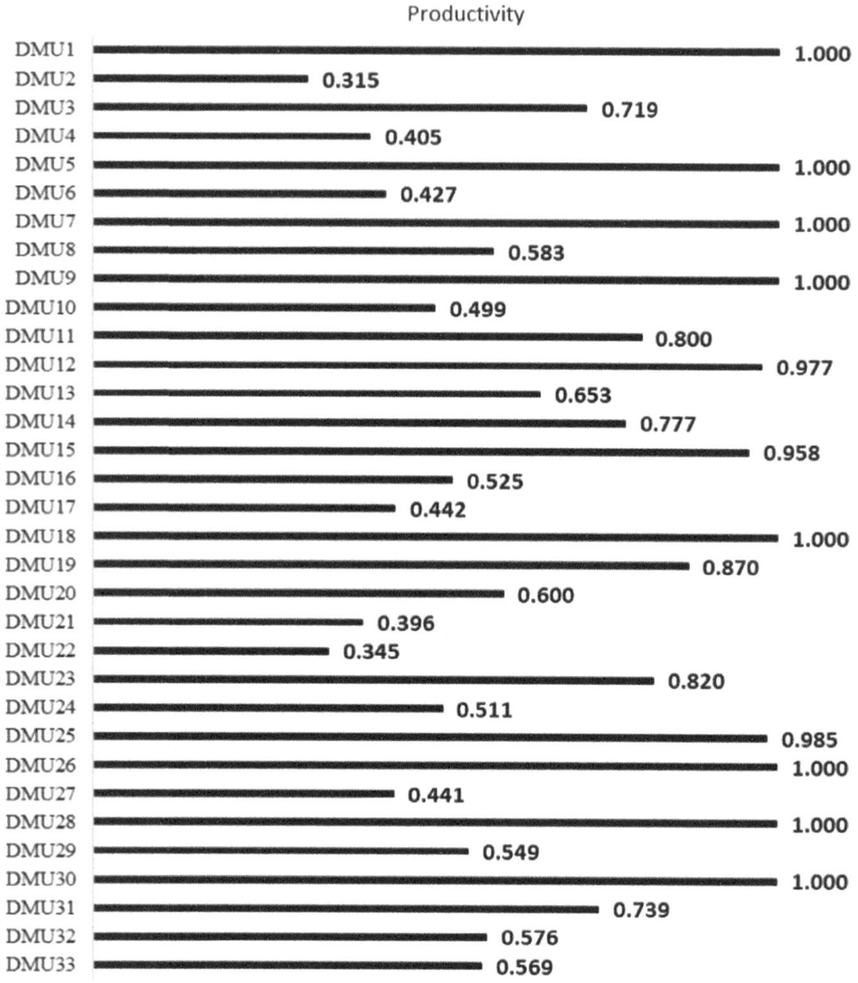

Fig. 2 Productivity chart

Bibliography

1. Girod, O. A., & Triantis, K. P. (1999). The evaluation of productive efficiency using a fuzzy mathematical programming approach: the case of the newspaper preprint insertion process. *IEEE Transactions on Engineering Management, 46*(4), 429–443.
2. Guo, P., & Tanaka, H. (2001). Fuzzy DEA: A perceptual evaluation method. *Fuzzy Sets and Systems, 119*(1), 149–160.
3. Lee, H. S., & Yeh, C. H. (2006). A multiple criteria decision making model based on fuzzy multiple objective DEA. In B. Gabrys, R. J. Howlett, & L. C. Jain (Eds.), *Knowledge-based intelligent information and engineering systems, Pt 2, proceedings* (pp. 917–921). Springer.

4. Angiz, L. M. Z., Emrouznejad, A., & Mustafa, A. (2010). Fuzzy assessment of performance of a decision making units using DEA: A non-radial approach. *Expert Systems with Applications, 37*(7), 5153–5157.
5. Chen, Y.-C., et al. (2013). The analysis of bank business performance and market risk—applying Fuzzy DEA. *Economic Modelling, 32*, 225–232.
6. Ghobadi, S., Jahanshahloo, G. R., Lotfi, F. H., & Rostamy-Malkhalifeh, M. (2014). Dynamic inverse DEA in the presence of fuzzy data. *Advances in Environmental Biology, 8*(24), 139–151.
7. Dotoli, M., Epicoco, N., Falagario, M., & Sciancalepore, F. (2015). A cross-efficiency fuzzy data envelopment analysis technique for performance evaluation of decision making units under uncertainty. *Computers & Industrial Engineering, 79*, 103–114.
8. Baeza-Sampere, I., Coll-Serrano, V., M'Zali, B., & Mendez-Rodriguez, P. (2016). A fuzzy data envelopment analysis model for evaluating the efficiency of socially responsible and conventional mutual funds. *Journal of Risk, 19*(1), 77–90.
9. Arya, A., & Yadav, S. P. (2018). Development of intuitionistic fuzzy super-efficiency slack based measure with an application to health sector. *Computers & Industrial Engineering, 115*, 368–380.
10. Peykani, P., Seyed Esmaeili, F. S., Hosseinzadeh Lotfi, F., & Rostamy-Malkhalifehm M. (2019). Estimating most productive scale size in DEA under uncertainty—The 11th Conference on Data Envelopment Analysis
11. Soleimani-Chamkhorami, K., Hosseinzadeh Lotfi, F., Jahanshahloo, G., & Rostamy-Malkhalifeh, M (2020). A ranking system based on inverse data envelopment analysis. *IMA Journal of Management Mathematics, 31*(3), 367–385.

A Generalized DEA Approach to Performance Assessment: Case Study of Northern Adriatic Ports

Mozhgan Mansouri Kaleibar and Evelin Krmac

Abstract Ports are an important interface between inland and maritime transport. They also play an important role in increasing trade and economic growth of the country. Therefore, improving the efficiency of ports is crucial as it can boost a country's competitiveness and economy. It also improves a country's access to global markets, leading to growth in trade and consequently higher incomes. Improving the efficiency of ports and the maritime transport system is important to increase the EU's competitiveness. Cargo volumes in the Adriatic ports of Central and Eastern European countries have increased in recent years. For this reason, the evaluation of ports in the Adriatic region is of particular importance. Current methodologies ignore the assessment in case of uncertainties. For this reason, the efficiency scores are not accurate enough to be used for improving port performance and, in general, they are not useful for port users. Therefore, the main objective of this paper is to assess and measure the efficiency of the studied ports over the last two years. In this way, efficient and inefficient ports as well as the causes of inefficiency of each port will be identified. The evaluation is based on the ports' operational capacity, which includes general cargo, vehicles, liquid bulk, containers and dry bulk. To assess the performance, this paper develops a model that takes into account undesirable inputs and outputs as well as uncertainties and estimates performance accordingly. In this paper, generalized non-classical DEA models are used. Therefore, the generalized fuzzy SBM model is used to evaluate the performance of the ports. The results of this project will help to improve the performance and efficiency of the studied ports. Policy makers can take the assessment results into account when deciding how to improve the European transport network and trade competitiveness.

Keywords Adriatic Sea ports · Fuzzy DEA · Generalized Data envelopment analysis (DEA) · Port efficiency · Slacks based measure model (SBM)

M. M. Kaleibar · E. Krmac (✉)
Faculty of Maritime Studies and Transport, University of Ljubljana, Pot pomorščakov 4, 6320 Portorož, Slovenia
e-mail: evelin.krmac@fpp.uni-lj.si

M. M. Kaleibar
e-mail: Mozhgan.Mansouri.Kaleibar@fpp.uni-lj.si

A. Emrouznejad et al. (eds.), *Advances in the Theory and Applications of Performance Measurement and Management*, Lecture Notes in Operations Research,
https://doi.org/10.1007/978-3-031-61597-9_11

1 Introduction

Ports are an important interface between inland and maritime transport. They also play an important role in increasing trade and economic growth of the country. Therefore, improving the efficiency of ports is crucial as it can boost a country's competitiveness and economy. As ports serve as engines of growth both locally and internationally, assessing their efficiency is of great importance to maritime stakeholders. Low port efficiency increases transport costs and reduces commercial growth and port competitiveness, thereby reducing revenues. Optimising port efficiency improves a country's access to global markets, leading to growth in trade and consequently higher incomes. As outlined in the EU Integrated Maritime Policy [1], improving of the efficiency of ports and the maritime transport system is important for increasing the EU's competitiveness.

The Adriatic ports were selected for the assessment because of their special importance for maritime traffic. The Adriatic ports can be considered as multi-purpose ports. These ports are also transit ports, usually serving more than 70% foreign hinterland markets and less than 30% domestic markets [2]. These ports are gradually gaining importance for the development of the economies in Central and Western Europe. The volume of cargo traffic in the Adriatic ports of Central and Eastern European countries increased from 36 to 48 million tonnes from 2015 to 2019 [3]. Given these conditions and the progress made by maritime transport in the region in recent years, port activities need to be maximised in order to achieve the best results for the regional and European economies. Therefore, solutions must be found to improve efficiency, which in turn will lead to a more efficient use of resources and ultimately to the development of ports. For this reason, the assessment of ports in the Adriatic region is of particular importance. Therefore, this study was conducted with the aim of evaluating the performance of the ports of the Northern Adriatic in years 2021 and 2022.

Data envelopment analysis is the best method for evaluating unit efficiency because it is a non-parametric method that takes into account most performance criteria and includes multiple inputs and outputs [4]. In order to conduct a successful performance evaluation, accurate data must be available. In most of the methods for evaluation of decision-making units (DMUs) with uncertain data that have been presented so far, the α-cut method has been used [5]. In these methods, the Sample DMUs [6] cannot be evaluated. Target DMUs in traditional models are limited to internal units that cannot evaluate SDMUs. As well, in such methods, several linear planning problems must be solved to estimate the membership function and only then the evaluation of the performance of a DMU can be done [7]. In this paper, the generalised DEA methods are used for uncertain data coverage, and includes models that can evaluate SDMUs [8]. The available data used in ports are sometimes inaccurate or vague. Due to the ambiguity of the data associated with ports, uncertainty must be considered and included in the models if we want to obtain realistic and useful estimates. This paper presents a new model that combines these two methods, fuzzy and non-classical generalised DEA. This approach therefore proposes a simple and accurate framework for assessing the efficiency of ports. The results of this paper will help to improve the performance and efficiency of the ports studied.

The paper is structured as follows. Section 2 provides a general overview of the literature on previous studies. The generalized method DEA is presented in Sect. 3. In Sect. 4, the proposed framework is applied to the efficiency of the Northern Adriatic ports in an empirical case. Finally, the conclusion is drawn in Sect. 5.

2 Literature Review

Review of the literature shows that some methods of efficiency analysis, including DEA, have been already used for the evaluation of ports. The first article on port efficiency, published in 1993, used DEA by Roll and Hayuth to evaluate Israeli ports. They showed that DEA efficiency assessments can be a useful tool for port managers and researchers, providing a deeper insight into port performance [9]. Six years later, Ibáñez et al. used DEA to evaluate all Spanish ports. From 1999 to 2009, most articles dealt with classical methods for assessing the efficiency of ports around the world [10]. Since then, many articles have also been written on measuring port performance using the DEA method have also been conducted. After that, other DEA models were used by the authors for the assessment. Since 2010, the number of articles on port efficiency and methods DEA has increased significantly. Hung et al. investigated the operational efficiency; scale efficiency targets and variability of DEA efficiency estimates of 31 Asian container ports [11]. Wu and Joh compared port operating efficiency in emerging markets (BRIC and the Next-11) with more advanced markets (G7) using CCR, BCC and Ap models [12]. Cullinane and Wang evaluated the available DEA panel data approaches by applying them to a sample of 25 leading container ports. The empirical results confirm the use of panel data and show that there is significant waste in container port production [13]. De Oliveira and Cariou assessed the efficiency of 122 iron ore and coal ports. The results aggregated at the country level show that national efficiency can be achieved either by a limited number of large ports or by combining smaller ports with complementary characteristics—national network effect [14]. Choi examined both traditional efficiency and environmental efficiency of the 20 largest container ports in the United States using DEA [15]. Niavis and Tsekeris studied the benchmarking, measurement and identification of the key determinants of technical efficiency of container ports in the Southeast Europe (SEE) region, including Italian ports [16]. Schøyen and Odeck assessed the technical efficiency of Norwegian container ports against a frontier consisting of the best performing ports and other comparable Nordic and UK ports [17]. Bray et al. evaluated the efficiency of transport systems under uncertainty using fuzzy sets theory DEA [18]. In their study, Lu and Park gain a variety of analytical results on the operational efficiency of the world's leading container ports by applying three types of DEA models [19]. Wanke and Barros evaluated the impact of public–private partnerships on large Brazilian public ports using 2 levels

DEA [20]. Morales-Fusco et al. studied the efficiency of Mediterranean ports using classical DEA models [21].

Since 2017, there has been an increase in the number of articles and, of course, in the number of articles assessing environmental performance. Schøyen and Odeck measured productivity changes in the six largest Norwegian container ports compared to 14 similar small and medium-sized ports in the Nordic countries and the UK [22]. Chang and Tovar measured the productivity of port terminals in Peru and Chile and evaluated the influence of certain explanatory variables that could explain the productivity differences [23]. Chang et al. evaluated the efficiency of Chinese dry ports using a two-stage approach based on panel data collected from eight dry ports of the Port of Ningbo, China [24]. Schøyen et al. studied container ports in six countries: Denmark, Finland, Iceland, Norway, Sweden and the United Kingdom [25]. Tovar and Wall estimated environmental efficiency for a cross-section of 28 Spanish port authorities [26]. Quintano et al. evaluate the efficiency of ten of Europe's leading ports using the BCC model [27]. Quintano et al. provided a comparative analysis of the eco-efficiency of twenty-four leading container ports in Europe [28]. Wang et al. evaluated the environmental efficiency of ports under conditions of environmental control, non-environmental control and particulate matter (PM) emissions through inter-port cooperation [29]. Quintano et al. evaluated the performance of 24 European ports in terms of eco-efficiency [30]. Kong and Liu constructed an evaluation framework to measure the sustainability of port cities by combining the Coupling Coordination Degree (CCD) model and the two-stage SBM-DDF model [31]. Nong studied the performance efficiency of Vietnam's ports using the Delphi technique and the DEA model. The theoretical contribution of this study was to propose an integrated technique (Delphi, KAMET and DEA) to deal with a variety of inputs and outputs and to evaluate the efficiency of ports [32].

In summary, there are various methods of performance evaluation, with DEA being the most preferred method by evaluators due to its effectiveness in performance evaluation [33]. In all the current methods, the evaluation is ignored in case of uncertainties. For this reason, the efficiency scores are not accurate enough to be used for improving port performance and in general they are not useful for port users. There is also no study in the literature on port performance evaluation in the Adriatic region.

3 Research Methodology

In order to evaluate the efficiency of the Adriatic ports, this study proposes a generalized DEA approach. After defining the purpose of the study, input and output variables are determined. Then, generalized SBM methods are used to measure the performance efficiency of the Northern Adriatic ports. Finally, the research results are analysed and discussed before management implications are proposed to improve the efficiency of Northern Adriatic ports.

3.1 SBM Model for Measuring Efficiencies

Assume that there are n DMUs to be evaluated, each consisting of m inputs and s outputs. $X_{ij},\ i = 1, ..., m$ and $Y_{rj},\ r = 1, ..., s$ denote the input and output values of DMUj ($j = 1, ..., n$), all of which are known and non-negative. Then, the production possibility set is defined as follows:

$$T = \left\{ (X, Y) | \sum_{j=1}^{n} \lambda_j X_j \le X,\ \sum_{j=1}^{n} \lambda_j Y_j \ge Y,\ \lambda_j \ge 0,\ j = 1, ..., n \right\} \quad (1)$$

T is a closed and convex set. Boundary points of T are defined as the efficient production frontier.

For direct assessment of the efficiency with slack values, an efficiency SBM was proposed by [34]. According to the concept of the efficient production frontier, the SBM model is defined as follows:

$$\begin{aligned} &\textit{Minimize} \quad \rho = \frac{1 - {}^{1}\!/_{m} \sum_{i=1}^{m} {}^{s_i^-}\!/_{x_{io}}}{1 + {}^{1}\!/_{s} \sum_{r=1}^{s} {}^{s_r^+}\!/_{y_{ro}}} \\ &\textit{subject to} \quad \sum_{j=1}^{n} \lambda_j x_{ij} + s_i^- = x_{io},\ i = 1, ..., m, \\ &\sum_{j=1}^{n} \lambda_j y_{rj} - s_r^+ = y_{ro},\ r = 1, ..., s, \\ &\lambda_j \ge 0,\ j = 1, ..., n,\ s_i^- \ge 0,\ i = 1, ..., m,\ s_r^+ \ge 0\ \ r = 1, ..., s. \end{aligned} \quad (2)$$

where DMUo denotes the DMU under evaluation. s_i^- $(i = 1, \ldots, m)$ and s_r^+ $(r = 1, \ldots, s)$ are called slacks. If $x_{io} = 0$, then the term ${}^{s_i^-}\!/_{x_{io}}$ is eliminated. If $y_{ro} = 0$, then it is replaced by a very small number, so that the term ${}^{s_r^+}\!/_{y_{ro}}$ has compensatory effect. Using scale transformation, model (2) can be converted into the following linear programming (LP) model:

$$
\begin{aligned}
&Minimize \quad \tau = t - {}^{1}\!/_{m}\sum_{i=1}^{m} {}^{s_i^-}\!/_{x_{io}} \\
&subject\ to \quad \sum_{j=1}^{n} \Lambda_j x_{ij} + s_i^- = t x_{io}\,,\ i = 1, \ldots, m\,, \\
&\sum_{j=1}^{n} \Lambda_j y_{rj} - s_r^+ = t y_{ro}\,,\quad r = 1, \ldots, s\,, \\
&1 = t + {}^{1}\!/_{s}\sum_{r=1}^{s} {}^{s_r^+}\!/_{y_{ro}}\,, \\
&\Lambda_j \geq 0\,, j = 1, \ldots, n,\ s_i^- \geq 0\,,\ i = 1, \ldots, m,\ s_r^+ \geq 0\,,\ r = 1, \ldots, s\,,\ t > 0\,.
\end{aligned} \tag{3}
$$

An important property of efficiency SBM is that τ is independent of the measurement unit used for inputs and outputs, and is monotonically decreasing in each input and output slack. If the optimal value of ρ occurs, i.e., $\rho^* = 1$, the respective DMU is called efficient, otherwise, it is called non-efficient.

One of the problems with some DEA models is that they are sensitive to the number of units and do not produce acceptable results with a small number of units. In this article, an improved DEA method is used to solve this problem. Similar to the KAM method [35], the improved model is as follows.

$$
\begin{aligned}
&Minimize \quad \tau = t - {}^{1}\!/_{m}\sum_{i=1}^{m} {}^{s_i^-}\!/_{x_{io}} \\
&subject\ to \quad \sum_{j=1}^{n} \Lambda_j x_{ij} + s_i^- = t x_{io} + \varepsilon_i^-\,,\ i = 1, \ldots, m\,, \\
&\sum_{j=1}^{n} \Lambda_j y_{rj} - s_r^+ = t y_{ro} - \varepsilon_r^+\,,\quad r = 1, \ldots, s\,, \\
&1 = t + {}^{1}\!/_{s}\sum_{r=1}^{s} {}^{s_r^+}\!/_{y_{ro}}\,, \\
&\Lambda_j \geq 0\,, j = 1, \ldots, n,\ s_i^- \geq 0\,,\ i = 1, \ldots, m,\ s_r^+ \geq 0\,,\ r = 1, \ldots, s\,,\ t > 0\,,
\end{aligned} \tag{4}
$$

where $\varepsilon^- = (\varepsilon_1^-, \ldots, \varepsilon_m^-)$ and $\varepsilon^+ = (\varepsilon_1^+, \ldots, \varepsilon_r^+)$.

3.2 Generalized SBM Model with Uncertain Data

In this section, we extend SBM models in uncertain environments to generalized models.

Definition 1 Suppose a DMU is a decision making unit in a decision making problem. All data that have the same input and output data as the DMU are called a sample decision making unit (SDMU) based on this decision making problem [35].

The differences between an SDMU and a DMU are illustrated below:

A DMU must be in the production possibility set, while an SDMU can be outside the production possibility set.

- The efficiency value of the DMU must be equal to or less than 1, while that of the SDMU can be equal to, less than or greater than 1.
- A DMU must occur in the constraints, while an SDMU can either occur in the constraints or not occur.

Some researchers have found that the fuzzy DEA models have an efficiency value greater than 1 [37–39]. This is because all constraints that affect the target DMU select worse units, while the target DMU selects better units for evaluation. Therefore, the evaluated DMU is an SDMU and not a DMU [36]. Sample DMUs also include normal DMUs.

SDMU can be any point of the domain including: Best DMU, Worst DMU, Max DMU, Min DMU, Center DMU, 1-Cut DMU and Vertex DMU.

When evaluating the target FDMU of the FSBM model, either the best or the worst DMU is selected. The remaining five special DMUs are never selected. In the proposed method, the special DMUs or any DMU of the domain that the decision maker prefers can be selected.

Definition 2 If the efficiency value of the SDMU is equal to or greater than 1 when the FSDMU is replaced by the worst SDMU and the FDMUs are replaced by the best DMUs, the FSDMU is said to be strongly efficient.

Definition 3 If the efficiency value of the DMU is equal to or greater than 1 when the FSDMU and FDMUs are replaced by one of the DMUs, the FSDMU is said to be efficient.

The generalized SBM model is shown in Eq. (5).

$$
\begin{aligned}
&Minimize \quad \tau = t - {}^{1}\!/_{m}\sum_{i=1}^{m} {}^{s_i^-}\!/_{x_{So}} \\
&subject\ to \quad \sum_{j=1}^{n} \Lambda_j x_{ij} + s_i^- = t x_{io} + \varepsilon_i^-,\ i = 1, \ldots, m, \\
&\sum_{j=1}^{n} \Lambda_j y_{rj} - s_r^+ = t y_{ro} - \varepsilon_r^+,\ r = 1, \ldots, s, \\
&1 = t + {}^{1}\!/_{s}\sum_{r=1}^{s} {}^{s_r^+}\!/_{y_{So}}, \\
&\Lambda_j \geq 0,\ j = 1, \ldots, n,\ s_i^- \geq 0,\ i = 1, \ldots, m,\ s_r^+ \geq 0,\ r = 1, \ldots, s, t > 0.
\end{aligned} \tag{5}
$$

The difference between models (4) and (5) is the sample fuzzy DMU. In the above model, (x_S, y_S) is FSDMU. The generalized SBM model is shown in Eq. (6).

$$
\begin{aligned}
&Maximize \quad \varphi = t + {}^{1}\!/_{m}\sum_{i=1}^{m} {}^{s_i^+}\!/_{x_{So}} \\
&subject\ to \quad \sum_{j=1}^{n} \Lambda_j x_{ij} - s_i^+ = t x_{io} + \varepsilon_i^-,\ i = 1, \ldots, m, \\
&\sum_{j=1}^{n} \Lambda_j y_{rj} + s_r^- = t y_{ro} - \varepsilon_r^+,\ r = 1, \ldots, s, \\
&1 = t - {}^{1}\!/_{s}\sum_{r=1}^{s} {}^{s_r^-}\!/_{y_{So}}, \\
&\Lambda_j \geq 0,\ j = 1, \ldots, n,\ s_i^+ \geq 0,\ i = 1, \ldots, m,\ s_r^- \geq 0,\ r = 1, \ldots, s, t > 0.
\end{aligned} \tag{6}
$$

The uncertain SDMU is replaced by one of the SDMUs and uncertain DMUi is replaced by one of the DMUs. Once the model has become a precise DEA model, it can be solved using the appropriate software. The selected SDMU or DMUi can be any point in the domain.

4 Empirical Study

In this section, we have evaluated the efficiency of North Adriatic ports using the proposed method. The performance of each port in 2021 and 2022 was analysed separately. The assessment is based on the ports' operational capacity, which includes general cargo, vehicles, liquid bulk, containers and dry bulk. In this analysis, models (5) and (6) are used to evaluate the ports based on indicators. The data is taken from the ports' websites.

4.1 Inputs and Outputs Selection

This study measures the performance efficiency of 5 ports in the Northern Adriatic, of which 3 ports in Italy (Trieste, Venice and Ravenna), one port in Croatia (Rijeka) and one port in Slovenia (Koper). Inputs and outputs were determined based on the results of the literature review and availability. The inputs and outputs are described below.

The **input indicators** used are the following:

- *Water depth*

Water depth refers to the depth of water in meters into which a vessel is submerged when fully loaded.

- *Berth length*

This factor refers to the total size of the berth in the terminal under study, expressed in meters.

- *Total terminal area*

This is the total usable area of the terminal including the storage area, measured in square meters.

The following indicators were used as **output indicators**:

- *Throughput*

Throughput is the total volume of cargo handled annually, measured in tonnes. Container terminals, solid and liquid bulk terminals and general cargo are taken into account.

- *Number of passengers*

It represents the number of handled passengers in the passenger terminal of the port under study in the year observed.

Tables 1 and 2 contain the data used in this analysis for input and output variables respectively.

4.2 Efficiency Scores

One of the inputs (water depth) is an interval. To capture this interval count, we use the generalised method. For this purpose, the DMUs are evaluated after selecting the

Table 1 Input data

	2021			2022		
Port name	Total terminal area	Water depth [min, max]	Berth length	Total terminal area	Water depth [min, max]	Berth length
Port of Koper	273.140	[10.10, 15.10]	598	273.140	[10.10, 15.10]	598
Port of Trieste	2,300,000	18	770	2,300,000	18	770
Port of Rijeka	122,200	[8, 18]		122,200	[8, 18]	628
Port of Venice	1,620,970	[7.1, 13.7]	500	1,620,970	[7.1, 13.7]	500
Port of Ravenna	987,000	[4.9, 10]	346	987,000	[4.9, 10]	346

Table 2 Output data

	2021		2022	
Port name	Throughput	Number of passengers	Throughput	Number of passengers
Port of Koper	17,821,177	4,450	23,237,584	72,753
Port of Trieste	48,279,658	134,823	50,181,687	448,763
Port of Rijeka	11,081,945	3000	11,668,009	4000
Port of Venice	24,242,354	102,613	24,613,092	371,486
Port of Ravenna	27,100,051	1946	27,389,886	19,397

different points (worst, best and centre). The generalized SBM model (6) is used to score the ports in 2020 and 2021. Lingo software is used to determine the different efficiency values of the DMUs. The efficiency scores of the ports for three points are shown in Table 3. Table 4 and Fig. 1 also show the ranking of the North Adriatic ports and the comparison of the efficiency of the ports in 2020 and 2021.

5 Discussion and Conclusion

The 2021 results show that the Port of Koper received the best efficiency score, and after the Port of Koper comes the Port of Ravenna. The port of Venice is in third place. The ports of Rijeka and Trieste are also in the next ranks. The port of Trieste has a worse result due to the high number of inputs.

Of course, the difference between the efficiency values of the ports is not significant; the lowest efficiency value is 0.8160263 and the highest is 0.8693516.

Table 3 Efficiency value of ports

	2021			2022		
Port name	Worst	Best	Centre	Worst	Best	Centre
Port of Koper	0.8620114	0.8651413	0.8693516	0.8733802	0.8708856	0.8787671
Port of Trieste	0.8160263	0.8180227	0.8195435	0.8275465	0.8286889	0.8279547
Port of Rijeka	0.8234654	0.8230545	0.8235353	0.8254568	0.8259895	0.8260345
Port of Venice	0.8507793	0.8409742	0.8528419	0.8555979	0.8499309	0.8530035
Port of Ravenna	0.8581853	0.8535403	0.8605291	0.8620213	0.8632570	0.8688558

Table 4 Ranking of ports

	2021		2022
Port name	Rank	Port name	Rank
Port of Koper	1	Port of Koper	1
Port of Trieste	5	Port of Trieste	4
Port of Rijeka	4	Port of Rijeka	5
Port of Venice	3	Port of Venice	3
Port of Ravenna	2	Port of Ravenna	2

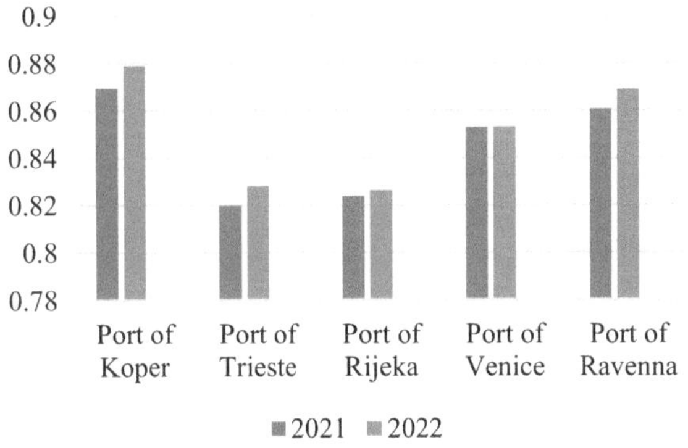

Fig. 1 Comparison of the efficiencies of the North Adriatic ports

In 2022, the port of Koper has again achieved better results and is in first place, followed by the ports of Ravenna and Venice. The results of the ports of Trieste and Rijeka have changed. The port of Trieste has risen to fourth place, which could be the reason for the increase in output (production), passenger numbers and throughput. However, the performance values of all ports have improved in 2022, which could be due to the reduction in the impact of the pandemic, which had led to a decline in 2021.

There are no significant differences in the efficiency scores in 2022. The lowest efficiency score is 0.8254568 and the highest score is 0.8787671. In general, none of the ports are efficient in the two years analysed, but they have achieved acceptable results.

Despite the fact that the evaluation was carried out in 3 different points, there was no change in the overall results of the list, and in general it can be said that the results are the same in all three items, the best, the worst and the centre point. One of the features of the generalized method is that it is able to determine performance scores in different situations without the need for complex non-deterministic methods.

The largest increase in efficiency score compared to 2021 and 2022 is recorded by the ports of Koper and Trieste, followed by the ports of Venice, Ravenna and Rijeka.The reason for this increase is that the port of Koper has the highest output (increase in production) in 2022 compared to the other ports in the region.

The port of Trieste also has a significant increase in output, but due to the size of the terminal, i.e., (one of the inputs), it could not reach a high rank, as it has a larger capacity. The port of Koper had obtained the best results in terms of desired inputs. The other three ports had a lower increase in output compared to 2021, with the port of Rijeka having the lowest tonnage increase.

There were restrictions on the procurement of data. It was not possible to evaluate all Adriatic ports from all aspects, as it was not possible to access the financial reports and the reports on emissions in all ports. The presented methodology can also be used for the evaluation and ranking of other transport systems and even other industries. For future studies, more data should be collected and studies should be conducted for all Adriatic ports in order to make a more accurate performance assessment. A comparison of the ports in this region with other ports in the Mediterranean and Europe is also possible.

References

1. Progress of the EU's Integrated Maritime Policy. (2012). Report from the Commission to the European Parliament, the Council, the European Economic and Social Committee and the Committee of Regions.
2. AdriaPorts. https://www.adriaports.com
3. Statista. https://www.statista.com/statistics
4. Charnes, A., Cooper, W. W., & Rhodes, E. (1978). Measuring the efficiency of decision making units. *European Journal of Operational Research, 2*(6), 429–444.
5. Jameel, A. F., Anakira, N. R., Alomari, A. K., Alsharo, D. M., & Saaban, A. (2019). New semi-analytical method for solving two point nth order fuzzy boundary value problem. *International journal of Mathematical Modelling and Numerical Optimisation, 9*(1), 21–31.
6. Wei, Q. L., & Yan, H. (2010). A data envelopment analysis (DEA) evaluation method based on sample decision making unit. *International Journal of Information Technology and Decision Making, 9*(4), 601–624.

7. Wang, Y. M., & Chin, K. S. (2009). A new approach for the selection of advanced manufacturing technologies: DEA with double frontiers. *International Journal of Production Research, 47*(23), 6663–6679.
8. Ashrafi, A., & Mansouri Kaleibar, M. (2019). Generalized fuzzy Slacks-Based measures of efficiency and its applications. *Jordan Journal of Mathematics and Statistics, 12*(2), 135–149.
9. Roll, Y., & Hayuth, Y. (1993). Port performance comparison applying data envelopment analysis (DEA). *Maritime Policy & Management, 20*(2), 153–161. https://doi.org/10.1080/03088839300000025
10. Ibáñez, M. N., Martínez, E., Diaz Armas, J., & Mesa, T. R. (1999). A study of the efficiency of Spanish port authorities using data envelopment analysis. *International Journal of Transport Economics, 26*(2), 235–252.
11. Hung, S. W., Lu, W. M., & Wang, T. P. (2010). Benchmarking the operating efficiency of Asia container ports. *European Journal of Operations Research, 203*(3), 706–713. https://doi.org/10.1016/j.ejor.2009.09.005
12. Wu, Y. J., & Goh, M. (2010). Container port efficiency in emerging and more advanced markets. *Transportation Research Part E, 46*, 1030–1042. https://doi.org/10.1016/j.tre.2010.01.002
13. Cullinane, K., & Wang, T. F. (2010). The efficiency analysis of container port production using DEA panel data approaches. *OR Spectrum, 32*(3), 717–738. https://doi.org/10.1007/s00291-010-0202-7
14. De Oliveira, G. F., & Cariou, P. (2011). A DEA study of the efficiency of 122 iron ore and coal ports and of 15/17 countries in 2005. *Maritime Policy & Management, 38*(7), 727–743. https://doi.org/10.1080/03088839.2011.625989
15. Choi, Y. (2011). The efficiency of major ports under logistics risk in northeast Asia. *Asia-Pacific Journal of Operations Research, 28*(1), 111–123. https://doi.org/10.1142/S0217595911003089
16. Niavis, S., & Tsekeris, T. (2012). Ranking and causes of inefficiency of container seaports in South-Eastern Europe. *European Transportation Research Review, 4*(4), 235–244. https://doi.org/10.1007/s12544-012-0080-y
17. Schøyen, H., & Odeck, J. (2013). The technical efficiency of Norwegian container ports: A comparison to some Nordic and UK container ports using Data Envelopment Analysis (DEA). *Maritime Economics & Logistics, 15*(2), 197–221. https://doi.org/10.1057/mel.2013.3
18. Bray, S., Caggiani, L., Dell'Orco, M., & Ottomanelli, M. (2014). Measuring transport systems efficiency under uncertainty by fuzzy sets theory-based Data Envelopment Analysis. *Procedia—Social and Behavioral Sciences, 111*, 770–779. https://doi.org/10.1016/j.sbspro.2014.01.111
19. Lu, B., & Park, N. K. (2015). The evaluation of operational efficiency of the world's leading container seaports. *Journal of Coastal Research, 73*, 248–254. https://doi.org/10.2112/SI73-044.1
20. Wanke, P. F., & Barros, C. P. (2015). Public-private partnerships and scale efficiency in Brazilian ports: Evidence from two-stage DEA analysis. *Socio-Economic Planning Science, 51*, 13–22. https://doi.org/10.1016/j.seps.2015.06.002
21. Morales-Fusco, P., Saurí, S., Lekka, A. M., & Karousos, I. (2016). Assessing customs performance in the Mediterranean ports. KPI selection and best practices identification as part of the MEDNET project. *Transportation Research Procedia, 18*, 374–383. https://doi.org/10.1016/j.trpro.2016.12.049
22. Schøyen, H., & Odeck, J. (2017). Comparing the productivity of Norwegian and some Nordic and UK container ports—An application of Malmquist productivity index. *Journal of Shipping and Transport Logistics, 9*(2), 234–256. https://doi.org/10.1504/IJSTL.2017.082526
23. Chang, V., & Tovar, B. (2017). Meta frontier analysis on productivity for West Coast of South Pacific terminals. *Transportation Research Part A, 103*, 118–134. https://doi.org/10.1016/j.tra.2016.12.012
24. Chang, Z., Yang, D., Wan, Y., & Han, T. (2018). Analysis on the features of Chinese dry ports: Ownership, customs service, rail service and regional competition. *Transport Policy, 31*(2), 231–251. https://doi.org/10.1016/j.tranpol.2018.06.008

25. Schøyen, H., Bjorbæk, C. T., Steger-Jensen, K., Bouhmala, N., Burki, U., Jensen, T. E., & Berg, O. (2018). Measuring the contribution of logistics service delivery performance outcomes and deep-sea container liner connectivity on port efficiency. *Research in Transportation Business & Management, 28*, 66–76. https://doi.org/10.1016/j.rtbm.2018.03.002
26. Tovar, B., & Wall, A. (2019). Environmental efficiency for a cross-section of Spanish port authorities. *Transportation Research Part D, 75*, 170–178. https://doi.org/10.1016/S0965-8564(99)00049-X
27. Quintano, C., Mazzocchi, P., & Rocca, A. (2020). A competitive analysis of EU ports by fixing spatial and economic dimensions. *Journal of Shipping and Trade, 5*(1), 1–18. https://doi.org/10.1186/s41072-020-00075-x
28. Quintano, C., Mazzocchi, P., & Rocca, A. (2020). Examining eco-efficiency in the port sector via non-radial data envelopment analysis and the response based procedure for detecting unit segments. *Journal of Cleaner Production, 259*(5). https://doi.org/10.1016/j.jclepro.2020.120979
29. Wang, Z., Wu, X., Guo, J., Wei, G., & Dooling, T. A. (2020). Efficiency evaluation and PM emission reallocation of China ports based on improved DEA models. *Transportation Research Part D, 82*. https://doi.org/10.1016/j.trd.2020.102317
30. Quintano, C., Mazzocchi, P., & Rocca, A. (2021). Evaluation of the eco-efficiency of territorial districts with seaport economic activities. *Utilities Policy, 71*. https://doi.org/10.1016/J.JUP.2021.101248
31. Kong, Y., & Liu, J. (2021). Sustainable port cities with coupling coordination and environmental efficiency. *Ocean and Coastal Management 205*(6). https://doi.org/10.1016/j.ocecoaman.2021.105534
32. Nong, T. N. (2023). Performance efficiency assessment of Vietnamese ports: An application of Delphi with Kamet principles and DEA model. *The Asian Journal of Shipping and Logistics, 39*, 1–12.
33. Krmac, E., & Mansouri Kaleibar, M. (2022). A comprehensive review of data envelopment analysis (DEA) methodology in port efficiency evaluation. *Maritime Economics & Logistics*. https://doi.org/10.1057/s41278-022-00239-5
34. Tone, K. (2001). A slacks-based measure of efficiency in data envelopment analysis. *European Journal of Operational Research, 130*(3), 498–509.
35. Khezrimotlagh, D. (2014). How to deal with numbers of decision making units and variables in data envelopment analysis. In: *13th International Conferences on Data Envelopment Analysis*.
36. Muren, Z., & Wei, C. (2014). Generalized fuzzy data envelopment analysis methods. *Applied Soft Computing, 19*, 215–225.
37. Guo, P., & Tanaka, H. (2001). DEA fuzzy, a perceptual evaluation method. *Fuzzy Sets and Systems, 119*(1), 149–160.
38. Wang, Y. M., Creatbanks, R., & Yang, J. B. (2005). Interval efficiency assessment using data envelopment analysis. *Fuzzy Sets and Systems, 153*(3), 347–370.
39. Angiz, L. M. Z., Emrouznejad, A., & Mustafa, A. (2012). Fuzzy data envelopment: A discrete approach. *Expert Systems with Applications, 39*(3), 2263–2269.

Development of Intuitionistic Fuzzy Revenue Efficiency Models in DEA: An Application to the Indian Public Sector Banks

Anjali Sonkariya and Shiv Prasad Yadav

Abstract Data envelopment analysis (DEA) is a non-parametric linear programming (LP) based technique to measure the relative efficiencies of homogeneous decision-making units (DMUs). The classical models of DEA rely on crisp input-output data, which may not always be available in real-life scenarios. Due to the existence of uncertainty and vagueness in real-life data, the concept of intuitionistic fuzzy (IF) has been introduced to handle imprecise data. Here, the conventional revenue efficiency models of DEA are extended to the IF environment. Also, the lower and upper revenue efficiency models are developed using α and β-cuts approaches. The input-output data and output prices are considered triangular intuitionistic fuzzy numbers (TIFNs). An application to the public sector banks of India is provided to illustrate the practicality of the proposed intuitionistic fuzzy revenue efficiency models (IFREMs). Data is collected from the official website of the Reserve Bank of India (RBI), Govt. of India, India.

Keywords α and β-cuts · Revenue efficiency · Data envelopment analysis · Intuitionistic fuzzy enviornment

1 Introduction

The CCR-DEA model was introduced by Charnes et al. [4] as a linear programming (LP) technique to evaluate the relative efficiencies of homogeneous DMUs. Zadeh [12] first introduced the theory of fuzzy sets (FS) to deal with uncertainty. Later, Sengupta in 1992 [10] proposed fuzzy data envelopment analysis (FDEA) by utilizing fuzzy regression, fuzzy entropy, and fuzzy mathematical programming techniques. The aim was to handle imprecise and uncertain inputs, outputs, and prior knowledge. Kao et al. [7] included an FDEA model using the α-cut method to the group of classic DEA models. They also used the intuitionistic fuzzy set (IFS) theory to address

A. Sonkariya (✉) · S. P. Yadav
Indian Institute of Technology Roorkee, Roorkee, India
e-mail: asonkariya1@ma.iitr.ac.in

A. Emrouznejad et al. (eds.), *Advances in the Theory and Applications of Performance Measurement and Management*, Lecture Notes in Operations Research,
https://doi.org/10.1007/978-3-031-61597-9_12

uncertainties in data. Atanassov [3] introduced IFS as a practical and applicable extension to FS theory, useful in solving real-world problems. The IFS considers not only the membership function but also the non-membership function. DMU's main objective is maximizing profit with minimum use of resources or expenses. Revenue efficiency maximizes output quantities while using the same level of inputs. Allocation DEA models estimate efficiency scores corresponding to different behavioral objectives to be achieved by the DMUs, i.e., cost reduction, revenue growth, and enhanced profit. The research conducted by Sufian et al. [11] involved an evaluation of revenue efficiency in the Islamic banking sector of Malaysia, covering the period from 2006 to 2010. They employed various measures of allocative efficiency, including cost, revenue, and profit. A discernible contrast is observed between the revenue efficiency of domestic and foreign Islamic banks, attributable to the discrepant levels of cost and profit efficiency. Puri and Yadav [9] developed fully fuzzy DEA models that incorporated undesirable outputs for assessing cost and revenue efficiencies. The models utilized a triad of inputs alongside two coveted outputs and one unfavorable output and were implemented in a fuzzy environment via data sourced from India's banking sector during the financial years of 2011–12 and 2012–13. Ashrafi [1] created models to assess cost, revenue, and profit efficiency, where fuzzy data for input, output, input cost, and output price were considered. Ghiyasi [5] proposed an inverse DEA model based on cost and revenue efficiency, which can be utilized when pricing data is available. The model preserves the technical efficiency and cost efficiency scores of DMUs, even when there is a perturbation in output. Kordrostami et al. [8] introduced a fully fuzzy DEA method to evaluate the revenue efficiency of sustainable supply chains and their components. Hussain et al. [6] analyzed the impact of country governance and other potential factors on the revenue efficiency of Islamic and conventional banks in countries across regions. The current research aims to establish a Revenue efficiency (RE) model using IF data. We employ the α and β-cut methods to convert the problem into a linear programming format and construct lower and upper RE models. We apply the model to evaluate the revenue efficiency of Indian public sector banks for 2021.

The study is structured subsequently: The foundational tenets of Intuitionistic Fuzzy (IF) theory are explicated in Sect. 2 through a comprehensive overview of the core concepts. Section 3 presents the DEA model for revenue efficiency, while Sect. 4 outlines the proposed IFREM. Section 5 describes the methodology for solving IFREM. Section 6 of the study demonstrates the model's implementation in the Indian banking sector, with the research culminating in Sect. 7.

2 Preliminaries

Definition 1 (*Intuitionistic fuzzy set (IFS)*) [3]: The IFS $\tilde{F}^I$ is expressed as $\tilde{F}^I = (w, \mu_{\tilde{F}^I}(w), \nu_{\tilde{F}^I}(w))$, where the entire domain of discussion is depicted in Ω. The membership function $\mu_{\tilde{F}^I} : \Omega \to [0, 1]$ and non-membership function $\nu_{\tilde{F}^I} : \Omega \to [0, 1]$ are the defining components of $\tilde{F}^I$.

In the context of IFS, the degree to which w belongs to $\tilde{F}^I$ is known as the membership grade and is represented by $\mu_{\tilde{F}^I}(w)$. Similarly, the degree to which w does not belong to $\tilde{F}^I$ is called the non-membership grade and is represented by $\nu_{\tilde{F}^I}(w)$. These grades are subject to the constraint that their sum must be between 0 and 1, i.e., $0 \leq \mu_{\tilde{F}^I}(w) + \nu_{\tilde{F}^I}(w) \leq 1$. The degree of hesitation or indeterminacy regarding the inclusion of an element w in the IFS $\tilde{F}^I$ is referred to as the hesitation degree and denoted by $\pi_{\tilde{F}^I}(w)$. This degree is precisely defined as follows:

$$\pi_{\tilde{F}^I}(w) = 1 - \mu_{\tilde{F}^I}(w) - \nu_{\tilde{F}^I}(w) \ \forall \ w \in \Omega, \quad 0 \leq \pi_{\tilde{F}^I}(w) \leq 1.$$

Definition 2 (*Intuitionistic fuzzy number (IFN)*) [2]: An IFS $\tilde{F}^I$ is considered to be an IFN if it is a convex set defined as $\tilde{F}^I = \{(w, \mu_{\tilde{F}^I}(w), \nu_{\tilde{F}^I}(w)) : w \in \mathbb{R}\}$ that satisfies the conditions of having a unique mean value $w_0 \in \mathbb{R}$ such that $\mu_{\tilde{F}^I}(w_0) = 1$, and $\exists$ a point $w_1 \in \mathbb{R}$ such that $\nu_{\tilde{F}^I}(w_1) = 1$. In other words, $\tilde{F}^I$ is a normal IFS, and w_0 represents the mean value of the set.

Definition 3 (*Triangular intuitionistic fuzzy number (TIFN)*) [2]: The IFN $\tilde{F}^I = \{(w, \mu^I_{\tilde{F}}(w), \nu^I_{\tilde{F}}(w)) : w \in \mathbb{R}\}$ is called a TIFN if

$$\mu_{\tilde{F}^I}(w) = \begin{cases} \dfrac{w - f^L}{f^M - f^L}, & f^L < w \leq f^M, \\ \dfrac{f^U - w}{f^U - f^M}, & f^M \leq w < f^U, \\ 0 & , otherwise. \end{cases} \qquad \nu_{\tilde{F}^I}(w) = \begin{cases} \dfrac{f^M - w}{f^M - f'^L}, & f'^L < w \leq f^M, \\ \dfrac{w - f^M}{f'^U - f^M}, & f^M \leq w < f'^U, \\ 1 & , otherwise, \end{cases}$$

where $f^L, f^M, f^U, f'^L, f'^U \in \mathbb{R}$ such that $f'^L \leq f^L \leq f^M \leq f^U \leq f'^U$. TIFN is denoted by $\tilde{F}^I = (f^L, f^M, f^U; f'^L, f^M, f'^U)$.

Definition 4 – The α-cut of $\tilde{F}^I$, for $\alpha \in (0, 1]$ is given by

$$(\tilde{F}^I)_\alpha = [(f)^L_\alpha, (f)^U_\alpha] = \left[\alpha f^M + (1-\alpha) f^L, \alpha f^M + (1-\alpha) f^U\right]$$

– The β-cut of $\tilde{F}^I$, for $\beta \in [0, 1)$ is given by

$$(\tilde{F}^I)_\beta = [(f)'^L_\beta, (f)'^U_\beta] = \left[\beta f'^L + (1-\beta) f^M, \beta f'^U + (1-\beta) f^M\right]$$

3 Revenue Efficiency (RE)

The revenue efficiency (RE) of a DMU is determined by the proportion of the actual revenue it generates to the maximum revenue it could potentially generate. Thus,

$$\text{RE} = \frac{\text{actual revenue}}{\text{max revenue}}$$

RE-DEA models estimate the revenue-maximizing level of output quantities while utilizing the same level of inputs. Let us consider a scenario where n DMUs exist, with each DMU having m inputs and s outputs. For any DMU_k, where $k = 1, 2, \ldots, n$, the RE can be defined in the following manner:

$$\text{RE}_k = \frac{\sum_{r=1}^{s} p_{rk} y_{rk}}{\sum_{r=1}^{s} p_{rk} y_r}$$

Model 1 (RE-DEA Model): For DMU_k,

$$\max \text{RE}_k = \frac{\sum_{r=1}^{s} p_{rk} y_{rk}}{\sum_{r=1}^{s} p_{rk} y_r}$$

subject to

$$\sum_{j=1}^{n} \lambda_j x_{ij} \leq x_{ik}, \qquad i = 1, 2, \ldots, m,$$

$$\sum_{j=1}^{n} \lambda_j y_{rj} \geq y_r, \qquad r = 1, 2, \ldots, s,$$

$$\lambda_j,\ y_r \geq 0,$$

where λ_j and y_r are variables, y_r = unknown quantity of the rth output, p_{rk} = output price of the rth output for DMU_k, x_{ij} = actual value of the ith input for DMU_j, y_{rj} = actual value of the rth output for DMU_j.

Definition 5 Let (y_r^*, λ_j^*) denote the optimal values of (y_r, λ_j) for $r = 1, 2, ..., s$ and $j = 1, 2, ..., n$, respectively, and let RE_k^* denote the optimal value of RE_k for DMU_k. A DMU_k is considered to be revenue-efficient if $\text{RE}_k^* = 1$, whereas if RE_k^* is less than 1, then DMU_k is classified as revenue-inefficient.

4 Proposed Intuitionistic Fuzzy Revenue Efficiency Model (IFREM)

The proposed IFREM- We take IF data for inputs, outputs, and output prices in this model. The IFREM (Model 2) is given as follows:

Model 2

$$\max \tilde{\text{RE}}_k{}^I = \frac{\sum_{r=1}^{s} \tilde{p}_{rk}^I \tilde{y}_{rk}^I}{\sum_{r=1}^{s} \tilde{p}_{rk}^I y_r}$$

subject to

$$\sum_{j=1}^{n} \lambda_j \tilde{x}_{ij}^{I} \leq \tilde{x}_{ik}^{I}, \qquad i = 1, 2, \ldots, m,$$

$$\sum_{j=1}^{n} \lambda_j \tilde{y}_{rj}^{I} \geq y_r, \qquad r = 1, 2, \ldots, s,$$

$$\lambda_j, \; y_r \geq 0.$$

Presume that IF input $\tilde{x}_{ij}^{I}$, IF output $\tilde{y}_{rj}^{I}$ and IF output price $\tilde{p}_{rk}^{I}$ are triangular intuitionistic fuzzy numbers (TIFNs).

The ensuing model obtained after the transformation of Model 2 is expressed as follows:

Model 3

$$\max \tilde{\mathrm{RE}}_k{}^{I'} = \frac{\sum_{r=1}^{s} \left(p_{rk}^{L}, p_{rk}^{M}, p_{rk}^{U}; p_{rk}^{\prime L}, p_{rk}^{\prime M}, p_{rk}^{\prime U}\right)\left(y_{rk}^{L}, y_{rk}^{M}, y_{rk}^{U}; y_{rk}^{\prime L}, y_{rk}^{\prime M}, y_{rk}^{\prime U}\right)}{\sum_{r=1}^{s} \left(p_{rk}^{L}, p_{rk}^{M}, p_{rk}^{U}; p_{rk}^{\prime L}, p_{rk}^{M}, p_{rk}^{\prime U}\right)\left(y_r, y_r, y_r; y_r, y_r, y_r\right)}$$

subject to

$$\sum_{j=1}^{n} \lambda_j \left(x_{ij}^{L}, x_{ij}^{M}, x_{ij}^{U}; x_{ij}^{\prime L}, x_{ij}^{\prime M}, x_{ij}^{\prime U}\right) \leq \left(x_{ik}^{L}, x_{ik}^{M}, x_{ik}^{U}; x_{ik}^{\prime L}, x_{ik}^{\prime M}, x_{ik}^{\prime U}\right), \; i = 1, 2, \ldots, m,$$

$$\sum_{j=1}^{n} \lambda_j \left(y_{rj}^{L}, y_{rj}^{M}, y_{rj}^{U}; y_{rj}^{\prime L}, y_{rj}^{\prime M}, y_{rj}^{\prime U}\right) \geq \left(y_r, y_r, y_r; y_r, y_r, y_r\right), \; r = 1, 2, \ldots, s,$$

$$\lambda_j, \; y_r \geq 0.$$

5 Methodology for Solving Proposed IFREM

Let $\mathrm{S}(\tilde{x}_{ij}^{I})$, $\mathrm{S}(\tilde{y}_{rj}^{I})$ and $\mathrm{S}(\tilde{p}_{rk}^{I})$ denote the supports of $\tilde{x}_{ij}^{I}$, $\tilde{y}_{rj}^{I}$ and $\tilde{p}_{rk}^{I}$ respectively.

$$\mathrm{S}(\tilde{x}_{ij}^{I}) = \left\{x_{ij} \mid \mu_{\tilde{x}_{ij}^{I}}(x_{ij}) > 0\right\} \quad \forall\, i, j$$

$$\mathrm{S}(\tilde{y}_{rj}^{I}) = \left\{y_{rj} \mid \mu_{\tilde{y}_{rj}^{I}}(y_{rj}) > 0\right\} \quad \forall\, r, j$$

$$\mathrm{S}(\tilde{p}_{rk}^{I}) = \left\{p_{rk} \mid \mu_{\tilde{p}_{rk}^{I}}(p_{rk}) > 0\right\} \quad \forall\, r, k$$

The α-cuts of $\tilde{x}_{ij}^{I}$, $\tilde{y}_{rj}^{I}$ and $\tilde{p}_{rk}^{I}$ are defined as

$$(\tilde{x}_{ij}^{I})_{\alpha} = \left\{x_{ij} \in \mathrm{S}(\tilde{x}_{ij}^{I}) \mid \mu_{\tilde{x}_{ij}^{I}}(x_{ij}) \geq \alpha\right\} = \left[(x_{ij})_{\alpha}^{L}, (x_{ij})_{\alpha}^{U}\right]$$
$$= \left[\min_{x_{ij}}\left\{x_{ij} \in \mathrm{S}(\tilde{x}_{ij}^{I}) \mid \mu_{\tilde{x}_{ij}^{I}}(x_{ij}) \geq \alpha\right\}, \max_{x_{ij}}\left\{x_{ij} \in \mathrm{S}(\tilde{x}_{ij}^{I}) \mid \mu_{\tilde{x}_{ij}^{I}}(x_{ij}) \geq \alpha\right\}\right],$$
$$\forall i, j \,\&\, 0 < \alpha \leq 1.$$

$$(\tilde{y}_{rj}^{I})_{\alpha} = \left\{y_{rj} \in \mathrm{S}(\tilde{y}_{rj}^{I}) \mid \mu_{\tilde{y}_{rj}^{I}}(y_{rj}) \geq \alpha\right\} = \left[(y_{rj})_{\alpha}^{L}, (y_{rj})_{\alpha}^{U}\right]$$
$$= \left[\min_{y_{rj}}\left\{y_{rj} \in \mathrm{S}(\tilde{y}_{rj}^{I}) \mid \mu_{\tilde{y}_{rj}^{I}}(y_{rj}) \geq \alpha\right\}, \max_{y_{rj}}\left\{y_{rj} \in \mathrm{S}(\tilde{y}_{rj}^{I}) \mid \mu_{\tilde{y}_{rj}^{I}}(y_{rj}) \geq \alpha\right\}\right],$$
$$\forall r, j \,\&\, 0 < \alpha \leq 1.$$

and

$$(\tilde{p}_{rk}^{I})_{\alpha} = \left\{p_{rk} \in \mathrm{S}(\tilde{p}_{rk}^{I}) \mid \mu_{\tilde{p}_{rk}^{I}}(p_{rk}) \geq \alpha\right\} = \left[(p_{rk})_{\alpha}^{L}, (p_{rk})_{\alpha}^{U}\right]$$
$$= \left[\min_{p_{rk}}\left\{p_{rk} \in \mathrm{S}(\tilde{p}_{rk}^{I}) \mid \mu_{\tilde{p}_{rk}^{I}}(p_{rk}) \geq \alpha\right\}, \max_{p_{rk}}\left\{p_{rk} \in \mathrm{S}(\tilde{p}_{rk}^{I}) \mid \mu_{\tilde{p}_{rk}^{I}}(p_{rk}) \geq \alpha\right\}\right],$$
$$\forall r, k \,\&\, 0 < \alpha \leq 1.$$

Further, using these α-cuts, IFREM can be transformed into a crisp model. Owing to the input and output data being intuitionistic fuzzy numbers, the efficiency score is also an intuitionistic fuzzy number denoted by $\tilde{\mathrm{RE}}_k{}^{I}$ with membership function $\mu_{\tilde{\mathrm{RE}}_k{}^{I}}$. Let $\mathrm{S}(\tilde{\mathrm{RE}}_k{}^{I})$ be the support of $\tilde{\mathrm{RE}}_k{}^{I}$.

$$\mathrm{S}(\tilde{\mathrm{RE}}_k{}^{I}) = \left\{\mathrm{RE}_k \mid \mu_{\tilde{\mathrm{RE}}_k{}^{I}}(\mathrm{RE}_k) > 0\right\} \quad \forall\, k$$

and α-cut is

$$(\tilde{\mathrm{RE}}_k{}^{I})_{\alpha} = \left\{\mathrm{RE}_k \in \mathrm{S}(\tilde{\mathrm{RE}}_k{}^{I}) \mid \mu_{\tilde{\mathrm{RE}}_k{}^{I}}(\mathrm{RE}_k) \geq \alpha\right\} = \left[(\mathrm{RE}_k)_{\alpha}^{L}, (\mathrm{RE}_k)_{\alpha}^{U}\right]$$
$$= \left[\min_{\mathrm{RE}_k}\left\{\mathrm{RE}_k \in \mathrm{S}(\tilde{\mathrm{RE}}_k{}^{I}) \mid \mu_{\tilde{\mathrm{RE}}_k{}^{I}}(\mathrm{RE}_k) \geq \alpha\right\}, \max_{\mathrm{RE}_k}\left\{\mathrm{RE}_k \in \mathrm{S}(\tilde{\mathrm{RE}}_k{}^{I}) \mid \mu_{\tilde{\mathrm{RE}}_k{}^{I}}(\mathrm{RE}_k) \geq \alpha\right\}\right],$$
$$\forall k \,\&\, 0 < \alpha \leq 1.$$

5.1 Model Based on α-Cut

Now using α-cut, we obtain the lower and upper bound RE-DEA models. Model 3 is reduced to Models 4 and 5 as follows:

Model 4

$$(\mathrm{RE}_k)_\alpha^L = \min_{\substack{(x_{ij})_\alpha^L \le x_{ij} \le (x_{ij})_\alpha^U \\ (y_{rj})_\alpha^L \le y_{rj} \le (y_{rj})_\alpha^U \\ (p_{rk})_\alpha^L \le p_{rk} \le (p_{rk})_\alpha^U}} \begin{cases} \max \mathrm{RE}_k = \dfrac{\sum_{r=1}^s p_{rk} y_{rk}}{\sum_{r=1}^s p_{rk} y_r} \\ \sum_{j=1}^n \lambda_j x_{ij} \le x_{ik}, & i = 1, 2, \ldots, m, \\ \sum_{j=1}^n \lambda_j y_{rj} \ge y_r, & r = 1, 2, \ldots, s, \\ \lambda_j, y_r \ge 0. \end{cases}$$

Model 5

$$(\mathrm{RE}_k)_\alpha^U = \max_{\substack{(x_{ij})_\alpha^L \le x_{ij} \le (x_{ij})_\alpha^U \\ (y_{rj})_\alpha^L \le y_{rj} \le (y_{rj})_\alpha^U \\ (p_{rk})_\alpha^L \le p_{rk} \le (p_{rk})_\alpha^U}} \begin{cases} \max \mathrm{RE}_k = \dfrac{\sum_{r=1}^s p_{rk} y_{rk}}{\sum_{r=1}^s p_{rk} y_r} \\ \sum_{j=1}^n \lambda_j x_{ij} \le x_{ik}, & i = 1, 2, \ldots, m, \\ \sum_{j=1}^n \lambda_j y_{rj} \ge y_r, & r = 1, 2, \ldots, s, \\ \lambda_j, y_r \ge 0. \end{cases}$$

Models 4 and 5 have been developed to assess the minimum and maximum thresholds of revenue efficiency using the α-cut method. Consequently, Model 6, presented below, is formulated as the RE-DEA model that gauges the lower limit of RE_k based on α-cut approach. In the objective function, the output price of DMU_k is modified to be at the lowest possible level. The constraints are tailored to be unfavorable for the evaluated DMU while favoring the other DMUs. The inputs are set at their upper limits and the outputs at their lower limits for DMU_k. Conversely, for other units, the inputs are set at their lower limits and the outputs at their upper limits. Thus, the mathematical expression for the lower bound RE-DEA model of DMU_k (Model 6) is as follows:

Model 6:

$$\max (\mathrm{RE}_k)_\alpha^L = \frac{\sum_{r=1}^s (p_{rk})_\alpha^L (y_{rk})_\alpha^L}{\sum_{r=1}^s (p_{rk})_\alpha^L y_r}$$

subject to

$$\sum_{\substack{j=1 \\ j \ne k}}^n \lambda_j (x_{ij})_\alpha^L + \lambda_k (x_{ik})_\alpha^U \le (x_{ik})_\alpha^U, \qquad i = 1, 2, \ldots, m,$$

$$\sum_{\substack{j=1\\j\neq k}}^{n} \lambda_j (y_{rj})_\alpha^U + \lambda_k (y_{rk})_\alpha^L \geq y_r, \qquad r = 1, 2, \ldots, s,$$

$$\lambda_j,\ y_r \geq 0.$$

Model 7, presented below, is a type of RE-DEA model that determines the upper limit of RE_k. In the objective function, the output price of DMU_k is adjusted to the highest level. The constraints are arranged in a way that benefits the other DMUs and penalizes the one being evaluated. For DMU_k, inputs are set at the lowest bounds and outputs are set at the highest bounds. On the other hand, for other units, inputs are set at the highest bounds, while outputs are set at the lowest bounds. Hence, the upper bound RE-DEA model of DMU_k (Model 7) is formulated as follows:

Model 7:

$$\max (\text{RE}_k)_\alpha^U = \frac{\sum_{r=1}^{s} (p_{rk})_\alpha^U (y_{rk})_\alpha^U}{\sum_{r=1}^{s} (p_{rk})_\alpha^U y_r}$$

subject to

$$\sum_{\substack{j=1\\j\neq k}}^{n} \lambda_j (x_{ij})_\alpha^U + \lambda_k (x_{ik})_\alpha^L \leq (x_{ik})_\alpha^L, \qquad i = 1, 2, \ldots, m,$$

$$\sum_{\substack{j=1\\j\neq k}}^{n} \lambda_j (y_{rj})_\alpha^L + \lambda_k (y_{rk})_\alpha^U \geq y_r, \qquad r = 1, 2, \ldots, s,$$

$$\lambda_j,\ y_r \geq 0.$$

Definition 6 Efficient and inefficient DMU based on α-cut:

- If $(\text{RE}_k)_\alpha^{L*} = (\text{RE}_k)_\alpha^{U*} = 1$ for any $\alpha \in (0, 1]$, then DMU_k is called α-based strongly revenue-efficient.
- If $(\text{RE}_k)_\alpha^{U*} = 1$ and $(\text{RE}_k)_\alpha^{L*} < 1$ for any $\alpha \in (0, 1]$, then DMU_k is called α-based weakly revenue-efficient.
- If $(\text{RE}_k)_\alpha^{U*} < 1$ and $(\text{RE}_k)_\alpha^{L*} < 1$ for any $\alpha \in (0, 1]$, then DMU_k is called α-based revenue-inefficient.

Axiom: $(\text{RE}_k)_\alpha^{L*} \leq (\text{RE}_k)_\alpha^{U*}\ \forall\ \alpha \in (0, 1]$, i.e., the lower bound revenue efficiency is less than or equal to the upper bound revenue efficiency of DMU under evaluation.

5.2 Model Based on β-Cut

Now using β-cut, we obtain the lower and upper bound RE. Model 3 is reduced to Model 8 and 9 as follows:

Model 8

$$(\mathrm{RE}_k)'^{L}_{\beta} = \min_{\substack{(x_{ij})'^{L}_{\beta} \le x_{ij} \le (x_{ij})'^{U}_{\beta} \\ (y_{rj})'^{L}_{\beta} \le y_{rj} \le (y_{rj})'^{U}_{\beta} \\ (p_{rk})'^{L}_{\beta} \le p_{rk} \le (p_{rk})'^{U}_{\beta}}} \begin{cases} \max \mathrm{RE}_k = \dfrac{\sum_{r=1}^{s} p_{rk} y_{rk}}{\sum_{r=1}^{s} p_{rk} y_r} \\ \sum_{j=1}^{n} \lambda_j x_{ij} \le x_{ik}, & i = 1, 2, \ldots, m, \\ \sum_{j=1}^{n} \lambda_j y_{rj} \ge y_r, & r = 1, 2, \ldots, s, \\ \lambda_j, y_r \ge 0. \end{cases}$$

Model 9

$$(\mathrm{RE}_k)'^{U}_{\beta} = \max_{\substack{(x_{ij})'^{L}_{\beta} \le x_{ij} \le (x_{ij})'^{U}_{\beta} \\ (y_{rj})'^{L}_{\beta} \le y_{rj} \le (y_{rj})'^{U}_{\beta} \\ (p_{rk})'^{L}_{\beta} \le p_{rk} \le (p_{rk})'^{U}_{\beta}}} \begin{cases} \max \mathrm{RE}_k = \dfrac{\sum_{r=1}^{s} p_{rk} y_{rk}}{\sum_{r=1}^{s} p_{rk} y_r} \\ \sum_{j=1}^{n} \lambda_j x_{ij} \le x_{ik}, & i = 1, 2, \ldots, m, \\ \sum_{j=1}^{n} \lambda_j y_{rj} \ge y_r, & r = 1, 2, \ldots, s, \\ \lambda_j, y_r \ge 0. \end{cases}$$

Models 8 and 9 are utilized to determine the lower and upper limits of revenue efficiency using the β-cut approach, respectively. Therefore, Model 10 presented below is the RE-DEA model, which measures the lower bound of revenue efficiency of DMU_k based on the β-cut approach. In the objective function, the output price of DMU_k is adjusted to the upper limit. In the constraints, the levels of inputs and outputs are tailored against the DMU under evaluation and in favor of the other DMUs. For DMU_k, inputs are fixed at upper bounds, and outputs are set at lower bounds. For the other units, inputs are arranged at lower bounds, and outputs are fixed at upper bounds. Thus, the lower bound RE-DEA model of DMU_k (Model 10) based on the β-cut is presented below:

Model 10:

$$\max (\mathrm{RE}_k)'^{L}_{\beta} = \frac{\sum_{r=1}^{s} (p_{rk})'^{L}_{\beta} (y_{rk})'^{L}_{\beta}}{\sum_{r=1}^{s} (p_{rk})'^{L}_{\beta} y_r}$$

subject to

$$\sum_{\substack{j=1\\j\neq k}}^{n} \lambda_j (x_{ij})'^{L}_{\beta} + \lambda_k (x_{ik})'^{U}_{\beta} \leq (x_{ik})'^{U}_{\beta}, \qquad i = 1, 2, \ldots, m,$$

$$\sum_{\substack{j=1\\j\neq k}}^{n} \lambda_j (y_{rj})'^{U}_{\beta} + \lambda_k (y_{rk})'^{L}_{\beta} \geq y_r, \qquad r = 1, 2, \ldots, s,$$

$$\lambda_j,\ y_r \geq 0.$$

Model 11, presented herewith, is the RE-DEA model that assesses the upper bound of RE_k based on β-cut. In the objective function, the output price of DMU_k is adjusted to the lower bound level. In the constraints, the levels of inputs and outputs are tailored to be unfavorable to the DMU under assessment and supportive of the other DMUs. The inputs are set at lower bounds, and outputs are set at upper bounds for DMU_k. Conversely, for other units, inputs are set at upper bounds, and outputs are set at lower bounds. Consequently, the upper bound RE-DEA model of DMU_k (Model 11) based on β-cut can be expressed as follows:

Model 11:

$$\max (\text{RE}_k)'^{U}_{\beta} = \frac{\sum_{r=1}^{s} (p_{rk})'^{U}_{\beta} (y_{rk})'^{U}_{\beta}}{\sum_{r=1}^{s} (p_{rk})'^{U}_{\beta}\, y_r}$$

subject to

$$\sum_{\substack{j=1\\j\neq k}}^{n} \lambda_j (x_{ij})'^{U}_{\beta} + \lambda_k (x_{ik})'^{L}_{\beta} \leq (x_{ik})'^{L}_{\beta}, \qquad i = 1, 2, \ldots, m,$$

$$\sum_{\substack{j=1\\j\neq k}}^{n} \lambda_j (y_{rj})'^{L}_{\beta} + \lambda_k (y_{rk})'^{U}_{\beta} \geq y_r, \qquad r = 1, 2, \ldots, s,$$

$$\lambda_j,\ y_r \geq 0.$$

Definition 7 Efficient and inefficient DMU based on β-cut:

- If $(\text{RE}_k)'^{L*}_{\beta} = (\text{RE}_k)'^{U*}_{\beta} = 1$ for any $\beta \in [0, 1)$, then DMU_k is called β-based strongly revenue-efficient.
- If $(\text{RE}_k)'^{U*}_{\beta} = 1$ and $(\text{RE}_k)'^{L*}_{\beta} < 1$ for any $\beta \in [0, 1)$, then DMU_k is called β-based weakly revenue-efficient.
- If $(\text{RE}_k)'^{U*}_{\beta} < 1$ and $(\text{RE}_k)'^{L*}_{\beta} < 1$ for any $\beta \in [0, 1)$, then DMU_k is called β-based revenue-inefficient.

Axiom: $(\text{RE}_k)'^{L*}_{\beta} \leq (\text{RE}_k)'^{U*}_{\beta}\ \forall\ \beta \in [0, 1)$, i.e., the lower bound revenue efficiency is less than or equal to the upper bound revenue efficiency of DMU under evaluation.

6 Application to the Indian Banking Sector

To demonstrate the practical relevance of the proposed models, an empirical study is conducted on the Indian banking industry. The dataset used in this study is obtained from the official website of the Reserve Bank of India for the year 2021. The study focuses on 12 public sector banks (DMUs), each having two input and two output variables. It should be noted that the choice of input-output variables is a subjective decision of the decision-maker. The chosen input-output variables for the analysis are presented below (Table 1).

The collected data is crisp in nature. To measure the revenue efficiencies of DMUs by the proposed models, the data is converted into IFNs, particularly TIFNs. The lower and upper bound revenue efficiencies of banks based on α-cut is calculated by Models 6 and 7 respectively for $\alpha = 0.25, 0.5, 0.75, 1$. The lower and upper bound revenue efficiencies of banks based on β-cut are calculated by Models 10 and 11, respectively, for $\beta = 0, 0.25, 0.5, 0.75$. The results are presented in Tables 2 and 3.

Table 1 Input-Output Data

Sr. No.	Input	Output	Output price
1.	Capital	Investment	Other income
2.	Borrowings	Advances	Interested earned

Table 2 The lower and upper revenue-efficiencies based on α-cuts

α	0.25	0.5	0.75	1
DMU$_k$	$[(RE_k)^L_{0.25}, (RE_k)^U_{0.25}]$	$[(RE_k)^L_{0.5}, (RE_k)^U_{0.5}]$	$[(RE_k)^L_{0.75}, (RE_k)^U_{0.75}]$	$[(RE_k)^L_1, (RE_k)^U_1]$
1. Bank of Baroda	[0.87,1]	[0.95,1]	[0.95,1]	[0.93,1]
2. Bank of India	[0.58,0.72]	[0.60,0.70]	[0.61,0.69]	[0.61,0.67]
3. Bank of Maharashtra	[0.54,1]	[0.59,0.94]	[0.64,0.85]	[0.68,0.78]
4. Canara Bank	[0.78,1]	[0.81,1]	[0.84,1]	[0.86,0.94]
5. Central Bank of India	[0.91,0.98]	[0.91,1]	[0.90,1]	[0.88,1]
6. Indian Bank	[0.78,1]	[0.80,1]	[0.83,0.97]	[0.86,0.92]
7. Indian Overseas Bank	[0.83,1]	[0.85,1]	[0.84,1]	[0.80,1]
8. Punjab and Sind Bank	[0.61,0.86]	[0.64,0.82]	[0.67,0.78]	[0.69,0.74]
9. Punjab National Bank	[0.91,1]	[0.91,1]	[0.91,1]	[0.89,1]
10. State Bank of India	[0.93,1]	[0.93,1]	[0.93,1]	[0.91,1]
11. UCO Bank	[0.22,0.35]	[0.24,0.34]	[0.25,0.32]	[0.25,0.31]
12. Union Bank of India	[0.59,0.71]	[0.60,0.70]	[0.61,0.68]	[0.61,0.67]

Table 3 The lower and upper revenue-efficiencies based on β-cuts

β	0	0.25	0.5	0.75
DMU ↓	$[(RE_k)_0^L, (RE_k)_0^U]$	$[(RE_k)_{0.25}^L, (RE_k)_{0.25}^U]$	$[(RE_k)_{0.5}^L, (RE_k)_{0.5}^U]$	$[(RE_k)_{0.75}^L, (RE_k)_{0.75}^U]$
1. Bank of Baroda	[0.93,1]	[0.95,1]	[0.88,1]	[0.75,1]
2. Bank of India	[0.61,0.67]	[0.59,0.71]	[0.56,0.74]	[0.52,0.78]
3. Bank of Maharashtra	[0.68,0.78]	[0.57,0.94]	[0.48,1]	[0.39,1]
4. Canara Bank	[0.86,0.94]	[0.78,1]	[0.71,1]	[0.67,1]
5. Central Bank of India	[0.75,1]	[0.90,1]	[0.88,0.98]	[0.72,0.98]
6. Indian Bank	[0.86,0.92]	[0.78,0.88]	[0.77,1]	[0.75,1]
7. Indian Overseas Bank	[0.80,1]	[0.84,1]	[0.77,0.98]	[0.67,0.97]
8. Punjab and Sind Bank	[0.69,0.74]	[0.63,0.83]	[0.57,0.98]	[0.50,0.97]
9. Punjab National Bank	[0.89,1]	[0.89,1]	[0.89,1]	[0.91,1]
10. State Bank of India	[0.91,1]	[0.93,1]	[0.93,1]	[0.93,1]
11. UCO Bank	[0.25,0.31]	[0.23,0.35]	[0.20,0.39]	[0.16,0.43]
12. Union Bank of India	[0.61,0.67]	[0.59,0.70]	[0.56,0.73]	[0.52,0.76]

By Definitions 6 and 7, we conclude that based on α, β-cuts, DMUs: Bank of India, Punjab and Sind Bank, UCO Bank and Union Bank of India are revenue-inefficient, and the rest of the DMUs are weakly revenue-efficient.

7 Conclusion

This paper's principal motivation is the expansion of the conventional revenue efficiency DEA model to the intuitionistic fuzzy environment in DEA. The conventional RE-DEA model is used for crisp data, but for all intents and purposes of real-life problems, we proposed IFREM. The proposed IFREM is an IF model that can be applied to all data types, i.e., crisp, fuzzy and IF. This study develops the lower and upper IFREMs based on α and β-cuts. The proposed models are developed by taking TIFNs. New models can be developed using trapezoidal fuzzy numbers (TrFNs), L-R type fuzzy numbers, or any other form of IF numbers to represent imprecision and hesitation in input-output and output price data.

Furthermore, the proposed IFRE-DEA approach is examined under constant returns to scale (CRS) in this study; however, it can be extended to the variable returns to scale (VRS) approach. The proposed IFREM is demonstrated in a real-world study, specifically in the Indian public sector banking sector. Future work may look at results from other DEA models extended in unfavorable settings like stochastic and robust. New models and approaches can be developed for the performance evaluation of DMUs based on cost and revenue efficiencies.

References

1. Ashrafi, A., & Kaleibar, M. (2017). Cost, revenue and profit efficiency models in generalized fuzzy data envelopment analysis. *Fuzzy Information and Engineering, 9*, 237–246.
2. Arya, A., & Yadav, S. (2019). Development of intuitionistic fuzzy data envelopment analysis models and intuitionistic fuzzy input-output targets. *Soft Computing, 23*, 8975–8993.
3. Atanassov, K. (1986). Intuitionistic fuzzy sets. *Fuzzy Sets and Systems, 20*(1), 87–96.
4. Charnes, A., Cooper, W., & Rhodes, E. (1978). Measuring the efficiency of decision making units. *European Journal of Operational Research, 2*, 429–444.
5. Ghiyasi, M. (2017). Inverse DEA based on cost and revenue efficiency. *Computers & Industrial Engineering, 114*, 258–263.
6. Hussain, H., Kamarudin, F., Mohamad Anwar, N., Nassir, A., Sufian, F., & Tan, K. (2020). Impact of country's governance dimensions on bank revenue efficiency: Overview on middle East, Southeast Asia, and South Asia countries. *Transformations in Business & Economics, 19*.
7. Kao, C., & Liu, S. (2000). Fuzzy efficiency measures in data envelopment analysis. *Fuzzy Sets and Systems, 113*, 427–437.
8. Kordrostami, S., & Jahani Sayyad Noveiri, M. (2019). Fuzzy revenue efficiency in sustainable supply chains. *International Journal of Applied Operational Research, 9*, 63–70
9. Puri, J., & Yadav, S. (2016). A fully fuzzy DEA approach for cost and revenue efficiency measurements in the presence of undesirable outputs and its application to the banking sector in India. *International Journal of Fuzzy Systems, 18*, 212–226.
10. Sengupta, J. (1992). A fuzzy systems approach in data envelopment analysis. *Computers & Mathematics with Applications, 24*, 259–266.
11. Sufian, F., Kamarudin, F., & Noor, N. (2014). Revenue efficiency and returns to scale in Islamic banks: Empirical evidence from Malaysia. *Journal of Economic Cooperation & Development, 35*, 47.
12. Zadeh, L. (1965). Fuzzy sets. *Information and Control, 8*, 338–353.

Environmental Performance of Coal-Based Thermal Generation Plants: A Two-Stage DEA Model with Undesirable Outputs

Shambhavi Mishra, Anoop Singh, and Sufia

Abstract The thermal plants in India operate within a regulated framework under the ambit of the Electricity Act, 2003. The tariff framework has evolved through various stages from a cost-of-service approach to a normative rate of return approach by introducing benchmark norms for various operational parameters. This study evaluates the operational and environmental performance of 20 coal/lignite-based Thermal Power Plants (TPPs) whose tariff is regulated by the Central Electricity Regulatory Commission (CERC). A Two-stage Data Envelopment Analysis (DEA) was applied to 102 DMUs over a period of three financial years, from 2014–15 to 2016–17. The first model estimates the operational performance of the plants, while the second model also includes the environmental indicators as undesirable outputs. Additionally, in the second stage, the efficiency scores from the first stage for both the models are regressed with external factors i.e., average unit size, vintage, and plant type to analyze their impact on the performance of TPPs. OLS regression results of the first model show that vintage is significant, while in the second model, both average unit size and plant type were significant.

Keywords Environmental performance · Two-stage DEA · Thermal power plants · India

1 Introduction

India, under the Paris Agreement, has committed to achieve net-zero emissions by 2070. This commitment includes meeting 50% of cumulative electric power installed capacity from non-fossil fuel-based energy resources by 2030 and reducing emissions intensity of its GDP by 45% by 2030 [1]. Alternatively, the surge in electricity demand has propelled India to become the third-largest global producer

S. Mishra (✉) · A. Singh · Sufia
Centre for Energy Regulation (CER), Department of Management Sciences, Indian Institute of Technology, Kanpur, India
e-mail: sbmishra@iitk.ac.in

A. Emrouznejad et al. (eds.), *Advances in the Theory and Applications of Performance Measurement and Management*, Lecture Notes in Operations Research,
https://doi.org/10.1007/978-3-031-61597-9_13

and consumer, boasting an installed capacity of 417.67 GW as of May 31, 2023, trailing behind only China and the United States. India's energy landscape comprises 41.4% (173.62 GW) from renewable sources, including large hydro, making it the second largest after coal, which still commands 49.1% (205.24 GW) [2]. Despite the increasing share of renewable energy, thermal generation continues to dominate India's power sector.

Though, there is decline in coal-based generation from 74% in 2018–19 to 50% in 2022–23, these plants provide a stable base load capacity, making them indispensable [3, 4]. The shift towards renewables, though promising, faces hurdles of intermittency leading to grid instability, due to its dependence on weather conditions. The conventional plants may continue to contribute towards new capacity addition and play significant role in the electricity supply mix, even in the context of growing renewable energy sources. However, Indian coal has a relatively low gross calorific value (GCV) and high ash content compared to the coal quality of other countries.

India's industrial heritage was built upon indigenous coal, which is the most essential and abundant fossil fuel. Approximately 50% of electricity production depends on coal-based thermal generating plants [5]. Thermal power plants have been categorized as one of the 17 highly polluting industries by the Central Pollution Control Board (CPCB) as they generate all kinds of pollution problems, including air pollution, water pollution, and hazardous wastes in India [6]. With a major share of emission, i.e., 60% of Particulate Matter, 45% of SO_2, 30% of NO_2, and 80% of Mercury, coal-based TPPs are contributing more towards polluting the environment than other industrial sectors [7]. Thus, to stringently regulate the emission norms, the Ministry of Environment, Forest and Climate Change (MoEFCC) has issued new revised emission standards in Environment (Protection) Amendment Rules, 2015 for TPPs of different vintages [8].

In this context, it is important not only to analyze the operational performance which is of regulators' interest but also environmental performance which is of government's interest, to meet the renewable energy and net zero emission targets. It is necessary to study the performance of existing coal-based TPPs, check the implementation of existing norms, and further encourage new norms that benefit both the beneficiaries and consumers in the electricity supply chain. This study assesses the environmental and operational performance of 16 central coal-based TPPs and 4 central lignite-based power plants for a period of 3 years from 2014–15 to 2016–17. Model 1 assesses the operational performance, while Model 2 assesses both operational and environmental performance of the TPPs.

The subsequent sections of this paper are structured as follows: In Sect. 2, a comprehensive literature review on the performance analysis of TPPs is presented. Section 3 outlines the methodology employed in this research, detailing the model used. In Sect. 4, the selection process for input and output indicators specific to TPPs is thoroughly explained. The results are presented in Sect. 5, where the factors influencing thermal performance are discussed in detail. Finally, the concluding remarks and future scope are provided in the last section.

2 Performance Analysis of Thermal Power Plants: A Literature Review

Thermal power plants have wide variation in design and operating conditions. In India, thermal plants use a mix of different qualities of coal, which leads to diversity in both coal quality and quantity between plants, resulting in large differences in emission factors. Mittal et al. developed a plant-specific methodology for the estimation of emissions (CO_2, SO_2, NO) from 86 Indian TPPs for the period of 2001–02 to 2009–10 using basic principles of combustion and operating conditions and showed future emission trends [9]. While Chakraborty et al. measured the emissions from coal-based TPPs using IMR 2800 flue gas analyzer. Emission rates from different plants vary due to variability and non-homogeneity in mixing ratios of different quality and quantity of coal, which varies with calorific value [10]. Environmental awareness and initiatives taken by the government towards sustainable development and net zero emissions have motivated researchers to think about environmental efficiency apart from only focusing on operational and economic benefits.

Data Envelopment Analysis (DEA), a linear programming-based performance benchmarking tool used to measure the relative efficiency score of Decision Making Units (DMUs), has been extensively used to do the performance analysis of firms [11–13]. There have been many other studies internationally that have used frontier models for analyzing the operational performance of TPPs [14–16].

Many studies have examined the impact of power sector restructuring, and policy reforms on the performance of TPPs. Malik et al. analyzed the impact of restructuring of the power sector on the operating efficiency of Indian coal-fired plants for a period from 1988 to 2009 by choosing operating reliability and thermal efficiency as output variables [17, 18]. Heshmati found negative impact of restructuring on Korean electricity generation [19]. Nakano and Manai found positive effects of regulatory reforms on Japanese electricity industry [20]. Li and Wang studied the impact of the operational environment due to policy reforms on the technical efficiency of electricity utilities in China to promote sustainable development [21].

Recent developments and policy reforms have emphasized more on the environmental performance of TPPs rather than only analyzing the operational performance. Bajpai and Singh carried out an efficiency analysis of Indian coal-fired plants using 3 different models by changing their input and output variables [22]. Murty et al. used the directional output distance function to determine the technical and environmental efficiency of some Indian coal-fired TPPs. Significant variation in shadow prices for bad outputs among firms was observed due to different environmental regulation compliance levels. Further pollution taxes is suggested to be implemented on firms for better compliance with environmental regulations [23]. Adnan et al. evaluated TPPs in Turkey, considering the cost of power generation and environmental effects, using two efficiency indices, one each for operational performance and environmental performance [24]. Chitkara applied DEA to determine operational inefficiencies in Indian power-generating units, mainly focusing on the causes of these inefficiencies. Research studies often neglect the external conditions that affect the efficient performance of plants. This paper covers all these non-discretionary factors like the vintage of the plant, number of outages, part loading, etc. [11]. Chandel et al. evaluated the operational performance of state-owned TPPs of Punjab and Haryana from 2009–14 and suggested efficient utilization of input resources for overall productive and operational improvement of plants [25].

Two-stage DEA has extensively been used to analyze efficiency of firms in different sectors, and industries. Izadikhah et al. proposed a model to measure sustainability and resilience of supply chains in the transport sector. Further to deal with undesirable data, a new chance constrained fuzzy two-stage DEA model was proposed [26]. Nemati et al. applied two-stage DEA approach to analyze overall and sub-unit efficiencies in refinery industries, considering partial impacts between inputs and outputs assuming that the inputs are separable [27].

Jing Fu applied two-stage DEA in the air quality improvement network in China. They extended the study by addressing disposability factors for desirable and undesirable inputs and outputs [28]. Based on the proportion of thermal power ratio, Xie et al. classified 30 provincial administrative regions in China into three categories and then applied a two-stage environment network DEA [29]. Zeng et al. evaluated the environmental efficiency of China's TPPs using a two-stage DEA model by considering both the production and abatement stages. Results show a negative impact of abatement efficiency on the overall environmental efficiency of a TPP [30]. Liu et al. analyzed two-stage DEA models with undesirable inputs and intermediate outputs using free disposal axioms [31].

Most studies evaluating thermal generator performance use electricity generated or PLF as an output indicator. The present research work analyses the performance of TPPs considering net electricity generated as output and primary production indicators as inputs similar to models found in literature. Most studies neglect the impact of external factors on the performance of thermal plants. Hence, considering these gaps in the literature, the present study encompasses these aspects and evaluates the performance of TPPs in India.

3 Methodology and Model Description

Data Envelopment Analysis (DEA) is a non-parametric and deterministic method based on linear programming (LP) principles. Additionally, alternate approaches within both parametric and non-parametric frameworks have been utilized to analyze operational efficiency. Stochastic Frontier Analysis (SFA) is a parametric approach in which the mathematical form of the production function needs to be pre-defined. The present study uses DEA to examine thermal power plants' operational and environmental efficiency. DEA method analyzes each decision-making unit (DMU) individually. DEA identifies entities that demonstrate best practices, establishing a frontier against which the relative efficiency of each Decision-Making Unit (DMU) is assessed. It calculates the efficiency scores and ranks the DMUs accordingly.

Charnes, Cooper and Rhodes (CCR) and Banker, Charnes and Cooper (BCC) proposed the traditional and most commonly used DEA models. The CCR model was based on constant returns to scale (CRS), while the BCC model accommodates variable returns to scale (VRS) [32]. Since the introduction of conventional DEA models, there have been several contributions to the DEA methodology according to the application required [33, 34]. Another important significance of using DEA as a methodology is its capability to treat multiple inputs and outputs, including non-discretionary and undesirable variables.

In the present research work, the first model investigates the operational efficiency of the TPPs. The second model is the extension of the first model by treating the desired output as uncontrollable along with undesirable outputs since, in the Indian context, the net electricity generated is not under the control of the generator and hence analyzes both operational and environmental performance. Different methods are divided into direct and indirect approaches to treat undesirable variables in DEA [35, 36].

A two-stage DEA analysis is used for this study. Two-stage DEA is an extension of basic DEA methodology in which the first stage is similar to conventional DEA, and the efficiency from the first stage is regressed with external factors in the second stage. Model 1 describes the purely operating efficiency of a TPP considering net electricity generated as output. At the same time, input variables chosen are plant capacity, quantity of coal and oil, and operation and maintenance cost. Input-oriented BCC model was used for the first operating efficiency model's first stage analysis. The VRS efficiency score for the 0th Decision-Making Unit (DMU) can be computed by solving the following Linear Programming (LP) problem:

$$\text{Min } \theta$$

Subject to

$$\sum_{j=1}^{N} Y_{rj}\lambda_j \geq Y_{r0}\forall r = 1, \ldots, m \tag{1}$$

$$\sum_{j=1}^{N} X_{ij}\lambda_j \leq \theta X_{io}\forall i = 1, \ldots, n \tag{2}$$

$$\sum\nolimits_{j=1}^{N} \lambda_j = 1 \tag{3}$$

$$\lambda_j \geq 0\forall j = 1, \ldots\ldots, N \tag{4}$$

where m is the number of outputs (Y), n is the number of inputs (X) and N is the number of DMUs.

Model 2 incorporates both environmental efficiency and operational efficiency of a plant. It considers SPM, SOx, and NOx concentrations from thermal plants as undesirable outputs, along with the desirable outputs from Model 1 and the same set of input variables. In the second model, output-oriented BCC model with undesirable outputs and non-controllable desired output was used to estimate the operational and environmental efficiency.

In the second stage, the efficiency scores were regressed with the external factors for both the models respectively using Ordinary Least Square (OLS) Regression. OLS regression has been preferred over Tobit as efficiency scores are fractional variables, not censored ones [37]. Various external factors can affect the performance of a plant, some of them considered in this study are the vintage of the plant, average unit size, and a dummy variable for sub-critical or super-critical plant type.

4 Selection of Output and Input Indicators for Thermal Power Plants

Identifying input and output variables for frontier-based models is the most essential part. There is a thumb rule to guarantee the discriminatory power of DEA models, i.e., the number of DMUs (n) must be greater than thrice the total number of inputs and outputs [38, 39]. Suppose the number of DMUs are insufficient as per the aforesaid rule, then the data can be categorized into operational and non-operational inputs, and integrated-DEA approach can be used to deal with lesser number of DMUs [39]. The selection of inputs and outputs should be done considering that they should explain the underlying process.

The research study comprises self-reported data from 20 NTPC ISGS plants ranging from 2014–15 to 2016–17, thus falling into tariff period 2014–19, which is collected from operational data by stakeholders published by CERC [40]. The data collected has been cross verified from Thermal Performance Review published by the Central Electricity Authority (CEA) [41]. Out of 20 plants, 16 plants are coal-based, while 4 plants are lignite-based. Since the plants come up in stages, and each stage can comprise one or more units of varying capacities, the 20 plants have been segregated stage-wise into 34 DMUs for each year respectively, thus summing up to 102 DMUs for three years.

The nature of a basic DEA model is to maximize its outputs by minimizing the inputs. But in some cases, it is required to minimize the undesirable outputs due to the production process. There are also some external factors that affect the performance of thermal plants. There are three types of variables considered in the present analysis. These are input variables, output variables, and external factors, which have been defined as follows:

Input Variables: The primary production indicators of power generation in a thermal power plant have been chosen as input variables.

i. Capacity (CAP) (MW)—It is the total capacity of all units in a particular stage of a plant.

ii. Operation and Maintenance Cost (O&M) (Rs Lakh)—It is the total annual operation and maintenance cost incurred for each plant stage.
iii. Quantity of Coal (Qc) (MT)—It is the total quantity of coal used for a particular stage of a plant to generate electricity annually.
iv. Quantity of Oil (Qo) (KL)—It is the total quantity of oil used by a particular stage of a plant in the process of generation of electricity annually.

Output Variables: Net electricity generated is the desired output variable. While emission parameters like SPM, SOx, and NOx, which are undesirable outputs of the power generation process and cause environmental pollution have also been considered in this study. SOx, and NOx cause acid rain and air pollution, while SPM affects visibility during winter and causes respiratory problems.

i. Net Electricity Generated (GEN) (MU)—It is actual net generation ex-bus i.e., the amount of gross electricity generated minus auxiliary power consumption.
ii. Suspended Particulate Matter (SPM) (mg/Nm^3)—Concentration of SPM or atmospheric aerosol particles that are suspended in the air, measured under standard temperature and pressure conditions from a thermal power plant.
iii. SOx (mg/Nm^3)—Pollutant Concentration of oxides of sulphur measured under standard temperature and pressure from a thermal power plant.
iv. NOx (mg/Nm^3)—Pollutant Concentration of oxides of nitrogen measured under standard temperature and pressure from a thermal power plant.

External Factors: Various external factors can affect the performance of a thermal plant. Here, average unit size, vintage of plant, and plant type (whether a plant is sub-critical or super-critical), were selected as external factors influencing the performance of thermal plants.

i. Average Unit Size (MW)—It is the average of capacities of all units in a particular stage of a plant.
ii. Vintage (years)—It is the age of a stage of TPP from the date of commissioning.
iii. Plant Type—This is a dummy variable which has been used for differentiating between a sub-critical and super-critical plant. Plants with units having capacity of 500 MW and above are the super-critical plants with value 1 while units with capacities below 500 MW are sub-critical TPPs which 0 value.

Summary statistics for input, output and environmental variables are presented in Table 1.

5 Factors Affecting Thermal Efficiency: Results and Discussions

Figure 1 displays the results of Stage 1 in Model 1. The relative efficiency scores for all the 102 DMUs vary from 0.56 to 1. The lignite plants, i.e., NLC TPS-I, NLC TPS-II Stage-I, and Stage-II, show the least efficiency score. The reason for

Table 1 Summary statistics for all the variables for the period 2014–15 to 2016–17

Variables	Unit	Max	Min	Mean	Median	Std. Dev
Input variables						
CAP	MW	2100	210	1002.8	1000	533.2
Qc	MT	12,228,643	877,004	5,130,178	4,850,176	2,921,778
Qo	KL	16,670.5	52.97	2942.8	2277.2	2509.95
O&M	Rs Lakh	59,823.6	6363.4	25,798.8	22,287.1	13,410.4
Desirable output						
GEN	MU	15,509.7	1102.7	6459.7	6258.5	4063.0
Undesirable outputs						
SPM	Mg/Nm^3	261.9	25	108.7503	89.5	66.0749
SOx	Mg/Nm^3	2908	334	870.0252	718.545	455.375
NOx	Mg/Nm^3	1150	46	416.0097	357	226.638
External factors						
Avg. unit size	MW	660	66.67	348.97	350	165.22
Vintage	Years	50.04	1.00	15.50	14.17	10.49

this inefficiency is the high consumption of coal and oil, with higher operation and maintenance costs. Also, it is observed that NLC TPS-I has very low net electricity generation. 12 DMUs had an efficiency score of 1. Most of the efficient plants are of super-critical type. Efficiency scores of 72 DMUs lie between 0.93 and 1, reflecting that the relative operational performance of most of NTPC ISGS plants is good.

Figure 2 shows the relative efficiency scores of Model 2 Stage 1. Model 2 considers both operational and environmental performance of plants and shows variation of relative efficiency scores from 0.3 to 1. 42 DMUs had an efficiency score of 1. This is because of the inclusion of three undesirable outputs, i.e., SPM, SOx, and NOx, along with the desirable outputs of Model 1. The increase in the number of variables in a DEA model gives some DMUs the advantage of getting efficient, which were earlier inefficient as their emissions are very low.

Here, it is observed that the worst performing DMUs of Model 1 were efficient as they were the least polluting plants. 7 DMUs, i.e., Sipat Stage-I, Unchahar Stage-III, Dadri Stage-II, Rihand Stage-III, NLC TPS-I Ext, and NLC TPS-II Stage-I and Stage-II, were consistently efficient during the three years of study. Due to higher emission parameters, Ramagundam Stage-I&II and Talcher TPS were the least efficient during the last two years of the study.

Table 2 presents the results of the second-stage regression. In both models, the efficiency scores from the first stage were regressed against average unit size, vintage, and plant type.

The coefficient of vintage is negative and significant for the first model, which signifies that operational efficiency decreases with the age of the plant. However, the sign of the coefficient of vintage is negative in the case of the second model and is not statistically significant. The coefficient of average unit size is positive in

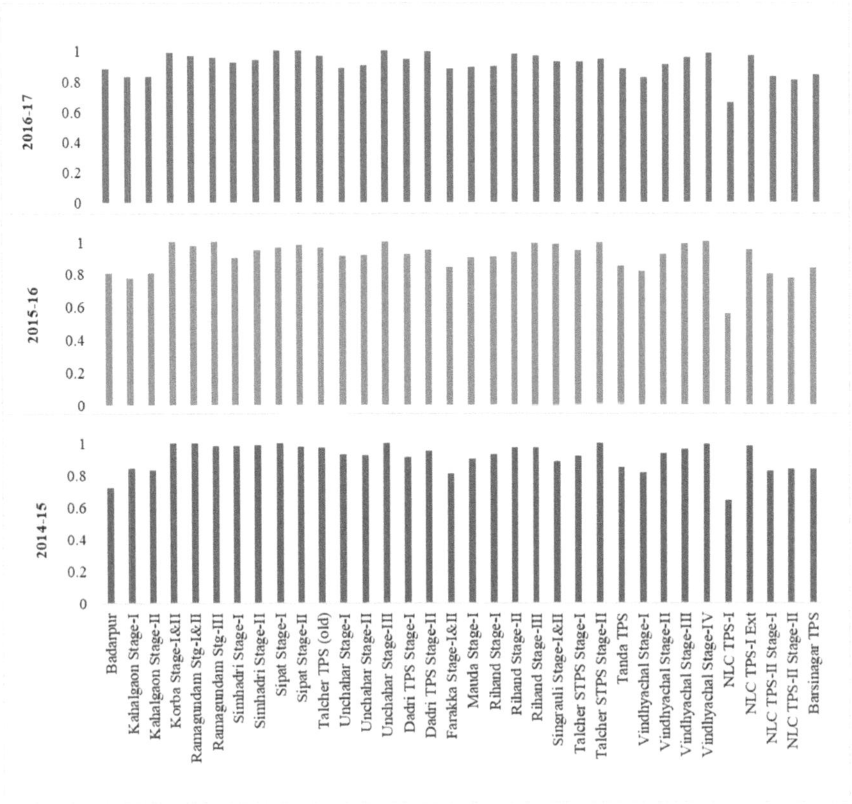

Fig. 1 Model 1 Results: Operational relative efficiency scores for thermal power plants

both models but statistically significant in the second model, implying an increase in efficiency with an increase in average unit size. An increase in efficiency score with increasing average unit size can be because of economies of scale. The coefficient of plant type in Model 1 represents that the super-critical plants are more efficient than the sub-critical ones. The combined operational and environmental efficiency of super-critical plants is found to be lower than that for the sub-critical plants and statistically significant. This contrasts with the general expectation that the super-critical plants are operationally and environmentally efficient.[1] Depending on data availability and a longer study duration, other variables such as plant life extension with higher vintage, installation of pollution control devices and regulatory changes might provide further insights.

[1] Stage-wise analysis of the plants was done, and more data will allow us to check whether data is robust or not.

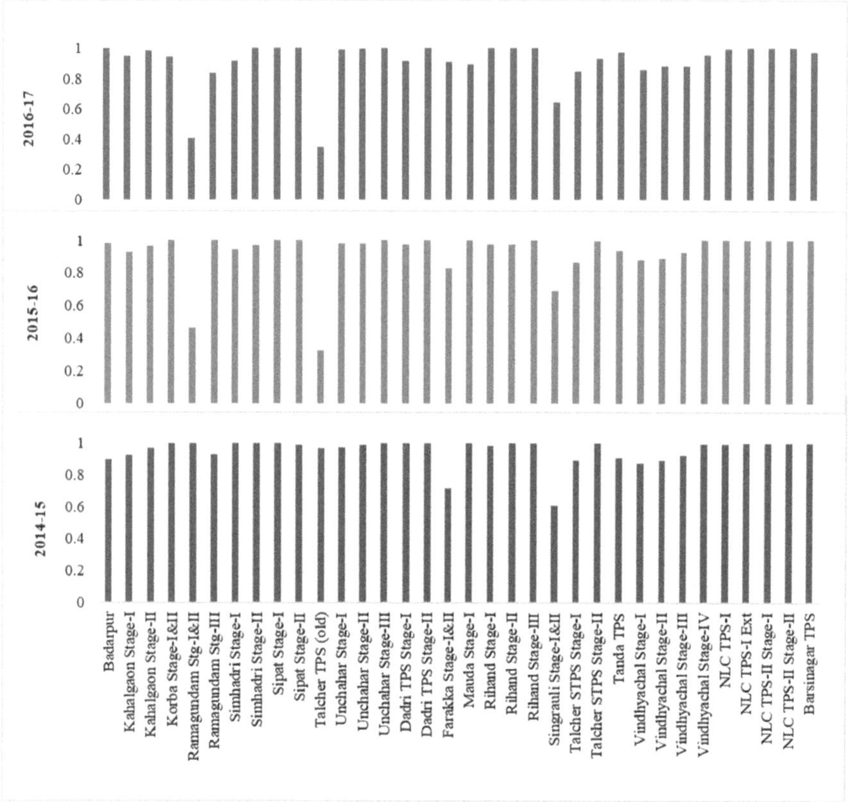

Fig. 2 Model 2 Results: Operational and environmental relative efficiency scores for thermal power plants

Table 2 Stage 2: Regression results of both the models

Regressor	Model 1: Op. Eff		Model 2: Op. and Env. Eff	
	Coef.	Std. Err.	Coef	Std. Err.
Avg. unit size (MW)	0.0001	0.0001	0.0006***	0.0002
Vintage (years)	−0.0026**	0.0008	−0.0019	0.0014
Plant type	0.0342	0.0274	−0.1893***	0.0475
Constant	0.8981***	0.0313	0.8737***	0.0543
R square	0.3565		0.2253	
Adj. R Square	0.3368		0.2015	
Std. error	0.0689		0.1196	
Observations	102		102	

Signif. codes: 0 '***', 0.001 '**'

6 Conclusions and Scope for Further Research

Coal-based plants belonging to NTPC were analyzed for this study, which is considered as an efficient performing Central Sector entity. The results reveal that over 70 DMUs had relative efficiency scores between 0.93 and 1. However, the results may portray a different picture if state level plants, which generally tend to be more inefficient, are included in the analysis.[2] Super-critical power plants, having higher thermal efficiency, place a lower carbon burden. However, results of the study reveal that in the context of local emissions, super-critical plants are found to be relatively inferior in terms of emission intensity.

Further investigation involving plants from the state sector as well as the private sector over an extended period can be interesting future work which can provide further useful insights. This would be useful for designing appropriate policy and regulatory framework for adoption of efficient and environmentally benign technologies for the thermal power plants.

References

1. India' s Updated First Nationally Determined Contribution Under Paris Agreement. (2022).
2. Government of India, M. of P. Power sector at a glance ALL INDIA. (2023). *Power sector at a glance ALL INDIA*. https://powermin.gov.in/en/content/power-sector-glance-all-india
3. Authority, C. E. (2023). National Electricity Plan. **I**.
4. Policy, E., Level, D., & Reform, G. (2021). National Electricity Policy, 2021 (Draft). **2021.**
5. Government of India, M. of C. Coal- Indian Energy Choice. *Major Statistics* https://coal.nic.in/en/major-statistics/coal-indian-energy-choice
6. CPCB_Latest_Final_Directions_highly polluting industries.pdf
7. Bhati, P., & Ramanathan, S. (2017). National clean energy fund to clean coal power plants (pp. 1–16). Centre for Science and Environment. New Delhi.
8. Powe, T. (1999). IPC – II Division Thermal Power Plants 1. Environmental Regulations.
9. Mittal, M. L. (2010). Estimates of emissions from coal fired thermal power plants in India. 39, 1–22.
10. Chakraborty, N., et al. (2008). Measurement of CO_2, CO, SO_2, and NO emissions from coal-based thermal power plants in India. *Atmospheric Environment, 42*, 1073–1082.
11. Chitkara, P. (1999). A data envelopment analysis approach to evaluation of operational inefficiencies in power generating units: A case study of indian power plants. *IEEE Transactions on Power Systems, 14*, 419–425.
12. Chowdhury, S. A., Aziz, S., & Hossan, M. B. (2022). Cost efficiency evaluation of thermal power plants in Bangladesh using a two-stage DEA model. *Economics of Energy & Environmental Policy, 11*, 145–163.
13. Sueyoshi, T., Goto, M., & Ueno, T. (2010). Performance analysis of US coal-fired power plants by measuring three DEA efficiencies. *Energy Policy, 38*, 1675–1688.
14. Ferdushi, K. F., Kamil, A. A., Ahmed, S., & Kawsar, L. A. (2020). The measures of efficiency of power generation plants in Sylhet of Bangladesh. *International Journal of Mathematics and Mathematical Sciences, 2020.*

[2] The study may also include IPPs if appropriate data is available. Such plants are generally of recent vintage and have higher unit size.

15. See, K. F., & Coelli, T. (2012). An analysis of factors that influence the technical efficiency of Malaysian thermal power plants. *Energy Economics, 34*, 677–685.
16. Sirasoontorn, P. (2011). Privatisation and technical efficiency of Thai electricity generation state owned enterprise. *SSRN Electronic Journal*, 1–37. https://doi.org/10.2139/ssrn.677070
17. Malik, K., Cropper, M., Limonov, A., & Singh, A. (2011). Estimating the impact of restructuring on electricity generation efficiency: The case of the Indian thermal power sector. *NBER Working Papers, 17383*, 17383.
18. Economics, E. (2019). The impact of electricity sector restructuring on coal-fired power plants in India author (s): Kabir Malik, Maureen Cropper, Alexander Limonov and Anoop Singh Source. *The Energy Journal, 36*(4), 287–312. (October 2015).
19. Heshmati, A. (2013). Efficiency and productivity impacts of restructuring 1. *Introduction and Background, 14*, 57–89.
20. Nakano, M., & Managi, S. (2008). Regulatory reforms and productivity: An empirical analysis of the Japanese electricity industry, *36*, 201–209.
21. Li, F., & Wang, W. (2019). Incentivizing sustainable development : The impact of a recent policy reform on electricity production efficiency in China, 1–11. https://doi.org/10.1002/sd.1942
22. Bajpai, V. K., & Singh, S. K. (2014). Measurement of operational and environmental performance of the coal-fired power plants in India by using data envelopment analysis, *8*, 3979–3988.
23. Surender, M. N. M. Æ., & Kishore, K. Æ. (2007). Measuring environmental efficiency of industry : A case study of thermal power generation in India, 31–50. https://doi.org/10.1007/s10640-006-9055-6
24. Sözen, A., Alp, I., & Özdemir, A. (2010). Assessment of operational and environmental performance of the thermal power plants in Turkey by using data envelopment analysis. *Energy Policy, 38*, 6194–6203.
25. Chandel, R., Singh, H., & Kumar, R. (2017). Performance evaluation of state-owned thermal power plants in northern India using DEA. *International Journal of Global Energy Issues, 40*, 380–399.
26. Izadikhah, M., Azadi, M., Toloo, M., & Hussain, F. K. (2021). Sustainably resilient supply chains evaluation in public transport: A fuzzy chance-constrained two-stage DEA approach. *Applied Soft Computing, 113*, 107879.
27. Nemati, M., Kazemi Matin, R., & Toloo, M. (2020). A two-stage DEA model with partial impacts between inputs and outputs: Application in refinery industries. *Annals of Operations Research, 295*, 285–312.
28. Fu, J. (2018). Two-stage data envelopment analysis with undesirable intermediate measures: An application to air quality improvement in China. *Central European Journal of Operation Research, 26*, 861–885.
29. Xie, B., Fan, Y., & Qu, Q. (2012). Does generation form influence environmental efficiency performance ? An analysis of China's power system. *Applied Energy, 96*, 261–271.
30. Zeng, X., Zhou, Z., Liu, Q., Xiao, H., & Liu, W. (2020). Environmental efficiency and abatement potential analysis with a two-stage DEA model incorporating the material balance principle. *Computers & Industrial Engineering, 148*, 106647.
31. Liu, W., Zhou, Z., Ma, C., Liu, D., & Shen, W. (2015). Two-stage DEA models with undesirable input-intermediate-outputs. *Omega (United Kingdom), 56*, 74–87.
32. Banker, R. D., Charnes, A., & Cooper, W. W. (1984). Some models for estimating technical and scale inefficiencies in data envelopment analysis. *Management Science, 30*, 1078–1092.
33. Toloo, M., & Mensah, E. K. (2019). Robust optimization with nonnegative decision variables: A DEA approach. *Computers & Industrial Engineering, 127*, 313–325.
34. Toloo, M., Mensah, E. K., & Salahi, M. (2022). Robust optimization and its duality in data envelopment analysis. *Omega (United Kingdom)*, 108, 102583.
35. Halkos, G., & Petrou, K. N. (2019). Treating undesirable outputs in DEA: A critical review. *Economic Analysis and Policy, 62*, 97–104.
36. Hua, Z., & Bian, Y., Chapter 6 DEA with undesirable factors.

37. McDonald, J. (2009). Using least squares and tobit in second stage DEA efficiency analyses. *European Journal of Operational Research, 197*, 792–798.
38. Friedman, L., & Sinuany-Stern, Z. (1998). Combining ranking scales and selecting variables in the DEA context: The case of industrial branches. *Computers & Operations Research, 25*, 781–791.
39. Mahmoudi, R., Emrouznejad, A., Khosroshahi, H., Khashei, M., & Rajabi, P. (2019). Performance evaluation of thermal power plants considering CO2 emission: A multistage PCA, clustering, game theory and data envelopment analysis. *Journal of Cleaner Production, 223*, 641–650.
40. CERC. Operational data received from stakeholders. https://cercind.gov.in/O&M_Data.html
41. Central, G. of I. M. of P. (2016). Review of performance of thermal power station 2016–17.

Does Income Diversification Affect Bank Efficiency, Effectiveness and Performance? Evidence from Indian Banking Industry

Mehak Gupta and Sunil Kumar

Abstract Banks are currently facing significant pressure to enhance their performance and bolster their financial stability. The strategy of diversifying income has emerged as a powerful tool to aid banks in improving their performance, particularly when their interest income is strained. The present study delves into the impact of income diversification on the efficiency, effectiveness, and overall performance of Indian banks. In this study, we conceptualize bank performance as an outcome of two distinct yet interconnected components: efficiency and effectiveness. To measure these facets of bank performance, we construct a two-stage performance evaluation model that accounts for undesirable outputs. To quantify bank performance and its components, we model a two-stage performance evaluation model with undesirable outputs and applied an extended version of Tone and Tsutsui (2009) network SBM model. To investigate the linear and non-linear impact of income diversification on efficiency, effectiveness, and performance, we employ panel LASSO technique. The findings highlight the presence of performance disparities among Indian banks, categorized by ownership groups. Remarkably, private banks outperform their counterparts in the public sector across all dimensions of performance. Interestingly, our study reveals a negative and significant influence of income diversification on bank efficiency, effectiveness, and overall performance.

Keywords Two-stage network DEA model · LASSO regression · Bank ownership · Bank performance · Income diversification

M. Gupta · S. Kumar (✉)
Faculty of Economics, South Asian University, New Delhi 110074, India
e-mail: skumar@econ.sau.ac.in

M. Gupta
e-mail: mehakgupta@students.sau.ac.in

A. Emrouznejad et al. (eds.), *Advances in the Theory and Applications of Performance Measurement and Management*, Lecture Notes in Operations Research,
https://doi.org/10.1007/978-3-031-61597-9_14

1 Introduction

Why do banks choose to diversify? It is an intriguing issue that arises in banking literature. The primary justifications for this strategic approach of banks are twofold. First, banks diversify their portfolios to reduce credit risk and improve the quality of their assets [20, 30]. Second, diversification promotes greater stability by reducing vulnerability to shocks and assuring a more consistent and stable revenue stream [13, 48]. In simplified terms, bank diversification encompasses a variety of activities, such as income generation, lending financial services, and non-lending financial services, which ultimately results in increased revenue and stability [4].

In recent years, banks have been minimizing their reliance on income derived from traditional activities and are instead moving towards diversifying their income sources through fee-based non-traditional activities. This trend has spawned voluminous literature on the effects of bank income diversification on various aspects of bank operations. Existing research can be broadly categorized into three groups. The first group comprises studies that analyse the effect of income diversification on bank performance, which is typically measured by accounting ratios such as ROA and ROE (see, for instance, [9, 18, 21] among others). The second group of studies examines the impact of income diversification on bank efficiency, often measured using frontier approaches (see for instance, [3, 12, 15, 17] among others). The third group directs their attention toward studying the effect of income diversification on risk using metrics such as Z-score (see, for example, [1, 29, 47, 52]).

Against this backdrop, the primary objective of the current study is to analyze the influence of income diversification on the operational efficiency, effectiveness, and overall performance of Indian banks. Furthermore, we seek to explore whether this influence holds uniformly across the various ownership groups that characterize the Indian banking industry. To gauge the metrics of efficiency, effectiveness, and performance, we adopt the conceptual framework formulated by [40]. This framework defines an entity's performance as the outcome of two distinct yet interrelated elements: efficiency and effectiveness. To derive empirical estimates for these metrics, we employ the extended version of slack-based network data envelopment analysis (NDEA) model proposed by [49] that incorporates undesirable outputs.

The present study makes a noteworthy contribution to the existing literature concerning the effects of income diversification on bank performance in several significant ways. Firstly, the present study will be perhaps the first that will analyze the impact of income diversification on efficiency, effectiveness, and performance of Indian banks across distinct ownership groups. Secondly, the methodological framework employed in this study is a cohesive and integrated approach that not only facilitates the computation of efficiency, effectiveness, and performance metrics, but also encompasses non-performing loans (NPLs) as an undesirable output. This unique aspect permits the computation of risk-adjusted performance indicators. By employing the Network Data Envelopment Analysis (NDEA) model, the inherent distortion in calculated scores is minimized, ensuring adequate and risk-adjusted estimates. Third, time horizon of this study spans over two decades. The extensive bank-level dataset employed over such a prolonged period engenders more robust findings that can be instrumental in the formulation of sound policies. Lastly, the study further enriches the existing literature by employing an advanced and more rigorous regularized machine learning technique known as Lasso regression. The Lasso regression is applied to dissect the repercussions of income diversification on bank performance and its constituent elements, while effectively managing a substantial number of control variables. To the best of our knowledge, this pioneering endeavor represents the first instance of incorporating Lasso regression into the empirical landscape of income diversification research in India.

This research endeavor sets out to address the following hypotheses:

(a) Does income diversification have a positive impact on the efficiency, effectiveness, and overall performance of Indian banks?
(b) Does the influence of diversification exhibit variations across distinct ownership groups within the Indian banking landscape?
(c) Has the global financial crisis left a lasting imprint on the long-term efficiency, effectiveness, and performance of Indian banks?

The findings derived from the Lasso regression analyses diverge from our initial expectations and do not uphold a positive relationship between income diversification and bank performance. This counterintuitive outcome extends to both the efficiency and effectiveness components of bank performance. Remarkably, private banks emerge as outperforming their public counterparts. Moreover, the lingering effects of the global financial crisis appear to have dissipated.

The subsequent sections of this paper are structured as follows: The second section delves into a brief review of the existing body of literature, examining the effects of income diversification on the various dimensions of bank performance, including efficiency and effectiveness. The third section delves into the specifics of data and variables used and expounds upon the methodological framework adopted for this study. In the fourth section, we present the empirical outcomes derived from our analysis. The final section encapsulates the study by summarizing the key findings and offers important policy implications.

2 Literature Review

This section offers a succinct overview of the existing literature on income diversification, encompassing both theoretical and empirical frameworks. The theoretical underpinnings play a pivotal role in shaping the empirical investigations. Within the theoretical discourse on income diversification, several key theories are illuminated. First, **market power theory** asserts a U-shaped relation between diversification and performance [28, 44]. Second, **resource-based view theory** underlines the long-run benefits of resource utilization due to income diversification [8, 22, 42]. Third,

theory of financial intermediation posits a non-linear relation between income diversification and performance coupled with a decrease in insolvency risk [32]. Fourth, **trade-off theory** advocates a reduced risk of failure in diversified banks as compared to undiversified banks [11, 39]. Fifth, modern portfolio theory contends the use of income diversification as a hedging mechanism, effectively mitigating insolvency risk [52, 53].

Following the theoretical framework, empirical inquiries into income diversification offer valuable insights. Firstly, only a limited number of researchers have delved into the impact of non-interest income components when assessing income diversification (evident in works such as [16, 34, 46]). Secondly, a common metric for quantifying income diversification remains absent from the existing literature (as observed in studies like [28, 36]), leading to a diversity of approaches. Thirdly, empirical investigations focusing on bank income diversification are notably scarce, particularly within developing financial economies such as India (as highlighted by [35, 50]). Fourth, the viewpoints on the impact of income diversification on performance or efficiency exhibit a degree of heterogeneity. Some studies portray income diversification as beneficial (e.g., [45]), while others characterize it as a burden on the financial system (e.g., [33]). Fifth, the integration of the non-linear impact of income diversification into empirical models has been relatively infrequent. Only a handful of researchers have hypothesized the U-shaped relationship between diversification and performance measures, acknowledging its potential significance (e.g., [3]).

The exploration of empirical literature can be categorized into two distinct strands. The first strand revolves around investigating the effects of income diversification on bank performance. The second strand delves into examining the relationship between income diversification and bank efficiency. A concise overview of both these research strands is presented in the subsequent sub-sections.

2.1 Bank Performance and Diversification

A thorough review of the literature pertaining to performance and income diversification offers valuable insights. Contemporary research frequently employs accounting ratios as the primary means to measure bank performance (as observed in studies such as [1, 27]). However, these metrics have limitations that hinder them from accurately capturing the intricate details of bank performance. First, the stability of financial ratio analysis has been called into question [43]. Second, financial ratios are computed under the assumption of normality, a condition that is not always met and can lead to erroneous results [38]. Third, the diversity in empirical methodologies adopted by various researchers has contributed to a wide array of results within the income diversification literature (as exemplified by [10, 27]). Fourth, he interchangeability of the terms "performance" and "profitability" has been prevalent in the literature. However, critical works on bank diversification emphasize that these terms should not be treated as interchangeable. Rather, they represent distinct aspects—"performance" encompasses both efficiency and effectiveness, whereas "profitability" pertains primarily to effectiveness (as highlighted in [24]). In alignment with this perspective, some researchers measure performance as a composite outcome of efficiency and effectiveness (see [26, 41]).

2.2 Bank Efficiency and Diversification

Numerous studies have been undertaken to investigate various dimensions of banks' efficiency, leading to several significant observations. First, a limited body of research has observed a non-linear correlation between efficiency and diversification [3]. Second, within the banking literature, bank efficiency is commonly measured using either Data Envelopment Analysis (DEA) or the Stochastic Frontier Approach (SFA). Among the studies reviewed, those employing the DEA approach outnumber those using SFA. Third, a notable trend within the DEA literature is the emergence of a new research streams such as Network DEA (NDEA) and Dynamic Network DEA (DNDEA) frameworks. The NDEA and DNDEA methodologies consider the interconnected activities within a bank, leading to a more comprehensive evaluation of its operational efficiency. The NDEA and DNDEA approaches have been applied in studies predominantly centered around financially developed economies. However, only a handful of studies have applied these approaches to developing countries like India (see, for example, [5, 7, 23, 31, 51]).

3 Data and Methodology

3.1 Data

The performance metrics and their components for Indian banks are derived from data sourced from the Reserve Bank of India (RBI) and the Indian Banks' Association (IBA). The data is extracted from various issues of Statistical Tables Related to Banks in India, published by the RBI. The sample period is 1998–1999 to 2021–2022. Notably, the years 2007–08 and 2008–2009 are treated as the years of the global financial crisis (GFC). As of March 2023, India's banking sector comprises 12 public sector banks, 21 private banks, and 45 foreign banks. However, this study focuses exclusively on banks under public and private ownership, which collectively constitute approximately 90% of banking operations in India. The research employs an unbalanced panel dataset, totalling 1179 observations for each variable after necessary adjustments. To enhance data quality, all variables in the dataset are standardized as values per branch. Additionally, any banks with missing data entries have been excluded from the analysis.

3.2 Methodology for Computing Bank Performance, Efficiency, and Effectiveness Score

Efficiency and effectiveness stand as central concepts when evaluating and gauging organizational performance [37]. In both profit-driven and non-profit setups, performance embodies a harmonious blend of efficiency and effectiveness. However, a certain inconsistency has emerged within the existing literature, muddling the usage of these terms within the context. While these terms might appear interchangeable to managers, each carries its distinct significance. Drucker [14] demarcated efficiency and effectiveness, relating efficiency to "doing things right" and effectiveness to "doing the right things." Under his framework, measuring efficiency revolves around evaluating an organization's capacity to achieve outputs using minimal inputs. This gauge does not measure marketplace triumph, but rather excellence in operational efficiency and resource utilization [2]. Precision in cost reduction and resource allocation across multiple uses encapsulates efficiency's essence.

On the other hand, [25] elucidated effectiveness as the extent to which an organization attains its predetermined objectives and goals. In essence, an organization's efficacy correlates directly with its goal realization [6]. In essence, effectiveness gauges the attainment of policy objectives. It's vital to note that although efficiency and effectiveness are distinct elements of overall performance, they can still exert influence on one another. To be more specific, efficiency can impact effectiveness, and vice versa, thereby influencing the overarching performance outcome [40]. With this interplay in mind, we define bank performance as the product stemming from the interrelation of bank efficiency and effectiveness i.e.,

$$\text{Performance} = \text{Efficiency} \times \text{Effectiveness}$$

This study employs a network framework to model bank performance, which is delineated into two discrete processes: (i) the Efficiency Process and (ii) the Effectiveness Process. The network architecture for calculating efficiency and effectiveness is illustrated in Fig. 1.

For the computation of efficiency, effectiveness, and performance metrics for individual banks, we extend the network SBM (Slacks-based Measure) model introduced

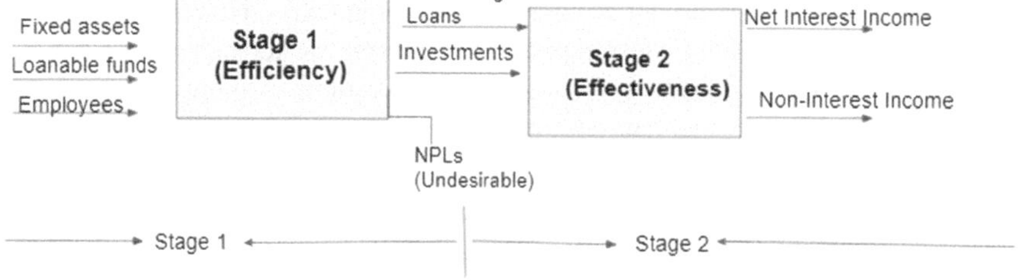

Fig. 1 A two-stage performance evaluation model with undesirable outputs

by [49]. This extension involves the inclusion of undesirable outputs. Let's consider a scenario where there are n banks ($j = 1, \ldots\ldots, n$) and each undergoing two distinct stages: Stage 1 and Stage 2. Withing this context, for each bank j, Stage 1 employs inputs $X_j(x_{j1}, \ldots, x_{jm})$ to produce the desirable outputs $Z_j(z_{j1}, \ldots, z_{jh})$ and undesirable outputs $U_j(u_{j1}, , \ldots, u_{jp})$, while Stage 2 uses desirable outputs $Z_j(z_{j1}, \ldots, z_{jh})$ from Stage 1 to generate desirable outputs $Y_j(y_{j1}, \ldots, y_{js})$. Here, $m, h, p,$ *and* s are the dimensions of vectors $X_j, Z_j, U_j,$ and Y_j,. The $PPS(X, Z, U, Y)$ of the system, given the assumption of strong disposability of undesirable outputs and variable returns to scale, can be defined as follows:

$$PPS(X, Z, U, Y) =$$

$$\sum_{j=1}^{n} \lambda_j^1 x_{ji} \le x_i \qquad i = 1, \ldots, m \tag{1a}$$

$$\sum_{j=1}^{n} \lambda_j^1 u_{jk} \le u_k \qquad k = 1, \ldots, p \tag{1b}$$

$$\sum_{j=1}^{n} \lambda_j^1 z_{jd} \le z_d \qquad d = 1, \ldots, h \tag{1c}$$

$$\sum_{j=1}^{n} \lambda_j^2 z_{jd} \ge z_d \qquad d = 1, \ldots, h \tag{1d}$$

$$\sum_{j=1}^{n} \lambda_j^2 y_{jr} \ge y_r \qquad \mathrm{r} = 1, \ldots, s \tag{1e}$$

$$\sum_{j=1}^{n} \lambda_j^1 = 1 \tag{1f}$$

$$\sum_{j=1}^{n} \lambda_j^2 = 1 \tag{1g}$$

$$\lambda_j^1, \lambda_j^2 \ge 0 j = 1, \ldots, n$$

where λ_j^1 and λ_j^2 are the intensity vector for Stage 1 and Stage 2, respectively. Drawing from the PPS, we formulate a two-stage network SBM model for the "o-th" bank as follows:

$$\text{Minimize } \rho_O = \frac{1 - \frac{1}{m}\sum_{i=1}^{m} \frac{s_i^-}{x_{oi}}}{1 + \frac{1}{s+p}\left(\sum_{r=1}^{s} \frac{s_r^+}{y_{or}} + \sum_{k=1}^{p} \frac{s_k^-}{u_{ok}}\right)} \tag{2a}$$

subject to

$$\sum_{j=1}^{n} \lambda_j^1 x_{ji} + s_i^- = x_{oi} \qquad i = 1, \ldots, m \tag{2b}$$

$$\sum_{j=1}^{n} \lambda_j^1 u_{jk} + s_k^- = u_{ok} \qquad k = 1, \ldots, p \tag{2c}$$

$$\sum_{j=1}^{n} \lambda_j^2 y_{jr} - s_r^+ = y_{or} \qquad \mathrm{r} = 1, \ldots, s \tag{2d}$$

$$\sum_{j=1}^{n} \lambda_j^1 z_{jd} \geq z_{od} \qquad d = 1, \ldots, h \tag{2e}$$

$$\sum_{j=1}^{n} \lambda_j^2 z_{jd} \leq z_{od} \qquad d = 1, \ldots, h \tag{2f}$$

$$\sum_{j=1}^{n} \lambda_j^1 = 1 \tag{2g}$$

$$\sum_{j=1}^{n} \lambda_j^2 = 1 \tag{2h}$$

$$\lambda_j^1, \lambda_j^2, s_i^-, s_k^-, s_r^+ \geq 0$$

The formulations (2e) and (2f) find their basis in the work of [19]. The constraints $\sum_{j=1}^{n} \lambda_j^1 = 1$ and $\sum_{j=1}^{n} \lambda_j^2 = 1$ ensures variable returns to scale. Within this framework, a slack-based measure of efficiency, derived from Stage 1, is defined as follows:

$$\rho_1^* = \frac{1 - \frac{1}{m} \sum_{i=1}^{m} \frac{s_i^-}{x_{oi}}}{1 + \frac{1}{h+p} \left(\sum_{d=1}^{h} \frac{s_d^{+*}}{z_{od}} + \sum_{k=1}^{p} \frac{s_k^{-*}}{u_{ok}} \right)} \tag{3}$$

A slack-based measure of effectiveness derived from Stage 2 is defined as:

$$\rho_2^* = \frac{1 - \frac{1}{h} \sum_{d=1}^{h} \frac{s_d^{+*}}{z_{od}}}{1 + \frac{1}{s} \left(\sum_{r=1}^{s} \frac{s_r^+}{y_{or}} \right)} \tag{4}$$

All computations have been conducted using MaxDEA Ultra software (version 7).

Diversification Index Computation

Diversification index of each bank is computed by using Herfindahl Hirschman index applied by [1] in his paper. The index is computed as follows:

$$DIV = 1 - \left[\left(\frac{\text{Net Interest income}}{\text{Net Operating Income}} \right)^2 + \left(\frac{\text{Non-interest income}}{\text{Net Operating income}} \right)^2 \right]$$

Index value lies between 0 and 0.5. A higher value of index implies a more diversified income mix.

Econometric Specification

To empirically investigate whether bank's income diversification impacts its performance, efficiency and effectiveness, this paper employs regularized regression methodology known as LASSO regression. LASSO regression outperforms OLS regression in terms of its better out of sample prediction. LASSO regression is suitable for handling the complex model like ours having too many control variables. LASSO regression selects only those variables from the list of control variables, which are significant in predicting dependent variable and make the coefficient of insignificant variables zero. Its econometric specification is as follows:

$$y_{it} = \beta_1 DIV_{it} + \beta_2 DIV_{it}^2 + \beta_3 OWN + \beta_4 CRISIS + \beta_i Z_{jit} + \varepsilon_{it} \quad (5)$$

y	efficiency or effectiveness or performance
DIV	Income Diversification Index
OWN	Ownership dummy
CRISIS	Global Financial Crisis dummy
Z_j	control variables (see Table 1).

To obtain β coefficients, the error term is minimized as:

$$\beta_{LASSO}(\lambda) = \arg\min \frac{1}{n} \sum_{i=1}^{n} (y_i - x_i{\prime}\beta)^2 + \frac{\lambda}{n} \sum_{j=1}^{p} \Psi_j |\beta_j| \quad (6)$$

The parameter λ controls overall penalty and Ψ_j represents the predictor specific penalty loadings. LASSO employs the L1 penalty norm, causing certain predictors to be entirely set to zero, thus facilitating model selection. Notably, this technique is sensitive to the scale of variables, unlike the Ordinary Least Squares (OLS) method. To mitigate this sensitivity, pre-standardization is employed during the application of LASSO.

3.3 *Results and Inferences*

Table 1 reports mean score of operational efficiency, effectiveness, and performance on yearly basis. Broadly, grand mean of overall efficiency and performance score stands at 65% and 68% while the effectiveness score stands at 75%. Contrastingly, public banks performance is at a lower edge in comparison to private banks. Results clearly demarcate the differences in mean score of public banks and private banks over the years.

The results of LASSO regression are reported in Tables 2, 3, and 4. Within each table, we present four distinct models. Model A considers income diversification as independent variable, coupled with control variables. To investigate potential non-linear relationships, Model B introduces both income diversification and its squared term. Model C effectively controls for the influence of the Ownership Dummy (Public or Private), alongside income diversification, on the dependent variables. Lastly, Model D incorporates the global financial crisis dummy to account for its effect. Joint significance of LASSO model is tested using W-statistics computed by sup-score. This joint significance test obtains p-value through multiplier bootstrap. Its score is compared with asymptotic 5% critical value (i.e., 3.75).

3.3.1 Efficiency as a Dependent Variable

Table 2 presents the findings stemming from the LASSO regression, focusing on the efficiency of banks as the dependent variable. The results draw attention to the substantial and significant negative influence of income diversification on efficiency. Notably, the relationship between diversification and efficiency seems to follow a linear pattern, and the inclusion of the squared term of diversification increases the joint significance of the lasso model. Additionally, the effect of income diversification varies among distinct ownership groups. A noteworthy observation is the adverse impact of diversification on efficiency for public sector banks, while private banks exhibit a contrasting positive effect on efficiency (as evidenced in Model C). This divergence is attributed to the adoption of modern technology and heightened competition, which contribute to the varying outcomes across ownership categories. Furthermore, the study indicates that the global financial crisis did not exert any discernible impact on bank efficiency. The control variables highlighted by the LASSO regression for the efficiency model reveal a positive influence of capital, non-interest income ratio, and leverage ratio, alongside a negative impact of the deposits/liability's ratio.

Table 1 Average efficiency, effectiveness and performance score

	Overall			Public			Private		
Year	Efficiency	Effectiveness	Performance	Efficiency	Effectiveness	Performance	Efficiency	Effectiveness	Performance
1999	0.680587	0.805286	0.721548	0.569463	0.791808	0.65357	0.771506	0.816314	0.777167
2000	0.694069	0.893946	0.772844	0.638189	0.867082	0.730477	0.741218	0.916612	0.808592
2001	0.697306	0.862319	0.761259	0.625274	0.838711	0.712143	0.760044	0.882881	0.804036
2002	0.703924	0.876825	0.767924	0.633069	0.863498	0.724873	0.767693	0.888819	0.806669
2003	0.669534	0.828804	0.731306	0.601444	0.807099	0.683197	0.730815	0.848338	0.774604
2004	0.62913	0.683559	0.63966	0.580371	0.673928	0.611352	0.673013	0.692227	0.665138
2005	0.640006	0.845309	0.719517	0.61191	0.861058	0.708802	0.668102	0.82956	0.730232
2006	0.565023	0.777185	0.646933	0.522739	0.755451	0.615268	0.607307	0.798919	0.678598
2007	0.621813	0.818186	0.694559	0.543994	0.761141	0.625139	0.708971	0.882076	0.772309
2008	0.635994	0.735376	0.670258	0.569946	0.640308	0.589825	0.716401	0.85111	0.768176
2009	0.67836	0.715846	0.687091	0.621486	0.599463	0.601009	0.74816	0.85868	0.792737
2010	0.597179	0.726508	0.64873	0.52179	0.619503	0.561991	0.589701	0.857833	0.755183
2011	0.610718	0.765385	0.669698	0.554522	0.671237	0.600441	0.680295	0.881948	0.755444
2012	0.638953	0.778857	0.692376	0.57126	0.676524	0.610631	0.726954	0.911889	0.798644
2013	0.657395	0.689444	0.655116	0.588803	0.589453	0.574108	0.746563	0.819431	0.760427
2014	0.645546	0.711643	0.664857	0.575352	0.573998	0.565487	0.736799	0.890582	0.794038
2015	0.613583	0.697852	0.645676	0.561266	0.621143	0.583718	0.684211	0.801409	0.729319
2016	0.644801	0.67272	0.649528	0.59429	0.566298	0.573154	0.709743	0.809548	0.747723
2017	0.639061	0.660309	0.639946	0.545528	0.54449	0.539632	0.754864	0.803704	0.764145
2018	0.680432	0.708521	0.682513	0.592468	0.565302	0.568465	0.768395	0.85174	0.796562
2019	0.710584	0.730625	0.704713	0.679451	0.602747	0.629769	0 738887	0.846877	0.772845

(continued)

Table 1 (continued)

Year	Overall			Public			Private		
	Efficiency	Effectiveness	Performance	Efficiency	Effectiveness	Performance	Efficiency	Effectiveness	Performance
2020	0.720523	0.666592	0.682557	0.6973	0.536781	0.602404	0.739524	0.772801	0.748136
2021	0.716095	0.707008	0.701612	0.636676	0.535587	0.573684	0.761478	0.804963	0.774713
2022	0.721324	0.698071	0.698605	0.684472	0.561709	0.611977	0.742383	0.775993	0.748107
Grand mean	**0.658831**	**0.752341**	**0.689534**	**0.596711**	**0.671847**	**0.618797**	**0.723876**	**0.837261**	**0.763481**

Table 2 Dependent variable: efficiency

Selected variables	Model A		Model B		Model C		Model D	
	LASSO	POST-estimation OLS	LASSO	POST-estimation OLS	LASSO	POST-estimation OLS	LASSO	POST-estimation OLS
Capital			0.00026	0.000179	0.000615	0.000452	0.0006151	0.000452
Leverage ratio	0.5247447	0.447974	0.011262	0.017622				
Spread	0.0099207	0.008296	0.011175	0.010655	0.006799	0.011014	0.0067832	0.011014
Loan/assets	−0.0225742	−0.13193						
Deposits/assets	−0.0035532	0.023936						
Car	0.0000714	0.004132	0.000742	0.004083				
Deposits/liabilities	−0.000094	−0.00085	−0.00049	−0.00141	−0.00069	−0.00199	−0.00068	−0.00199
Non-interest income/assets	0.0077768	0.035787	0.048368	0.070968	0.03079	0.064351	0.0306667	0.064351
Return on advances	0.0115836	0.024253	0.007187	0.018342				
Performing loans	0.0000414	0.000323						
Investment			1.86E-05	0.000334				
DIV	−0.1464435	−0.17341	−0.4579	−0.54334	−0.39956	−0.57405	−0.398921	−0.57405
DIV^2			−0.09968	−0.12052	−0.08743	−0.13562	−0.087256	−0.13562

(continued)

Table 2 (continued)

Selected variables	Model A		Model B		Model C		Model D	
	LASSO	POST-estimation OLS	LASSO	POST-estimation OLS	LASSO	POST-estimation OLS	LASSO	POST-estimation OLS
Ownership dummy								
0					−0.03271	−0.08305	−0.032654	−0.08305
1					0.034437	–	0.0344345	–
Sup-score W statistics (P-value)	6.70 (0.00)		7.99 (0.00)		12.35 (0.00)		12.35 (0.00)	

Note Among ownership dummy, 0 is used for public banks and 1 is used for private banks. Critical value for W-statistics is 3.75 at 5% significance level

Table 3 Dependent variable: effectiveness

Selected variables	Model A		Model B		Model C		Model D	
	LASSO	Post-estimation OLS	LASSO	Post-estimation OLS	LASSO	Post-estimation OLS	LASSO	Post-estimation OLS
Leverage ratio	0.5889024	0.507616	0.267524	0.323993	0.234145	0.576417	0.2328875	0.576417
Spread	0.0109916	0.014975	0.01165	0.014955	0.009481	0.010339	0.0094782	0.010339
NIM			0.222928	0.54835	0.494066	1.251127	0.4912852	1.251127
NPA/advance	−0.0035852	−0.00885	−0.00307	−0.00774	−0.00274	−0.00746	−0.002726	−0.00746
Loan/assets					−0.00721	−0.14079	−0.006721	−0.14079
Car	0.0029032	0.006128	0.002856	0.005389	0.002759	0.002995	0.0027579	0.002995
Deposits/liabilities	−0.0016179	−0.00175	−0.00211	−0.00218	−0.00213	−0.00251	−0.002128	−0.00251
Interest income/assets	0.0028294	0.010479			0.003555	0.017582	0.0035039	0.017582
Non-interest income/assets	0.060529	0.079706	0.104898	0.120657	0.098658	0.120725	0.0985767	0.120725
Intermediation cost/assets	0.026477	0.048059	0.022163	0.040518	0.02105	0.030317	0.0210159	0.030317
Return on advances	0.0155273	0.014425	0.012219	0.015339	0.003823	−0.00154	0.0038428	−0.00154
DIV	−0.0962987	−0.15046	−0.46798	−0.53505	−0.45062	−0.5276	−0.450338	−0.5276
DIV^2			−0.12198	−0.12947	−0.11874	−0.12869	−0.118702	−0.12869

(continued)

Table 3 (continued)

Selected variables	Model A		Model B		Model C		Model D	
	LASSO	Post-estimation OLS	LASSO	Post-estimation OLS	LASSO	Post-estimation OLS	LASSO	Post-estimation OLS
Ownership dummy								
0					−0.01484	−0.04497	−0.014849	−0.04497
1					0.029714	–	0.0297038	–
Sup-score W statistics (P-value)	12.68 (0.00)		14.23 (0.00)		14.50 (0.00)		14.50 (0.00)	

Note Among ownership dummy, 0 is used for public banks and 1 is used for private banks. Critical value for W-statistics is 3.75 at 5% significance level

Table 4 Dependent variable: performance

Selected variables	Model A		Model B		Model C		Model D	
	LASSO	POST-estimation OLS	LASSO	POST-estimation OLS	LASSO	POST-estimation OLS	LASSO	POST-estimation OLS
Leverage ratio	0.5674877	0.445107	0.391994	0.381844	0.153397	0.002295	0.153397	0.002295
Spread	0.010395	0.014973	0.012321	0.014604	0.010657	0.011973	0.010657	0.011973
NIM			0.062627	0.630224	0.110612	0.899024	0.1106117	0.899024
Loan/assets	−0.0254097	−0.08252	−0.01372	−0.10469	−0.01853	−0.08771	−0.018533	−0.08771
Car	0.0024678	0.005987	0.001975	0.004082	0.002569	0.004771	0.0025691	0.004771
Deposits/liabilities	−0.0012639	−0.00145	−0.00135	−0.00192	−0.00155	−0.00231	−0.00155	−0.00231
Non-interest income/assets	0.0296703	0.05031	0.07321	0.096279	0.062763	0.088295	0.0627625	0.088295
Intermediation cost/assets	0.0047625	0.026417	0.001646	0.016935	0.003358	0.016517	0.0033582	0.016517
Return on advances	0.0188789	0.02263	0.013504	0.016461	0.006335	0.007423	0.0063351	0.007423
DIV	−0.1121869	−0.14223	−0.47334	−0.56846	−0.44557	−0.55681	−0.445567	−0.55681
DIV^2			−0.11726	−0.1406	−0.11165	−0.13943	−0.111646	−0.13943
Ownership dummy								
0					−0.02362	−0.06187	−0.02362	−0.06187
1					0.031815	–	0.0318153	–
Sup-score W statistics (P-value)	9.68 (0.00)		11.73(0.00)		15.29 (0.00)		15.29 (0.00)	

Note Among ownership dummy, 0 is used for public banks and 1 is used for private banks. Critical value for W-statistics is 3.75 at 5% significance level

3.3.2 Effectiveness as a Dependent Variable

Table 3 presents the results of the LASSO regression with effectiveness as the dependent variable. In this analysis, we also observe an indirect linkage between effectiveness and income diversification. Notably, the inclusion of the squared term for diversification, in conjunction with the ownership dummy, notably amplifies the model's joint significance to a value of 14.50. Within the sphere of effectiveness, private banks demonstrate superior performance when contrasted with their public sector counterparts. Intriguingly, the enduring impact of the financial crisis seems to be limited. It is worth mentioning that these findings harmonize well with the insights derived from the efficiency model. Examining the chosen control variables, Table 2 underscores their direct associations with variables such as the leverage ratio, spread, Net Interest Margin (NIM), Capital Adequacy Ratio (CAR), interest income to assets ratio, non-interest income to assets ratio, returns on advances, and intermediation cost. Additionally, a reverse connection is observable between effectiveness and metrics like asset quality, loan to assets ratio, and deposits to liabilities ratio.

3.3.3 Performance as a Dependent Variable

The LASSO regression results, as depicted in Table 4, elucidate that (i) income diversification exerts a detrimental impact on bank performance, and (ii) private banks demonstrating superior performance compared to their public counterparts. Notably, the model incorporating diversification and ownership as independent variables, in conjunction with control variables, increases the joint significance value to 15.29 from the initial 9.68, as depicted in Model C of Table 3. Regarding the selected control variables, we note that loan to assets and deposits to liabilities ratios exert a negative influence, whereas indicators like leverage ratio, spread, Net Interest Margin (NIM), Capital Adequacy Ratio (CAR), return on advances, non-interest income ratio, and intermediation cost ratio have a positive effect. Intriguingly, the global financial crisis appears to have no discernible adverse impact on bank performance.

4 Conclusion and Limitations

The paper endeavors to conduct an empirical analysis of the impact of income diversification on the performance, efficiency, and effectiveness of Indian banks. This investigation utilizes panel LASSO regression techniques on a bank-level dataset spanning the period from 1999 to 2022. The computation of bank-level efficiency, effectiveness, and performance scores is accomplished through a network DEA model incorporating undesirable outputs. The findings highlight a pronounced asymmetry among individual banks concerning their performance throughout the years. The results derived from the LASSO approach indicate a lack of positive and significant influence of income diversification on bank efficiency, effectiveness, and performance. This can be attributed to the prevalent reliance on conventional sources of income and the utilization of obsolete technologies within the Indian banking sector. Empirical evidence distinctly establishes the superior performance of private banks in comparison to public banks. Notably, private banks exhibit a positive influence of diversification on performance and its components, while public banks display a negative impact. Moreover, the global financial crisis (GFC) of 2007–09 does not emerge as a significantly influential factor on bank performance.

We also observe noteworthy results concerning the influence of control variables on bank performance, as well as its constituent components of efficiency and effectiveness. The findings indicate that banks with a higher capital adequacy ratio (indicative of a lower-risk portfolio) exhibit superior performance. Conversely, banks with lower asset quality (higher non-Performing assets) experience weaker performance. The observed patterns in efficiency, effectiveness, and performance align with our expectations. Nonetheless, there remains substantial room for progress in the realm of diversification towards non-traditional income sources. This shift has the potential to optimize resource utilization and enhance overall performance. As our investigation has focused on a specific aspect of this expansive subject, opportunities for further exploration emerge in the following directions. First, while our study has employed the LASSO technique for model selection and prediction, the exploration of inferences utilizing methods such as ridge and elastic net regression lies beyond the scope of this current paper. Secondly, our examination has primarily encompassed the influence of the global financial crisis. Extending this research to encompass more recent endogenous and exogenous crises in India, such as the impacts of demonetization and the COVID-19 pandemic, would contribute to a more comprehensive understanding of the dynamics at play.

References

1. Abuzayed, B., Al-Fayoumi, N., & Molyneux, P. (2018). Diversification and bank stability in the GCC. *Journal of International Financial Markets, Institutions and Money, 57*, 17–43.
2. Achabal, D., Heineke, J. M., & McIntyre, S. H. (1984). Issues and perspectives on retail productivity. *Journal of Retailing, 60*(3), 107–127.
3. Alhassan, A. L. (2015). Income diversification and bank efficiency in an emerging market. *Managerial Finance, 41*, 1318–1335.
4. Ali, M., Gan, C., & Nadeem, M. (2023). A CEO's expertise power and bank diversification. *Accounting & Finance*.
5. Anand, N., & Kumar, D. A. (2020). Efficiency mapping of private banks in India through II-stage network DEA efficiency scores. *International Journal of Scientific & Technology Research, 9*(01), 3651–3657.
6. Asmild, M., Paradi, J. C., Reese, D. N., & Tam, F. (2007). Measuring overall efficiency and effectiveness using DEA. *European Journal of Operational Research, 178*(1), 305–321.
7. Bansal, P., Mehra, A., & Kumar, S. (2022). Dynamic metafrontier Malmquist-Luenberger productivity index in network DEA: An application to banking data. *Computational Economics, 59*(1), 297–324.

8. Brahmana, R., Kontesa, M., & Gilbert, R. E. (2018). Income diversification and bank performance: Evidence from Malaysian banks. *Economics Bulletin, 38*, 799–809.
9. Brighi, P., & Venturelli, V. (2013). The effect of revenue and geographic diversification on bank performance. *CEFIN, 33*.
10. Calmès, Christian Theoret & Raymond. (2012). *The rise of market-oriented banking and the Hidden Benefits of Diversification. 5*, 88–125.
11. Chiorazzo, V., Milani, C., & Salvini, F. (2008). Income diversification and bank performance: Evidence from Italian banks. *Journal of Financial Services Research, 33*(3), 181–203.
12. Curi, C., Lozano-Vivas, A., & Zelenyuk, V. (2015). Foreign bank diversification and efficiency prior to and during the financial crisis: Does one business model fit all? *Journal of Banking & Finance, 61*, S22–S35.
13. Doerr, S., & Schaz, P. (2019). *Bank loan supply during crises: The importance of geographic diversification.*
14. Drucker, P. F. (1977). *An introductory view of management*. Harper College Press.
15. Duho, K. C. T. (2019). Bank diversification and performance in an emerging market. *International Journal of Managerial Finance, 19*.
16. Elsas, R., Hackethal, A., & Holzhaeuser, M. (2006). *The anatomy of bank diversification.*
17. Elyasiani, E., & Wang, Y. (2012). Bank holding company diversification and production efficiency. *Applied Financial Economics, 22*(17), 1409–1428.
18. Flamini, V., VFlamini@imf.org, Schumacher, L., LSchumacher@imf.org, McDonald, C. A., & CMcDonald@imf.org. (2009). The determinants of commercial bank profitability in Sub-Saharan Africa. *IMF Working Papers, 09*(15), 1.
19. Fukuyama, H., & Mirdehghan, S. M. (2012). Identifying the efficiency status in network DEA. *European Journal of Operational Research, 220*(1), 85–92.
20. Gelman, M., Goldstein, I., & MacKinlay, A. (2022). Bank diversification and lending resiliency. *SSRN Electronic Journal.*
21. Githaiga, P. N. (2020). Human capital, income diversification and bank performance–an empirical study of East African banks. *Asian Journal of Accounting Research, 6*(1), 95–108.
22. Goddard, J., McKillop, D., & Wilson, J. O. S. (2008). *The diversification and financial performance of US credit unions* 14.
23. Gulati, R., & Kumar, S. (2017). Analysing banks' intermediation and operating efficiencies using the two-stage network DEA model: The case of India. *International Journal of Productivity and Performance Management, 66*(4), 500–516.
24. Kakwezi, P., & Nyeko, S. (2019). Procurement processes and performance: Efficiency and effectiveness of the procurement function. *International Journal of Social Sciences Management and Entrepreneurship (IJSSME), 3*(1), Article 1.
25. Keh, H. T., Chu, S., & Xu, J. (2006). Efficiency, effectiveness and productivity of marketing in services. *European Journal of Operational Research, 170*(1), 265–276.
26. Khan, A., Hassan, M. K., Maroney, N., Boujlil, R., & Ozkan, B. (2020). Efficiency, diversification, and performance of US banks. *International Review of Economics & Finance, 67*, 101–117.
27. Kim, J., & Kim, Y.-C. (2020). Heterogeneous patterns of income diversification effects in U.S. bank holding companies. *International Review of Economics & Finance, 69*, 731–749.
28. Lee, C.-C., Hsieh, M.-F., & Yang, S.-J. (2014). The relationship between revenue diversification and bank performance: Do financial structures and financial reforms matter? *Japan and the World Economy, 29*, 18–35.
29. Liang, Q., Xu, P., & Jiraporn, P. (2013). Board characteristics and Chinese bank performance. *Journal of Banking & Finance, 37*(8), 2953–2968.
30. Lin, H. -C., Huang, J. -C., & You, C. -F. (2022). Bank diversification and financial constraints on firm investment decisions in a bank-based financial system. *Sustainability, 14*(17), Article 17.
31. Martínez-Campillo, A., Wijesiri, M., & Wanke, P. (2020). Evaluating the double bottom-line of social banking in an emerging country: How efficient are public banks in supporting priority and non-priority sectors in India? *Journal of Business Ethics, 162*(2), 399–420.

32. Meng, X., Cavoli, T., & Deng, X. (2018). Determinants of income diversification: Evidence from Chinese banks. *Applied Economics, 50*(17), 1934–1951.
33. Mercieca, S., Schaeck, K., & Wolfe, S. (2007). Small European banks: Benefits from diversification? *Journal of Banking & Finance, 31*(7), 1975–1998.
34. Meslier, C., Tacneng, R., & Tarazi, A. (2013). *Is bank income diversification beneficial? Evidence from an emerging economy.*
35. Mostak Ahamed, M. (2017). Asset quality, non-interest income, and bank profitability: Evidence from Indian banks. *Economic Modelling, 63*, 1–14.
36. Moudud-Ul-Huq, S., Ashraf, B. N., Gupta, A. D., & Zheng, C. (2018). Does bank diversification heterogeneously affect performance and risk-taking in ASEAN emerging economies? *Research in International Business and Finance, 46*, 342–362.
37. Mouzas, S. (2006). Efficiency versus effectiveness in business networks. *Journal of Business Research, 59*(10), 1124–1132.
38. Nenide, B., Pricer, R. W., & Camp, S. M. (2003). The use of financial ratios for research: Problems associated with and recommendations for using large databases. *National Business and Economics Society.*
39. Nepali, S. R. (2018). Income diversification and bank risk-return trade-off in the Nepalese commercial banks. *Asian Economic and Financial Review, 8*(2), 279–293.
40. Ozcan, Y. A. (2014). *Health care benchmarking and performance evaluation-an assessment using data envelopment analysis (DEA)* (2nd ed., Vol. 210). Springer.
41. Pinprayong, B., & Siengthai, S. (2012). Restructuring for organization efficiency in the banking sector in Thailand: A case study of SIAM Commercial Bank. *Far East Journal of Psychology and Business, 8*(2).
42. Pisedtasalasai, A., & Edirisuriya, P. (2020). Diversification and performance of Sri Lankan banks. *The Journal of Asian Finance, Economics and Business, 7*(9), 1–10.
43. Salmi, T., & Martikainen, T. (1994). A review of the theoretical and empirical basis of financial ratio analysis. In *News Group* (_001; News Group).
44. Sang, N. M. (2017). *Income Diversification and Bank Efficiency in Vietnam, 19*, 16.
45. Sawada, M. (2013). How does the stock market value bank diversification? Empirical evidence from Japanese banks. *Pacific-Basin Finance Journal, 25*, 40–61.
46. Schwarze, S., & Zeller, M. (2005). Income diversification of rural households in Central Sulawesi. *Indonesia, 1*, 13.
47. Sharma, S., & Anand,. (2018). Income diversification and bank performance: Evidence from BRICS nations. *International Journal of Productivity and Performance Management, 67*, 00–00.
48. Shim, J. (2019). Loan portfolio diversification, market structure and bank stability. *Journal of Banking & Finance, 104*, 103–115.
49. Tone, K., & Tsutsui, M. (2009). Network DEA: A slacks-based measure approach. *European Journal of Operational Research, 197*(1), 243–252.
50. Trivedi, S. R. (2015). Banking Innovations and new income streams: Impact on banks' performance. *Vikalpa: The Journal for Decision Makers, 40*(1), 28–41.
51. Wanke, P., Skully, M., Wijesiri, M., Walker, T., & dalla Pellegrina, L. (2022). Does ownership structure affect firm performance? Evidence of Indian bank efficiency before and after the global financial crisis. *International Transactions in Operational Research, 29*(3), 1842–1867.
52. Wyss, L. (2018). *Essays on bank income diversification.*
53. Zhou, K. (2014). The effect of income diversification on bank risk: Evidence from China. *Emerging Markets Finance and Trade, 50*(sup3), 201–213.

Health Efficiency Measurement of Turkish Provinces Using DEA, TOPSIS and EDAS Methods

Fuad Selamzade and Yusuf Ersoy

Abstract Epidemic diseases and increasing disease types in recent years once again reveal the importance of the health sector. For this reason, efficiency evaluation has become popular in the health sector, as in every sector. This study aimed to evaluate the health services performance of 81 provinces in Türkiye. The input-oriented CCR model was used for efficiency measurement. The input-oriented SE-CCR model was used to evaluate the efficient provinces among themselves, and TOPSIS and EDAS methods were used to rank them. TOBIT analysis was conducted to determine the factors affecting the efficiency. In the study, analyses were carried out using three inputs and three outputs. According to the results of the input-oriented CCR model, 7 provinces were found to be efficient, while the remaining Decision Making Units (DMUs) were inefficient. Using TOBIT analysis, the effects of input and output variables on efficiency scores were determined. Efficient DMUs were evaluated among themselves by comparing the results of the CCR Super Efficiency (CCR-SE) model, TOPSIS and EDAS method. The relationship between the CCR-SE model, TOPSIS and EDAS methods was determined using Spearman's rho method. The study results reveal that using the DEA method and MCDM methods together is very useful for comparing and ranking efficient DMUs.

Keywords Efficiency measurement · Health efficiency · Data envelopment analysis · TOPSIS · EDAS

1 Introduction

Health, as a person's physical, mental and social well-being, is seen as a fundamental human right and also as a prerequisite for the economic and social development of countries [1]. With the effect of the rapid development of technology, especially communication technology, the advancement of health technology, the rapid change

F. Selamzade · Y. Ersoy (✉)
Mus Alparslan University, 49250 Mus, Türkiye
e-mail: y.ersoy@alparslan.edu.tr

A. Emrouznejad et al. (eds.), *Advances in the Theory and Applications of Performance Measurement and Management*, Lecture Notes in Operations Research,
https://doi.org/10.1007/978-3-031-61597-9_15

in the economic situations of countries and their impact on each other have increased health expenditures. Research has shown that higher-income countries generally have higher health expenditures than lower-income countries [2].

As a result of the negative impact of the COVID-19 pandemic on the economies and health systems of developed and developing countries, the inadequacies in the health systems of the countries have come to light. Following the decrease in the production of countries and economic stagnation due to the impact of the COVID-19 pandemic, the inflation problem arose. In countries such as Türkiye, where the social responsibility of the state is more prominent, expenditures on the health sector and health service provision have also been affected. As a matter of fact, before the economic recession, Türkiye concentrated on health investments with the establishment of city hospitals with high technologies, and the share of similar investments decreased after the economic recession [3].

Solving the increasing health problems in Türkiye, as in the rest of the world, due to both demographic changes and the resulting economic problems, is one of the primary goals of establishing health systems correctly in achieving the country's growth and development goals. In order for the health system to achieve the desired results, it is necessary to establish a health system that meets the expectations of the entire society. Comparing the performance of regions' health systems is seen as an effective way to determine whether the health system is achieving the desired results. Nowadays, parametric and non-parametric methods are used to compare the performance and efficiency levels of health systems. The most widely used method for performance measurement in the healthcare industry is the Data Envelopment Analysis (DEA) method. Although the DEA method makes a relative comparison of the performance values of decision-making units, it reveals whether a health system has achieved its goals or not [4].

There are many studies in the literature on calculating the efficiency of health systems using both Data Envelopment Analysis and other methods. As examples of these studies, Kutlar and Salamov (2016), Carrillo and Jorge (2017), Mardani et al. (2017), Debnath et al. (2017), Bhole and Deshmukh (2018), 9. Sival-ingam et al. (2022), Samut (2023), Jung et al. (2023), Antunes et al. (2023), Cinaroğlu (2023), and Selamzade et al. (2023) can be cited [4–14].

In this study, the health service delivery performance of 81 provinces was evaluated for 2021, using the data obtained from the Health Statistics Yearbook published every year by the Ministry of Health of the Republic of Türkiye. Input-oriented CRS with constant returns to scale methods of non-parametric Data Envelopment Analysis (DEA), which are frequently used in the literature to calculate performances, and Super Efficiency, TOPSIS and EDAS methods to determine which of the efficient provinces is more efficient. used. Then, Tobit analysis was performed to determine the factors affecting the effectiveness. Two different models were used to conduct efficient analyses.

The rest of the study is organized as follows. In the second part of the study, the methodology of the study is included. The third section contains information about the model of the study and the data used in the study. The fourth section contains the findings of the study. In the fifth chapter, a general evaluation of the study was made.

2 Methodology

In this study, the health efficiency of 81 provinces was evaluated using the DEA method. The input-oriented CCR model and the input-oriented CCR super efficiency (SE)-DEA model were used in the study. SE-DEA model results are widely used in the literature to compare efficient decision-making units (DMUs) among themselves according to the DEA results [15–19]. As explained in the study of Banker and Chang (2006), the SE-DEA model is insufficient to rank efficient DMUs among themselves [20]. Some researchers have also used MCDM methods together with the SE-DEA model as [21–27]. In some studies, MCDM methods were used to rank the DMUs that were efficient as a result of DEA analysis [14, 28, 29]. MCDM methods are widely used in different fields such as health, energy, defense, education, supplier selection, human research selection, equipment selection, place selection and tourism [30–35]. TOPSIS and EDAS methods are some of the commonly used MCDM methods [14, 28, 35–38].

It is believed that MDCM methods will be combined with the DEA method so that a full ranking of DMUs can be made. Some researchers believe that the DEA method can be considered as a MCDM method [39–41].

In this study, TOPSIS and EDAS methods were used and compared with the CCR-SE model results. The method developed by Tobin in 1958 is used to determine the factors affecting the efficiency scores obtained as a result of DEA. This approach is often called the "Tobit model" [42]. In the research, the E.M.S 1.3.0 package program was used for DEA analyses, the Eviews 7 program was used for TOBIT analyses, and the Microsoft Excel 2016 program was used for TOPSIS and EDAS method analyses. The methods used in the research and the data of the research are given below.

2.1 Data Envelopment Analysis

The basis of Data Envelopment Analysis (DEA) was created by Farrell's study titled "The Measurement of Productive Efficiency" in 1957 [43]. The constant returns to scale DEA method was developed by Charnes, Cooper and Rhodes in 1978 [44]. This method has been used in the literature as the CCR or CRS method. In 1984, the variable returns to scale method was developed by Banker, Charnes and Cooper, which was later used as the BBC or VRS method [45].

A mathematical programming model has been applied to evaluate the relative effectiveness of different decision-making units (DMU) in DEA. The main idea of the DEA model is to consider each de-rated unit as a DMU, different DMUs form the rated group. The evaluation process is carried out by taking the weight of each DMU input and output index as a variable after a comprehensive analysis of the input and output coefficients. The relative efficiency of a DMU is determined by comparing the distance between the DMU and its production frontiers. Since there is no need to adopt functions and distribution formats and expand input and output possibilities through the application of DEA, the impact of subjective judgments and errors is reduced [46].

In the analysis, for n number of DMUs, it represents the order of the jth DMU with input vector quantity x_j and output vector quantity $y_i : x_j = (x_{1j}, x_{2j}, \ldots, x_{mj})^T$, $y_j = (y_{1j}, y_{2j}, \ldots, y_{sj})^T$, $j = 1, 2, \ldots, n$. The efficiency score of each DMU can be calculated with the following CCR model [47].

$$\theta_{CCR} = \frac{\sum_{i=1}^{n} \lambda_j y_{rj}}{\sum_{i=1}^{n} \lambda_j x_{ij}}$$
$$s.t. : \sum_{i=1}^{n} \lambda_j x_{ij} \le \theta x_o; \sum_{i=1}^{n} \lambda_j y_{rj} \ge y_o; \lambda_j \ge 0; j = 1, 2, \ldots, n \tag{1}$$

In both models used in our study, the input variable is $x_j =$ (Doctors, Nurses, Bed)^{2021}. Output variables: $Y_j =$ (Visits, Inpatients, Surgeries)^{2021} was used in Model 1, $y_j =$ (Visits, Inpatients, Surgeries, Bed Turnover Rate)^{2021} was used in Model 2. λ_j is the coefficient of DMU portfolio j; θ is the efficiency score of KVB_0 obtained by dividing the weighted outputs by the weighted inputs. When $\theta = 1$ the DMU is active, otherwise it is inactive.

The BCC model with variable returns to scale was created by adding the convexity constraint $\left(\sum_{i=1}^{n} \lambda_j = 1\right)$ to the CCR model [45]. The scale efficiency of DMUs is obtained by dividing the efficiency score obtained by the CCR model by the efficiency score obtained by the BCC model [48].

The Super Efficiency Analysis model was created by Seiford and Thrall in 1990 to determine the most effective DMUs as a result of calculations made with DEA models. In this model, DMUs can take values of 1 and above [49–51]. The following formula is applied to calculate the Super Efficiency model with the CCR method:

$$P_{CCR} = \frac{\sum_{\substack{j=1 \\ j \neq 0}}^{n} \lambda_j y_j}{\sum_{\substack{j=1 \\ j \neq 0}}^{n} \lambda_j x_j}$$
$$s.t. : \sum_{\substack{j=1 \\ j \neq 0}}^{n} \lambda_j x_j \le p x_o; \sum_{\substack{j=1 \\ j \neq 0}}^{n} \lambda_j y_j \ge y_o; p, \lambda_j \ge 0; j \neq 0; \tag{2}$$

3 Models and Data

3.1 Models

Two different models were used to analyze the efficiency of the health systems of the 81 provinces of the Republic of Türkiye for 2021. Number of Doctors, Number of Nurses, and Number of Bed were used as input variables in both models. As output variables, Number of Visits, Number of Inpatients, and Number of Surgeries were used in Model 1, and Bed Turnover Rate was additionally used in Model 2.

3.2 Data and Correlation

Table 1 presents the general statistics of the input and output variables used in the study and the correlation between the data.

When the correlation relationship between the data presented in Table 1 is examined, it is seen that there is a high correlation between all variables except Bed Turnover rate. It has been observed that there is a very low, almost 0, correlation between Bed Turnover rate and other variables. In this case, it can be said that there is no relationship between Bed Turnover rate and other variables.

Table 1 Statistics of input and output variables and the correlation between them

	Inputs			Outputs			
	Doctors	Nurses	Bed	Visits	Inpatients	Surgeries	Bed turnover rate
Max	42,268	42,410	46,960	921,840	119,744,208	2,045,691	83.1
Min	151	230	150	674	562,195	2771	18.5
Mean	2266	2870	3142	58,075	8,341,385	145,500	45.8
Std. dev	5291.7	5316.5	5743.8	115,954.6	14,633,431.9	253,493.3	11.0
Correlation							
	Doctors	Nurses	Bed	Visits	Inpatients	Surgeries	Bed turnover rate
Doctors	1						
Nurses	0.995	1					
Bed	0.989	0.997	1				
Visits	0.982	0.990	0.994	1			
Inpatients	0.979	0.990	0.995	0.996	1		
Surgeries	0.987	0.995	0.997	0.995	0.996	1	
Bed turnover rate	0.014	0.025	0.026	0.069	0.085	0.054	1

4 Results

Two different models were used to conduct effectiveness analyses. The efficiency of 81 provinces for both models was evaluated according to the CRS DEA and CRS-SE DEA models and the DMUs with an efficiency score of "1" were determined. According to the DEA Analysis results, it was understood that the Bed Turnover Rate output variable did not have a great effect on the effectiveness, hence we removed this variable from the DEA model. The results of the DEA, TOBIT, TOPSIS and EDAS methods used in the study are given below, respectively.

4.1 Results of DEA

Data envelopment analysis results are shown in Table 2.

It can be seen from Table 2 that 7 DMUs are efficient according to the CCR model. According to the CCR-SE model, the efficiency scores of efficient DMUs are greater than "1".

Table 2 Data envelopment analysis results

Province	Model 1		Model 2	
	CCR	CCR-SE	CCR	CCR-SE
Adana	0,94	0,94	0,94	0,94
Adıyaman	0,69	0,69	0,69	0,69
Afyonkarahisar	0,79	0,79	0,79	0,79
Ağrı	0,81	0,81	0,81	0,81
Amasya	0,56	0,56	0,54	0,54
Ankara	0,78	0,78	0,78	0,78
Antalya	0,83	0,83	0,83	0,83
Artvin	0,58	0,58	0,55	0,55
Aydın	0,83	0,83	0,83	0,83
Balıkesir	0,75	0,75	0,75	0,75
Bilecik	0,64	0,64	0,57	0,57
Bingöl	0,64	0,64	0,62	0,62
Bitlis	0,76	0,76	0,76	0,76
Bolu	0,69	0,69	0,69	0,69
Burdur	0,67	0,67	0,62	0,62
Bursa	0,81	0,81	0,81	0,81
Çanakkale	0,61	0,61	0,60	0,60

(continued)

Table 2 (continued)

Province	Model 1		Model 2	
	CCR	CCR-SE	CCR	CCR-SE
Çankırı	0,67	0,67	0,55	0,55
Çorum	0,67	0,67	0,67	0,67
Denizli	1,00	1,13	1,00	1,13
Diyarbakır	0,82	0,82	0,82	0,82
Edirne	0,55	0,55	0,55	0,55
Elazığ	0,81	0 ,81	0,81	0,81
Erzincan	0,72	0,72	0,72	0,72
Erzurum	0,73	0,73	0,73	0,73
Eskişehir	0,82	0,82	0,82	0,82
Gaziantep	1,00	1,04	1,00	1,04
Giresun	0,80	0,80	0,80	0,80
Gümüşhane	0,47	0,47	0,36	0,36
Hakkâri	0,72	0,72	0,72	0,72
Hatay	0,92	0,92	0,92	0,92
Isparta	1,00	1,03	1,00	1,03
Mersin	0,81	0,81	0,81	0,81
İstanbul	0,76	0,76	0,76	0,76
İzmir	0,90	0,90	0,90	0,90
Kars	0,52	0,52	0,49	0,49
Kastamonu	0,58	0,58	0,55	0,55
Kayseri	0,86	0,86	0,86	0,86
Kırklareli	0,64	0,64	0,64	0,64
Kırşehir	0,55	0,55	0,55	0,55
Kocaeli	0,88	0,88	0,88	0,88
Konya	0,86	0,86	0,86	0,86
Kütahya	0,54	0,54	0,54	0,54
Malatya	0,74	0,74	0,74	0,74
Manisa	0,74	0,74	0,74	0,74
Kahramanmaraş	0,90	0,90	0,90	0,90
Mardin	0,89	0,89	0,89	0,89
Muğla	0,74	0,74	0,74	0,74
Muş	0,73	0,73	0,73	0,73
Nevşehir	0,79	0,79	0,71	0,71
Niğde	0,65	0,65	0,65	0,65
Ordu	0,87	0,87	0,87	0,87

(continued)

Table 2 (continued)

Province	Model 1		Model 2	
	CCR	CCR-SE	CCR	CCR-SE
Rize	0,64	0,64	0,64	0,64
Sakarya	0,78	0,78	0,78	0,78
Samsun	0,83	0,83	0,83	0,83
Siirt	0,95	0,95	0,91	0,91
Sinop	0,57	0,57	0,56	0,56
Sivas	0,60	0,60	0,60	0,60
Tekirdağ	0,91	0,91	0,91	0,91
Tokat	0,81	0,81	0,81	0,81
Trabzon	0,76	0,76	0,76	0,76
Tunceli	0,46	0,46	0,26	0,26
Şanlıurfa	1,00	1,01	1,00	1,01
Uşak	0,85	0,85	0,83	0,83
Van	0,75	0,75	0,75	0,75
Yozgat	0,57	0,57	0,57	0,57
Zonguldak	0,73	0,73	0,73	0,73
Aksaray	0,69	0,69	0,68	0,68
Bayburt	0,56	0,56	0,45	0,45
Karaman	0,98	0,98	0,87	0,87
Kırıkkale	0,72	0,72	0,68	0,68
Batman	0,86	0,86	0,86	0,86
Şırnak	0,85	0,85	0,85	0,85
Bartın	0,56	0,56	0,51	0,51
Ardahan	0,80	0,80	0,55	0,55
Iğdır	1,00	1,59	1,00	1,12
Yalova	1,00	1,10	1,00	1,08
Karabük	0,76	0,76	0,76	0,76
Kilis	1,00	1,06	0,90	0,90
Osmaniye	1,00	1,11	1,00	1,06
Düzce	0,94	0,94	0,94	0,94
Mean	0,77	0,78	0,75	0,75

4.2 *Results of TOPSIS*

Table 3 shows the ranking of the DMUs according to the TOPSIS method.

According to the ranking in Table 3, it was understood that the best alternative is Şanlıurfa, second is Osmaniye and fourth is Iğdır.

Table 3 Ranking of the alternatives according to the TOPSIS method

Province	S_i^*	S_i^-	C_i^*	Rank
Denizli	0,1372	0,1365	0,4988	5
Gaziantep	0,1927	0,1914	0,4983	6
Isparta	0,1574	0,1425	0,4750	7
Şanlıurfa	0,1413	0,1556	0,5242	1
Iğdır	0,1914	0,1927	0,5017	4
Yalova	0,1781	0,1811	0,5041	3
Osmaniye	0,1600	0,1668	0,5104	2

4.3 Results of EDAS

Table 4 shows the ranking of the DMUs according to the EDAS method.

According to the ranking in Table 4, it was understood that the best alternative is Şanlıurfa, second is Osmaniye and fourth is Gaziantep.

4.4 Results of Tobit Analysis

According to the Tobit analysis results, the effects of the input variables in Model 1 on CCR and SE-CCR scores are given in Table 5. Tobit analysis was performed using the Censored Normal (TOBIT) (Quadratic hill climbing) method. According to the Tobit analysis results, it was determined that all three input variables affected both CCR and SE-CCR scores at approximately the same rate. It was determined that the number of doctors and the number of beds had a positive significant effect on the efficiency scores, approximately 0.072%, and the number of nurses approximately 0.076%. According to these results, it can be concluded that the efficiency scores will increase by approximately 0.07% depending on the 1% increase in the input variables.

According to the Tobit analysis results, the effects of the input variables in Model 1 on CCR and SE-CCR scores are given in Table 6. It was determined that the number of doctors by approximately 0.049% and 0.040%, the number of nurses

Table 4 Ranking of the alternatives according to the EDAS method

Province	SP_i	SN_i	NSP_i	NSN_i	AS_i	Rank
Denizli	0,129	0,140	0,192	0,793	0,492	6
Gaziantep	0,673	0,673	1,000	0,000	0,500	4
Isparta	0,090	0,169	0,133	0,749	0,441	7
Şanlıurfa	0,384	0,315	0,571	0,531	0,551	1
Iğdır	0,425	0,431	0,632	0,359	0,496	5
Yalova	0,356	0,352	0,529	0,477	0,503	3
Osmaniye	0,257	0,233	0,383	0,653	0,518	2

Table 5 Effect of input variables on efficiency scores model 1 (dependent variables CCR and SE-CCR)

Variable	CCR coeff	SE CCR coeff	CCR coeff	SE CCR coeff	CCR coeff	SE CCR coeff
C	0.248**	0.260**	0.190*	0.205*	0.21*	0.227*
LnDoctors	0.072***	0.071***				
LnNurses			0.076***	0.074***		
LnBed					0.072***	0.071***
Error distribution						
SCALE:C(3)	0.137***	0.15***	Avg, log likelihood		0.566	0.479
Mean dep.var	0.747	0.752	S,D, dependent var		0.158	0.169
S,E, of reg	0.139	0.152	Akaike info criterion		−1.059	−0.884
Sum^2 resid	1.528	1.82	Schwarz criterion		−0.97	−0.795
Loglikelihood	45.878	38.798	Hannan-Quinn criter		−1.023	−0.848

Here, "p" is considered as an indicator of the existence of a statistically significant relationship. *** $p \leq 0.001$; **$p \leq 0.05$; * $p \leq 0.1$

by approximately 0.021% and 0.074%, and the number of beds by approximately 0.048% and 0.071% had a positive significant effect on the CCR and SE-CCR efficiency scores respectively. According to these results, it can be concluded that efficiency scores will increase as a result of an 1% increase in input variables.

According to the Tobit analysis results, the effects of the input variables in Model 1 and Model 2 on BCC [4, 45] and SE-BCC [11, 51] scores are given in Table 7. It was determined that the effects of the input variables on the BCC and Super BCC

Table 6 Effect of input variables on efficiency scores model 2 (dependent variables CCR and SE CCR)

Variable	CCR coeff	SE CCR coeff	CCR coeff	SE CCR coeff	CCR coeff	SE CCR coeff
C	0.421***	0.497***	0.388***	0.205*	0.41***	0.227*
LnDoctors	0.049***	0.040**				
LnNurses			0.021***	0.074***		
LnBed					0.048***	0.071***
Error distribution						
SCALE:C(3)	0.131***	0.170***	Avg, log likelihood		0.615	0.352
Mean dep.var	0.765	0.778	S,D, dependent var		0.142	0.177
S,E, of reg	0.132	0.172	Akaike info criterion		−1.156	−0.630
Sum^2 resid	1.386	2.346	Schwarz criterion		−1.068	−0.541
Loglikelihood	49.832	28.5	Hannan-Quinn criter		−1.121	−0.594

Here, "p" is considered as an indicator of the existence of a statistically significant relationship. *** $p \leq 0.001$; **$p \leq 0.05$; * $p \leq 0.1$

Table 7 Effect of input variables on efficiency scores model 1 (dependent variables BCC and SE BCC)

Variable	BCC coeff	SE BCC coeff	BCC coeff	SE BCC coeff	BCC coeff	SE BCC coeff
C	0.669***	0.818***	0.678***	0.863***	0.705***	0.912***
LnDoctors	0.022	0.005				
LnNurses			0.020	−0.002		
LnBed					0.016	−0.008
Error distribution						
SCALE:C(3)	0.121***	0.176***	Avg, log likelihood		0.687	0.318
Mean dep.var	0.825	0.851	S,D, dependent var		0.124	0.177
S,E, of reg	0.122	0.178	Akaike info criterion		−1.300	−0.561
Sum² resid	1.177	2.483	Schwarz criterion		−1.212	−0.472
Loglikelihood	56.434	25.395	Hannan-Quinn criter		−1.265	−0.525

Here, "p" is considered as an indicator of the existence of a statistically significant relationship. *** $p \leq 0.001$; ** $p \leq 0.05$; * $p \leq 0.1$

efficiency scores obtained using both models were not significant. Based on these results, it can be concluded that input variables do not affect BCC efficiency scores (Table 8).

According to the Tobit analysis results, the effects of the input variables in Model 1 and Model 2 on CCR and SE-CCR, BCC and SE-BCC scores are given in Table 9. When the effect of the output variables on the efficiency scores according to Model 1 was examined, it was determined that the number of visits had a negative significant

Table 8 Effect of input variables on efficiency scores model 2 (dependent variables BCC and SE-BCC)

Variable	BCC coeff	SE BCC coeff	BCC coeff	SE BCC coeff	BCC coeff	SE BCC coeff
C	0.704***	0.818***	0.678***	0.863***	0.705***	0.912***
LnDoctors	0.016	0.005				
LnNurses			0.020	−0.002		
LnBed					0.016	−0.008
Error distribution						
SCALE:C(3)	0.121***	0.176***	Avg, log likelihood		0.687	0.318
Mean dep.var	0.825	0 .851	S,D, dependent var		0.124	0.177
S,E, of reg	0.123	0.178	Akaike info criterion		−1.300	−0.561
Sum² resid	1.200	2.483	Schwarz criterion		−1.212	−0.472
Loglikelihood	55.642	25.395	Hannan-Quinn criter		−1.265	−0.525

Here, "p" is considered as an indicator of the existence of a statistically significant relationship. *** $p \leq 0.001$; ** $p \leq 0.05$; * $p \leq 0.1$

effect on the CCR and SE-CCR efficiency scores, and a positive significant effect on the number of inpatients and surgeries.

It was observed that the output variables had no significant effects on the BCC and Super BCC efficiency scores for the same model. When the effect of the output variables on the efficiency scores obtained from Model 2 was examined, it was determined that the number of visits had a negative significant effect on the efficiency scores, while the inpatients, surgeries, and bed turnover rate variables had a positive significant effect.

4.5 Discussion

The results of the efficient DMUs according to the CRS-SE model, TOPSIS and EDAS methods were compared and shown in Table 10.

When Table 10 is analyzed, Denizli ranks 1st according to the results of the CRS-SE model. While Şanlıurfa is in the last place according to the results of the CRS-SE model, it is in the first place according to the TOPSIS, EDAS method results. While Osmaniye is in the fourth place according to the results of the CRS-SE model, it is in the second place according to the TOPSIS and EDAS method results. According to all method results, it is possible to say that Şanlıurfa is in first place, Osmaniye is in second place and Isparta is in last place. Besides, the correlation between the results of the TOPSIS, EDAS methods and CCR-SE models has been examined with the Spearman Correlation approach. The correlation results can be seen in Table 11.

It is possible to say from Table 11 that there is a strong positive relationship between the TOPSIS and EDAS methods and the results obtained.

5 Conclusion

The health sector is an important sector that directly affects the continuation and quality of human life. Therefore, it is vital to evaluate the efficiency of the health sector. In this study, the efficiency of health services in 81 provinces in Türkiye was measured. The input-oriented CCR model was used for efficiency analysis in the research. The input-oriented SE-CCR model was used to compare the effective DMUs among themselves. TOPSIS and EDAS methods were used to rank the efficient DMUs more clearly. According to the CCR model results, 7 DMUs were efficient and 74 DMUs were not. In the discussion part of the study, the results of the CCR-SE model and the TOPSIS, EDAS method were compared. According to the results of three methods, Şanlıurfa is in first place, Osmaniye is in second place and Isparta is in last place. According to the sperm correlation test results, it was determined that there was a positive relationship between the TOPSIS and EDAS method results.

Table 9 Effect of output variables on efficiency scores model 1 and model 2 (dependent variables CCR, SE CCR, BCC and SE BCC)

Variables	Model 1			
	CCR 3 coeff	SE CCR 3 coeff	BCC 3 coeff	SE BCC 3 coeff
C	1.762***	1.879***	0.995**	0.887
LnVisits	−0.302***	−0.324***	−0.030	0.023
LnInpatients	0.224***	0.248***	−0.064	−0.107
LnSurgeries	0.109*	0.106*	0.100	0.080
Error distribution				
SCALE:C(5)	0.064***	0.097***	0.108***	0.121***
Mean dependent var	0.765	0.778	0.747	0.752
S,E, of regression	0.066	0.1	0.111	0.124
Sum squared resid	0.332	0.757	0.95	1.179
Log likelihood	107.699	74.287	65.137	56.354
Avg, log likelihood	1.33	0.917	0.804	0.696
S,D, dependent var	0.142	0.177	0.158	0.169
Akaike info criterion	−2.511	−1.686	−1.485	−1.268
Schwarz criterion	−2.334	−1.509	−1.337	−1.12
Hannan-Quinn criter	−2.44	−1.615	−1.426	−1.209
	Model 2			
Variables	CCR 4 coeff	SE CCR 4 coeff	BCC 4 coeff	SE BCC 4 coeff
C	−0.044	−0.379	−0.151	−0.475
LnVisits	−0.074*	−0.073	0.078	0.162
LnInpatients	−0.232***	−0.244***	−0.264**	−0.396**
LnSurgeries	0.276***	0.268***	0.171**	0.203**
LnBedTurnoverRate	0.465***	0.606***	0.270***	0.324***
Error distribution				
SCALE:C(5)	0.106***	0.159***	0.118***	0.175***
Mean dependent var	0.825	0.847	0.825	0.851
S,E, of regression	0.109	0.165	0.121	0.18
Sum squared resid	0.910	2.007	1.131	2.45
Log likelihood	66.882	32.972	58.057	25.923
Avg, log likelihood	0.826	0.417	0.717	0.324
S,D, dependent var	0.124	0.175	0.124	0.177
Akaike info criterion	−1.503	−0.683	−1.31	−0.523
Schwarz criterion	−1.326	−0.503	−1.162	−0.374
Hannan-Quinn criter	−1.432	−0.611	−1.251	−0.463

Here, "p" is considered as an indicator of the existence of a statistically significant relationship. *** $p \leq 0.001$; ** $p \leq 0.05$; * $p \leq 0.1$

Table 10 Efficiency comparisons of efficient DMUs

Province	CRS-SE	TOPSIS	EDAS
Denizli	1	5	6
Gaziantep	5	6	4
Isparta	6	7	7
Şanlıurfa	7	1	1
Iğdır	2	4	5
Yalova	3	3	3
Osmaniye	4	2	2

Table 11 The correlation between the results of the TOPSIS, EDAS methods and CCR-SE models

Correlations

			CCR-SE	TOPSIS	EDAS
Spearman's rho	CCR-SE	Correlation coefficient	1.000	−0.107	−0.357
		Sig. (2-tailed)		0.819	0.432
		N	7	7	7
	TOPSIS	Correlation coefficient	−0.107	1.000	0.893**
		Sig. (2-tailed)	0.819		0.007
		N	7	7	7
	EDAS	Correlation coefficient	−0.357	0.893**	1.000
		Sig. (2-tailed)	0.432	0.007	
		N	7	7	7

** Correlation is significant at the 0.01 level (2-tailed)
Source Author's calculation in the SPSS24 statistics software

The research results reveal that it is useful to use MCDM methods such as TOPSIS and EDAS to make a clearer ranking among the efficient DMUs according to the DEA method analysis.

Like many other studies, this study also has some limitations. Using 3 input and 3 output variables in the study is one of the limitations of the study. Another limitation is that the study was conducted in Türkiye and was limited to 2021 data.

In the future, studies on this subject can be carried out in the health sector or different sectors using different input and output variables. Efficient DMUs can be ranked using different MCDM methods. Since the number of studies conducted in the health sector using DEA and MCDM methods is limited, the study can contribute to the literature in this respect.

References

1. Chai, P., Zhang, Y., Zhou, M., Liu, S., & Kinfu, Y. (2019). Technical and scale efficiency of provincial health systems in China: A bootstrapping data envelopment analysis. *British Medical Journal Open, 2019*(9), 1–9. https://doi.org/10.1136/bmjopen-2018-027539
2. Wang, F. (2018). The roles of preventive and curative health care in economic development. *PLoS ONE, 13*(11), 1–12. https://doi.org/10.1371/journal.pone.0206808
3. Selamzade, F., & Özdemir, Y. (2020). Evaluation of the effectiveness of OECD countries against COVID-19 with DEA. *Turkish Studies, 15*(4), 977–991.
4. Kutlar, A., & Salamov, F. (2016). Evaluation of effectiveness of Azerbaijan public hospitals with data envelopment analysis. *KOSBED, 31*, 1–17.
5. Carrillo, M., & Jorge, J. M. (2017). DEA-like efficiency ranking of regional health systems in Spain. *Social Indicators Research, 133*, 1133–1149.
6. Mardani, A., Nilashi, M., Zakuan, N., Loganathan, N., Soheilirad, S., Saman, M. Z. M., & Ibrahim, O. (2017). A systematic review and meta-Analysis of SWARA and WASPAS methods: Theory and applications with recent fuzzy developments. *Applied Soft Computing, 57*, 265–292.
7. Debnath, A., Roy, J., Kar, S., Zavadskas, E. K., & Antucheviciene, J. (2017). A hybrid MCDM approach for strategic project portfolio selection of agro by-products. *Sustainability, 9*(8), 1302.
8. Bhole, G. P., & Deshmukh, T. (2018). Multi-criteria decision making (MCDM) methods and its applications. *International Journal for Research in Applied Science & Engineering Technology (IJRASET), 6*(5), 899–915.
9. Sivalingam, V., Poogavanam, G., Natarajan, Y., & Sun, J. (2022). Optimization of atomized spray cutting fluid eco-friendly turning of Inconel 718 alloy using ARAS and CODAS methods. *International Journal of Advanced Manufacturing Technology, 120*, 4541–4564. https://doi.org/10.1007/s00170-022-09047-w
10. Samut, P. K. (2023). Efficiency determinants in healthcare: A systematic review with an integrated canonical correlation analysis–data envelopment analysis/assurance region model. *Journal of Healthcare Management, 68*(5), 356–375.
11. Jung, S., Son, J., Kim, C., & Chung, K. (2023). Efficiency measurement using data envelopment analysis (DEA) in public healthcare: Research trends from 2017 to 2022. *Processes, 11*(3), 811–817.
12. Antunes, J., Hadi-Vencheh, A., Jamshidi, A., Tan, Y., & Wanke, P. (2023). TEA-IS: A hybrid DEA-TOPSIS approach for assessing performance and synergy in Chinese health care. *Decision Support Systems, 171*, 113916. https://doi.org/10.1016/j.dss.2022.113916
13. Cinaroglu, S. (2023). Fuzzy efficiency estimates of the turkish health system: A comparison of interval, bias-corrected, and fuzzy data envelopment analysis. *International Journal of Fuzzy Systems, 25*, 2356–2379. https://doi.org/10.1007/s40815-023-01519-9
14. Selamzade, F., Ersoy, Y., Ozdemir, Y., & Celik, M. Y. (2023). Health efficiency measurement of OECD countries against the COVID-19 pandemic by using DEA and MCDM methods. *Arabian Journal for Science and Engineering*, 1–18. https://doi.org/10.1007/s13369-023-08114-y
15. Yang, L., Ouyang, H., Fang, K., Ye, L., & Zhang, J. (2015). Evaluation of regional environmental efficiencies in China based on super-efficiency-DEA. *Ecological Indicators, 51*, 13–19. https://doi.org/10.1016/j.ecolind.2014.08.040
16. Chitnis, A., & Mishra, D. K. (2019). Performance efficiency of Indian private hospitals using data envelopment analysis and super-efficiency DEA. *Journal of Health Management, 21*(2), 279–293. https://doi.org/10.1177/0972063419835120
17. Pan, Z., Wang, Y., Zhou, Y., & Wang, Y. (2020). Analysis of the water use efficiency using super-efficiency data envelopment analysis. *Applied Water Science, 10*(139). https://doi.org/10.1007/s13201-020-01223-1
18. Li, L. L., Seo, Y. -J., & Ha, M. -H. (2021). The efficiency of major container terminals in China: Super-efficiency data envelopment analysis approach *6*(2), 173–187. https://doi.org/10.1108/MABR-08-2020-0051

19. Li, M., Wang, X., Agyeman, F. O., Gao, Y., & Sarfraz, M. (2023). Efficiency evaluation and the impact factors of sustainable forestry development in China: Adoption of Super-efficiency data envelopment analysis and malmquist index methods. *Forests, 14*(5), 1–27. https://doi.org/10.3390/f14050909
20. Banker, R. D., & Chang, H. (2006). The super-efficiency procedure for outlier identification, not for ranking efficient units. *European journal of operational research, 175*(2), 1311–1320.
21. Lotfi, F. H., Fallajnejad, R., & Navidi, N. (2011). Ranking efficient units in DEA by using TOPSIS method. *Applied Mathematical Sciences, 5*(1), 805–815. Retrieved May 10, 2022, from http://www.m-hikari.com/ams/ams-2011/ams-17-20-2011/lotfiAMS17-20-2011-1.pdf
22. Çelen, A., & Yalçın, N. (2012). Performance assessment of Turkish electricity distribution utilities: An application of combined FAHP/TOPSIS/DEA methodology to incorporate quality of service. *Util. Policy, 23*, 59–71.
23. Lai, P.-L., Potter, A., Beynon, M., & Beresford, A. (2015). Evaluating the efficiency performance of airports using an integrated AHP/DEA-AR technique. *Transport Policy, 42*, 75–85. https://doi.org/10.1016/j.tranpol.2015.04.008
24. Babaee, S., Bagherikahvarin, M., Sarrazin, R., Shen, Y., & Hermans, E. (2015). Use of DEA and PROMETHEE II to assess the performance of older drivers. *Transportation Research Procedia, 10*, 798–808. https://doi.org/10.1016/j.trpro.2015.09.033
25. Chitnis, A., & Vaidya, O. S. (2016). Efficiency ranking method using DEA and TOPSIS (ERM-DT): Case of an Indian bank. *Benchmarking International Journal, 23*(1), 166–182. https://doi.org/10.1108/bij-09-29013-0093F
26. Su, W., Wang, D., Xu, L., Zeng, S., & Zhang, C. (2020). A nonradial super efficiency Dea framework using a MCDM to measure the research efficiency of disciplines at Chinese universities. *IEEE Access, 8*, 86388–86399. https://doi.org/10.1109/ACCESS.2020.2993108
27. Stevic, Z., Miskic, S., Vojinovic, D., Huskanovic, E., Stankovic, M., & Pamucar, D. (2022). Development of a model for evaluating the efficiency of transport companies: PCA–DEA–MCDM model. *Axioms, 11*(3), 1–33. https://doi.org/10.3390/axioms11030140
28. Ersoy, Y. (2021). Equipment Selection for an e-commerce company using entropy-based TOPSIS, EDAS and CODAS methods during the covid-19. *LogForum, 17*(3), 341–358. https://doi.org/10.17270/J.LOG.2021.603
29. Wang, C. N., Dang, T. T., Nguyen, N. A. T., & Wang, J. W. (2023). A combined Data Envelopment Analysis (DEA) and Grey Based Multiple Criteria Decision Making (G-MCDM) for solar PV power plants site selection: A case study in Vietnam. *Energy Reports, 8*, 1124–1142. https://doi.org/10.1016/j.egyr.2021.12.045
30. Mardani, A., Jusoh, A., Nor, K. M. D., Kahalifah, Z., Zakwan, N., & Valipour, A. (2015). Multiple criteria decision-making techniques and their applications—A review of the literature from 2000 to 2014. *Economic Research-Ekonomska Istraživanja, 28*(1), 516–571. https://doi.org/10.1080/1331677X.2015.1075139
31. Rezaei, J. (2015). A systematic review of multi-criteria decision-making applications in reverse logistics. *Transportation Research Procedia, 10*, 766–776. https://doi.org/10.1016/j.trpro.2015.09.030
32. Emrouznejad, A., & Marra, M. (2017). The state of the art development of AHP (1979–2017): A literature review with a social network analysis. *International Journal of Production Research, 55*(22), 6653–6675. https://doi.org/10.1080/00207543.2017.1334976
33. Chowdhury, P., & Paul, S. K. (2020). Applications of MCDM methods in research on corporate sustainability: A systematic literature review. *Management of Environmental Quality: An International Journal, 31*(2), 385–405. https://doi.org/10.1108/MEQ-12-2019-0284
34. Soba, M., Ersoy, Y., Altınay, A. T., Erkan, B., & Şik, E. (2020). Application of multiple criteria decision-making methods in assignment place selection. *Mathematical Problems in Engineering*, 1–13. https://doi.org/10.1155/2020/6748342
35. Ersoy, Y. (2021). Performance evaluation in distance education by using data envelopment analysis (DEA) and TOPSIS Methods. *Arabian Journal for Science and Engineering, 46*, 1803–1817. https://doi.org/10.1007/s13369-020-05087-0

36. Hou, H., & Zhao, C. (2022). A novel D-SCRI–EDAS method and its application to the evaluation of an online live course platform. *Systems, 10*(5), 1–17. https://doi.org/10.3390/systems10050157
37. Hajduk, S., & Jelonek, D. (2021). A Decision-making approach based on TOPSIS method for ranking smart cities in the context of urban energy. *Energies, 14*(9), 1–23. https://doi.org/10.3390/en14092691
38. Oliveria, M. S., Steffen, V., Francisco, A. C., & Trojan, F. (2023). Integrated data envelopment analysis, multi-criteria decision making, and cluster analysis methods: Trends and perspectives. *Decision Analytics Journal, 8*, 1–54. https://doi.org/10.1016/j.dajour.2023.100271
39. Somogyi, R. M. (2011). Ranking efficient and inefficient decision making units in data envelopment analysis. *International Journal for Traffic and Transport Engineering, 1*(4), 245–256. http://ijtte.com/uploads/2011-12-19/d4c8811d-2d60-b00a245-256.pdf
40. Azadfallah, M. (2018). Multi-criteria decision making for ranking decision making units. *International Journal of Productivity Management and Assessment Technologies (IJPMAT), 6*(1), 17–36. https://doi.org/10.4018/IJPMAT.2018010102
41. Taherdost, H., & Madanchian, M. (2023). Multi-Criteria Decision Making (MCDM) methods and concepts. *Encyclopedia, 3*(1), 77–87. https://doi.org/10.3390/encyclopedia3010006
42. Tobin, J. (1958). Estimation for relationships with limited dependent variables. *Econometrica, 26*(1), 24–36.
43. Farrel, M. J. (1957). The measurement of productive efficiency. *Journal of the Royal Statistical Society, Series A (General), 120*(3), 253–290.
44. Charnes, A., Cooper, W. W., & Rhodes, E. (1978). Measuring the efficiency of decision making units. *European Journal of Operational Research, 2*, 429–444.
45. Banker, R. D., Charnes, A., & Cooper, W. W. (1984). Some models for estimating and scale inefficiencies in data envelopment analysis. *Management Science, 30*(9), 1078–1092.
46. Li, Q. Y., & Su, Z. (2018). Evaluation on efficiency of innovation and entrepreneurship education in a university based on DEA. In *International Conference On Computer, Communication and Network Technology (CCNT 2018)* (pp. 710–714).
47. Cooper, W. W., Seiford, L. M., & Tone, K. (2010). Data envelopment analysis, Boston: Kluwer Academic Publishers.
48. Selamzade, F., & Yüksel, O. (2021). Examination of health performance of provinces in Azerbaijan with Malmquist Index. *Journal of International Health Sciences and Management, 7*(14), 53–61.
49. Seiford, L. M., & Thrall, R. M. (1990). Recent developments in DEA: The mathematical programming approach to frontier analysis. *Journal of Econometrics, 4*, 7–38.
50. Coelli, T. J. et al. (1998). *An Introduction to Efficiency and Productivity Analysis* (2nd ed.). Springer.
51. Yesilyurt, O., & Salamov, F. (2017). Evaluation of efficiency and factors influencing the efficiency in the health systems of Turkic states with super-efficiency and Tobit models. *Balkan and Near Eastern Journal of Social Sciences, 3*(2), 128–138.

Performance Evaluation of University Faculty Members Using DEA Models with Categorical Variables

Tamás Koltai and Katalin Gerákné Krasz

Abstract Data Envelopment Analysis (DEA) is used for efficiency evaluation of decision-making units (DMU) based on input and output information. DMUs are generally for-profit or non-profit organizations. It is possible, however, to consider employees as DMUs, and in this case DEA can be used for personal performance appraisal. This paper shows, how performance of university faculty members can be evaluated with DEA. Output information is determined by the achievement of faculty members in teaching, research, and service areas, and the inputs are formed by the elements of the faculty compensation system. Input-oriented DEA models are used to explore how the compensation of a faculty member is justified by the achievements. To refine the analysis, categorical variables are applied to distinguish faculties being in different university positions. Weight restrictions are also applied to express the different expectations of management in three fundamental areas of performance: teaching, research, and service. The suggested method is illustrated with the case of the School of Economic and Social Sciences of the Budapest University of Technology and Economics. Output and input data are collected through the performance evaluation system of the school. The presented results show how DEA can support individual performance appraisal and may enrich the application experience of DEA in this special area.

Keywords Data envelopment analysis · Performance appraisal · Linear programming · Higher education

T. Koltai (✉)
Department of Management and Business Economics, Budapest University of Technology and Economics, Pf. 91., 1521 Budapest, Hungary
e-mail: koltai.tamas@gtk.bme.hu

K. G. Krasz
Department of Ergonomics and Psychology, Budapest University of Technology and Economics, Pf. 91., 1521 Budapest, Hungary
e-mail: krasz.kata@gtk.bme.hu

A. Emrouznejad et al. (eds.), *Advances in the Theory and Applications of Performance Measurement and Management*, Lecture Notes in Operations Research,
https://doi.org/10.1007/978-3-031-61597-9_16

1 Introduction

The efficient utilization of constrained resources is a primary objective in for-profit and non-profit organizations as well. Data envelopment analysis (DEA) is a widely used method for identifying the sources of inefficient operations, and to indicate improvement possibilities. Several papers reviewed the exponentially increasing application areas of DEA in the recent decades (e.g., [1] and [2]). Applications mostly concentrate on efficiency analysis of organizational units. In this paper, however, DEA is applied to individuals, instead of organizations and the possibility of performance appraisal of employees with DEA in higher education is explored.

Performance appraisal (PA) is a strategic and integrated approach to increase the effectiveness of organizations by improving the performance of employees. PA can be used for developing the capabilities of teams and individual contributors as well. The purpose of PA is to provide information for managers to support administrative decisions such as hiring, firing, promoting, replacing, transferring, terminating, rewarding. PA also provides feedback for employees about how well they performed and what they need to improve [3, 4].

A special area of PA is higher education, where a constantly argued balance of teaching, research and service must be established at organizational level and at individual level as well. Most of the currently available performance measures in universities mainly focus on the research performance of their academic staff, and no or little attention has been paid to the evaluation of teaching performance and of other service activities [5]. In addition to research, teaching and service performance should also be evaluated through the PA process [6]. A method is required which can evaluate results achieved in all relevant areas. DEA might be an appropriate tool, because it aggregates results of different areas and indicators measured in different units.

Only a limited number of papers discuss the application of DEA for individual PA. In PA, information about the achievements of employees must be contrasted with the resources used, with the working conditions prevailing, and with the rewards obtained. In some reported cases, performance indicators and information about the prevailing conditions have been collected with questionnaires in IT companies [7], in agricultural companies [8] and in a pipeline construction business [9]. Nurses at an intensive care unit have been evaluated with a DEA model based on an indicator system consisting of 45 elements. The indicators were clustered and grouped to get an acceptable number for the inputs and for the outputs [10] as well.

A complex DEA based appraisal of professors is presented in [11]. Self-evaluation was prepared by a modified CRS model, pear evaluation was generated by cross efficiency analysis, and subjective evaluation was provided by the Dean. The three results were combined and analyzed. The applied DEA models, however, allowed different weights for the same inputs, and outputs at different professor. A similar model was applied by Oukil [12] for the PA of university professors, and he recommended that professors with different university ranks must be analyzed independently.

Based on the spears literature in this area it can be concluded that DEA is a promising tool for employee's performance appraisal, but reliable and consistent data collection must be solved, and the application possibility of the obtained results must be further explored. The objective of the presented paper is to provide an example of PA of faculty members and to explore some theoretical problems which may facilitate the successful application of DEA in this area.

The paper is structured as follows. In Sect. 2, the application environment and the data collection process are introduced. In Sect. 3, the applied DEA models are described and in Sect. 4 the main results are presented. Finally, in Sect. 5, the conclusions are summarized, and future research possibilities are outlined.

2 Data Collcction

The presented research is based on the performance evaluation system of the School of Economic and Social Sciences of the Budapest University of Technology and Economics. Data related to the three main target areas of performance (research, teaching and service) are collected from four sources:

- *Teaching*: Teaching load of each faculty member is obtained from the official study management system of the university. The teaching load aggregates the weakly hours professors spend in personal contacts with students. Depending on the type of courses (lecture, practice, project consultation) and on the language of instruction (English, Hungarian) different weights are used when weakly teaching load is calculated individually. The weighted sum of teaching ours provides a score which characterizes the teaching activity of each faculty member.
- *Publication*: Publication data are collected from the publication database of the Hungarian Academy of Sciences which contains all publications of Hungarian university professors and scientists. This database contains Web of science and Scopus publications as well as those works which are not indexed internationally. Publications are evaluated according to a weighting scale approved by the doctoral council of the school. Applying these weights, a publication score is generated.
- *Other scientific activities*: There are several scientific activities which are part of the professional life of faculty members and pertains to research activity (e.g., scientific committee memberships, pear review activity). An extended list of these activities is available in an on-line questioner which must be filled by the faculty members. Each activity is weighted on a commonly accepted evaluation scale and the weighted sum of these activities provide a score. This score quantitatively characterizes the activity of each faculty member in this category.
- *Service*: Activities which may not necessarily contribute to research and to teaching but indispensable for the operation of a higher education system must also be considered (e.g., managerial positions, committee memberships, administrative duties). An on-line questioner containing these activities helps to estimate the time spent by each faculty member on these activities in the evaluation period.

The total sum of this time characterizes the activity of each faculty member in this category.

Note, that research is split into two categories (publication and other scientific activity) as a consequence of the two different sources and types of information.

Performance of faculty members is evaluated based on the listed four categories in each calendar year. The assessment period starts in April. By then all education and publication data of the previous calendar year are collected. Next, the collected information is checked, and the on-line questioners are filled. Having all the data, an individual personal appraisal meeting is organized with the head of department. In this meeting, performance data are evaluated, corrected, and evaluation of the past year, furthermore, plans for the next year are discussed. As a result of this meeting, performance data are approved and officially registered in the PA database. Salary of each faculty member is also imported from the human resource management module of the university information system to the PA database.

A method is needed which compares the result of activities of faculty members and the total yearly income obtained for it. If faculty members are considered as decision making units (DMUs), outputs are measured by the PA system in the above mentioned four categories and inputs are the total yearly salary, then DEA can be applied and several characteristics of the performance of faculty members can be explored.

3 DEA Models Applied for PA

DEA is a linear programming-based performance evaluation method which compares inputs and outputs of organizations or individuals which are universally called decision making units (DMU).

In this section, the mathematical models applied in the presented research are summarized. Notation used in the paper are listed in Table 1.

Assume that we have I number of inputs, each denoted by index i ($i = 1,\ldots,I$) and R number of outputs, each denoted by index r ($r = 1,\ldots,R$). We also have J number of DMUs, each denoted by index j ($j = 1,\ldots, J$). The input value of input i in DMU j is x_{ij} and input matrix X contains all x_{ij} values. Similarly, the output value of output r in DMU j is y_{rj} and output matrix Y contains all y_{rj} values. Let u_r be the weight of output r and vector u contains all output weights. Similarly, let v_i be the weight of input i and vector v contains all input weights. Input-oriented efficiency is calculated as the ratio of weighted outputs over weighted inputs for reference DMU 0, that is, uY_0/vX_0. The optimal value of each variable is distinguished by notation * (e.g., u^*, v^*). If the weighted input is normalized, and we assume that the ratio of weighted outputs and weighted inputs is less than or equal to 1 for each DMU, then the input oriented primal model of DEA is obtained as follows,

Table 1 Notation

Indices	
j	Index of decision-making units (DMU), $j = 1, \ldots, J$
i	Index of inputs, $i = 1, \ldots, I$
r	Index of outputs, $r = 1, \ldots, R$
h	Index of categories, $h = 1, \ldots, H$
k	Running index of inputs and outputs
Parameters	
y_{rj}	Value of output r of DMU j
Y	Matrix containing all output values of DMUs
Y_0	Output vector of reference DMU 0
x_{ij}	Value of input i of DMU j
X	Matrix containing all input values of DMUs
X_0	Input vector of the reference DMU
l_{ki}	Lower bound of the input weight ratio of DMUs k and i
u_{ki}	Upper bound of the input weight ratio of DMUs k and i
L_{kr}	Lower bound of the output weight ratio of DMUs k and r
U_{kr}	Upper bound of the output weight ratio of DMUs k and r
P	Matrix containing the coefficients of input weight restrictions
Q	Matrix containing the coefficients of output weight restrictions
Variables	
u_r	Weight of output r
u	Vector containing output weights
v_i	Weight of input i
v	Vector containing input weights
λ_j	Dual variable of DMU j
λ	Dual variable vector of DMU constraints in the primal model
θ	Relative efficiency score

$$\begin{aligned} &\text{Max } (uY_0) \\ &uY - vX \le 0 \\ &vX_0 = 1 \\ &u, v \ge 0 \end{aligned} \tag{1}$$

Model (1) is solved for each DMU, consequently J number of efficiency scores (u^*Y_0) are obtained. The optimal solution for each DMU also contains the optimal weights of inputs (v^*) and outputs (u^*).

Any input and/or output weight can be zero in the optimal solution of (1). In this case the corresponding input or output values are ignored in the efficiency evaluation. To avoid such solutions, weight restrictions (WR) can be applied. DEA models with weight restrictions are also called assurance region (AR) models [13]. A frequent way of avoiding zero weights is the application of pairwise weight restrictions (e.g., [14] and [15]), which can be formulated as follows,

$$l_{ki} \leq u_i/u_k \leq u_{ki} \tag{2}$$

$$L_{kr} \leq v_r/v_k \leq U_{kr} \tag{3}$$

In (2) and (3), l_{ki} and L_{kr} are lower bounds of the input and output weight pairs, while u_{ki} and U_{kr} are upper bounds of the input and output weight pairs. Constraint (2) and (3) for all inputs and outputs can be summarized in matrix form as follows,

$$uP \leq 0 \tag{4}$$

$$vQ \leq 0 \tag{5}$$

Adding constraints (4) and (5) to model (1), the input-oriented CRS model with weight restrictions is obtained.

The dual form of model (1) may provide further insight for management about the patterns efficient DMUs follow. Introducing dual variable λ_j for DMU j and θ for the input normalization equation the dual of model (1) is obtained as follows,

$$\begin{aligned} &\text{Min } (\theta) \\ &Y\lambda \geq Y_0 \\ &X\lambda \leq \theta X_0 \\ &\lambda \geq 0 \end{aligned} \tag{6}$$

The optimal solution of (6) also provides the efficiency score of DMU 0 and it is equal to the efficiency value obtained by model (1) as a consequence of the strong duality theorem ($u^*Y_0 = \theta^*$) [16]. The optimal value of λ_j provide information about those DMUs whose technology must be followed by the inefficient DMUs. The right-hand-side of (6) shows, that outputs of the reference DMU are not changed, however, inputs must be reduced by θ. If the reference DMU is efficient ($\theta^* = 1$), inputs are not changed. If, however, the reference DMU is inefficient ($\theta^* < 1$), inputs must be decreased in order to achieve efficient operation.

Several other models exist in DEA which change model orientation (output-oriented, non-oriented), introduce variable return to scale (VRS) properties, depart from the radial approach, etc. [13]. In this paper, however, only the listed models are used for evaluating the performance of faculty members.

4 Analysis of the Results

DEA models presented in Sect. 3 are applied to explore the relationship between the achievements and the total salary of faculty members. Data of 111 professors have been collected for the 2022 calendar year. Each professor has a category identifier. Ranking of professors (full professor, associate professor, assistant professor and lecturer) are identified with R1, R2, R3 and R4 accordingly. Name of faculty members is substituted by codes from P1 to P111.

Performance data in the four activity categories presented in Sect. 2 used as outputs, and total yearly salary used as input. Input oriented constant return to scale models are applied to explore what salary would be appropriate for the performance of each faculty member. Several conditions were, however, also considered to refine results and to provide a better insight for the management about the correlation of performances and financial compensation of professors.

Table 2 summarizes some parts of the results of the applied models. The table lists information only about associate professors (category R2) to satisfy paper length requirement. The first column identifies the faculty members. Column 2 shows efficiency results when faculty members of all ranks (R1, …, R4) are evaluated together, however, only the results of professors with rank R2 are listed. Column 3 shows the efficiency results obtained when only associate professors (R2) are considered in the analysis. Column 4 shows the modified efficiency scores of associate professors when weight restrictions are applied. In column 5, associate professors (R2) are evaluated together with full professors (R1) while in column 6, associate professors are evaluated together with assistant professors (R3) and lecturers (R4). Finally, columns 7–10 list the optimal output weights (u^*) obtained for associate professors (R2) with model (1).

4.1 Categorical Problem

Evaluating all faculty members together, independently of their university position, is not appropriate because different positions provide different input and output possibilities. If DMUs operate under different operating conditions, they have different possibilities for efficient operations. One possibility to cope with this problem is the application of hierarchical categorical variables. In our case, the different university ranks are the different categories. The algorithm proposed in [13] avoids the evaluation of a DMU together with DMUs which have better possibilities for efficient operation. In the case of faculty members, however, it is not clear which category provides better possibilities for efficient operation.

From the point of view of input (salary), lower ranked professors are in more advantageous situation. At lower ranks, salary is lower, consequently, lower input results in higher efficiency at equal output values. From the point of view of output, higher ranked professors are in more advantageous situation. Higher ranks generally

Table 2 Efficiency scores and output weights of associate professors (R2)

1	2	3	4	5	6	7	8	9	10
DMU	Total	R2	R2 WR	R1 + R2	R2 + R3 + R4	Weights publ.	Weights prof.	Weights service	Weights teaching
P07	26.02	56.05	53.06	56.05	20.01	0	0	0.0079	0.0153
P13	74.22	43.35	37.75	42.39	73.67	0.0143	0	0.0011	0.0068
P15	64.18	51.99	44.32	51.49	100	0.0155	0	0.0076	0.0106
P19	31.35	70.73	61.03	67.13	100	0.0126	0	0.0040	0.0015
P20	40.43	60.99	60.82	60.99	34.20	0	0.0019	0	0
P21	34.92	82.81	47.13	71.11	24.15	0.0292	0	0	0.0191
P23	24.12	52.49	46.69	51.92	26.02	0.0144	0	0.0070	0.0099
P24	79.90	32.7	29.00	32.7	74.22	0	0.0028	0	0.0329
P25	29.60	89.39	79.06	89.39	64.18	0	0	0.0072	0.0140
P28	36.19	31.81	26.42	31.81	31.35	0	0	0.0052	0.0100
P29	24.81	100	70.41	72.19	40.43	0.0219	0.0006	0	0.0039
P30	45.25	56.39	51.59	56.39	34.92	0	0	0.0063	0.0122
P31	35.63	100	74.38	81.61	24.12	0.0212	0	0.0009	0.0105
P32	31.19	100	100	100	79.90	0	0	0.0065	0.0126
P35	36.42	38.61	35.38	38.24	29.60	0.0112	0	0.0055	0.0077
P36	27.41	68.35	59.67	64.62	36.19	0.0168	0	0.0054	0
P37	29.59	68.39	66.62	68.39	24.81	0	0.0005	0.0054	0.0093
P38	17.48	100	100	100	45.25	0	0.0007	0.0067	0.0114
P40	45.77	54.11	41.41	46.78	35.63	0.0393	0.0001	0.0016	0.0192
P44	51.78	74.37	69.33	74.37	31.19	0	0	0.0160	0.0311

(continued)

Table 2 (continued)

1	2	3	4	5	6	7	8	9	10
DMU	Total	R2	R2 WR	R1 + R2	R2 + R3 + R4	Weights publ.	Weights prof.	Weights service	Weights teaching
P46	35.80	89.84	87.90	89.77	36.42	0.0152	0	0.0075	0.0104
P47	17.38	100	51.72	71.05	29.59	0.0318	0	0	0.0208
P49	45.53	57.3	49.25	57.3	17.48	0	0.0004	0	0.0561
P50	27.33	73.47	66.43	73.47	45.77	0	0	0.0048	0
P53	49.98	66.43	47.97	64.12	51.78	0.0340	0	0	0.0222
P55	41.71	62.37	44.79	51.85	35.80	0.0381	0.0009	0	0.0080
P61	60.34	45.36	36.73	42.27	17.38	0.0182	0	0.0057	0.0021
P62	56.08	100	94.72	100	45.53	0.0522	0	0.0041	0.0248
P66	21.14	62.02	51.89	58.09	27.33	0.0235	0	0.0074	0.0028
P71	30.03	44.39	36.49	39.89	49.98	0.0294	0.0011	0	0
P72	33.72	25.53	24.05	25.37	41.71	0.0103	0	0.0051	0.0071
P73	56.59	20.22	16.37	20.22	60.34	0	0	0.0047	0.0091
P74	79.01	42.68	37.95	41.97	56.08	0.0271	0	0.0012	0.0134
P75	27.95	16.07	7.98	13.93	21.14	0.0471	0	0	0.0308
P76	30.40	34.93	29.58	34.93	30.03	0	0.0002	0	0.0301
P81	35.71	46.76	24.43	46.76	33.72	0.0021	0	0	0.0559
P88	24.15	70.35	58.72	69.8	56.59	0.0137	0	0.0067	0.0094
P106	35.23	65.78	65.37	65.78	79.01	0	0.0006	0.0030	0
P107	13.98	100	72.92	100	27.95	0.0017	0	0	0.0455

imply more publication possibilities, more engagements in other scientific activities (e.g., committee work) and more service tasks (e.g., managerial positions), consequently, it is easier to obtain higher output values.

In order to evaluate the effect of categories on efficiency, different DEA models with professors in different categories are solved. First, faculty members in R2 category are evaluated together with higher ranked faculty members (R1). Next, faculty members in R2 category are evaluated together with lower ranked faculty members (R3 and R4). The related efficiency scores are listed in column 5 and 6 of Table 2. In Fig. 1, efficiency scores of the two cases are compared.

To see the effect of categorical variables, professors are listed according to a monotone decreasing order of efficiency obtained when only R2 category is applied. These scores are hided in the figure for better clarity of the presentation. If efficiency scores deviate differently from a monotone decreasing trend in the two presented cases, then the effect of the rank of professors on efficiency is contradictory. This is justified by Fig. 1.

The figure shows, that when higher (R1 and R2) or lower (R2, R3 and R4) ranked professors are added to the calculation efficiency is not monotone decreasing and deviation from the monotone decreasing trend is different in the two presented cases. For example, the efficiency of faculty members P32, P31 and P29 are monotone decreasing in this order when R1 and R2 ranks are evaluated together. The efficiency of P31 is, however, higher than the efficiency of the two neighboring professors when R2, R3 and R4 ranks are evaluated together. In the case of professors P25, P21 and P44 the opposite can be seen. In this order, efficiency is monotone decreasing when R2, R3 and R4 ranks are evaluated together, but the efficiency of P21 is lower than the efficiency of the two neighboring professors when R1 and R2 ranks are evaluated together.

Several similar patterns can be discovered in the figure. Statistical analysis may provide further details about the effect of the two different calculations on efficiency,

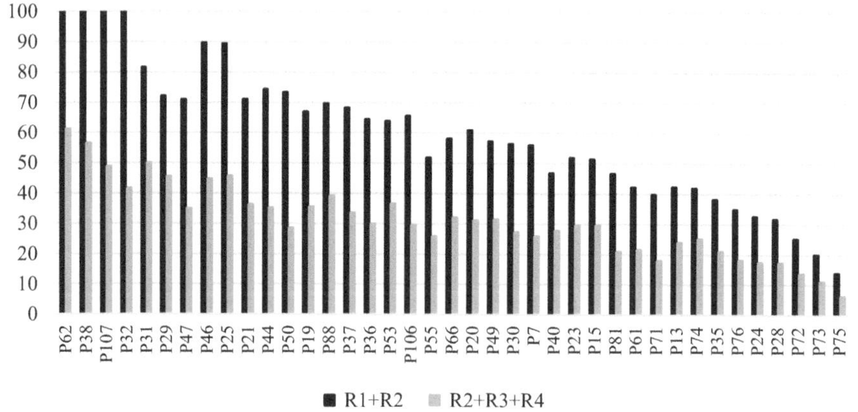

Fig. 1 Efficiency with categorical variables

but even based on Fig. 1 we may conclude, that the results of the two different approaches are contradictory. Consequently, it is recommended to evaluate efficiency of faculty members within the same category independently from professors of other categories.

4.2 *Effect of Weight Restrictions*

When model (1) is solved, weights of some inputs will be zero in the optimal solution. The analysis of the pattern of zero weights may highlight some important behavioral patterns of faculty members. The optimal output weights for associate professors provided by model (1) can be seen in column 8–11 of Table 2. To avoid zero weights, symmetrical pairwise weight restrictions are used for the inputs according to (3). The weight of any input cannot be 10 times higher than the weight of any other input, that is, $L_{kr} = 0.1$ and $U_{kr} = 1/L_{kr} = 10$ are applied in (3). Efficiency scores of associate professors (R2) calculated with weight restriction are listed in column 4 of Table 2. The effect of weight restrictions on efficiency is illustrated with Fig. 2. Faculty members are ordered in monotone decreasing order of efficiency calculated without weight restrictions. The R2_WR efficiency values are not monotone decreasing, which indicates the diverse effect of weight restrictions on efficiency. In some cases, significant differences can be seen between the two efficiency scores (R2 and R2_WR).

P47, for example, efficient without weight restriction, and a significant drop off efficiency can be seen when weight restrictions are applied. According to the optimal weights in Table 2, this professor has good results in publication and teaching, but other professional activity and service are excluded from the analysis when zero

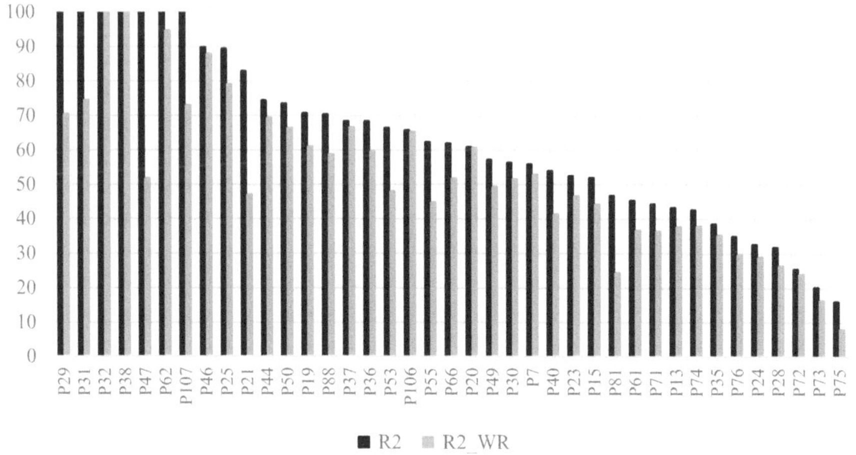

Fig. 2 The effect of weight restrictions

weights are allowed. This weight pattern can be seen at many other faculty members. Either efficient operation turns into inefficiency or the efficiency score of an inefficient operation drops significantly as a consequence of the poor performance in the area of other scientific activity and service (e.g., P21 and 81).

The optimal weights in Table 2 also show, that several faculty members have zero weights for the publication and for the other scientific activity outputs. These colleagues have good results in teaching and service but have a relatively poor performance in research. When weight restrictions are applied, their efficiency score drops, but not as significantly as in the above-mentioned cases (e.g., P7, 25 or P28).

Looking at the zero weights, some general conclusions can also be drawn. The highest number of zero values can be seen in column 9 (other professional activity), and the second highest number is in column 8 (publication). On the other hand, a significantly smaller number of zero values can be found in column 11. This pattern of 0 weights indicates, that most faculty members have better results is teaching, than in research (publication and other professional activity). This result may call for a stronger encouragement and motivation of colleagues in the scientific area.

Weights, however, must be interpreted carefully, because the optimal solution of model (1) is generally degenerate, alternative optimum may exist for the input and for the output weights. The different, equally optimal solutions may have different management interpretation. Selecting those optimal solutions, which correspond to some predetermined secondary management objective may solve this problem. For example, selecting waits which have the maximum support by the dataset, and maximize the minimum relative value of the corresponding inputs and outputs may provide unique optimum for model (1). The implementation of the algorithm proposed in [17] may provide unique solution and unique interpretation of the 0 weights.

5 Conclusions

A possible way of applying DEA for individual performance evaluation has been outlined in this paper. Input-oriented constant return to scale models have been used to evaluate the performance of university faculty members. Considering the limited length of the paper only results for associate professors have been presented, but similar techniques can be applied for other university positions. The official performance evaluation system of the school provided the output data in three classical activity area of higher education: research, teaching, and service. The total salary of professors was used as input, which was provided by the university accounting system. All data necessary for the analysis are available at the university independently of the application of DEA and are used for many other purposes.

The large number of results generated by DEA can only partly be presented in this paper, but two general conclusions can be drawn based on the results.

- Faculty members with different university ranks have different salaries and different output generation possibilities. Input favors lower ranked professors while output favors higher ranked professor. Using hierarchical categorical variables, we illustrated that, it is not possible to characterize universally advantageous or disadvantageous categories. Consequently, it is recommended to evaluate similarly ranked professors together, and independently from other ranks.

- School management wants to encourage faculty members to participate in research, teaching and service as well. Excellence in one or two areas cannot be an excuse for neglecting other areas. Weight restrictions can assist to introduce this policy into performance appraisal. Weight restriction parameters determine how much attention should be paid to the less favored areas by faculty members. The change of the efficiency scores as a consequence of weight restrictions shows the significant effect of the introduced policy in cases when some areas are neglected by the DMUs.

There are several ways to refine the presented DEA results. The introduction of non-controllable inputs may help to consider the fixed and variable elements of the faculty compensation system. The application of output orientation may provide information for target settings and for getting minimum expected values for research, teaching and service. The proper selection of the optimal weights in case of the degenerate solution of model (1) may clarify the management interpretation of 0 weights and opens new possibilities for the interpretation of reference sets.

Finally, the introduction of categorical variables for different scientific areas of faculty members may help to express scientific area differences in efficiency analysis. These questions are topics for our further research.

References

1. Cook, W. D., & Seiford, L. (2009). Data envelopment analysis (DEA)—Thirty years on. *European Journal of Operational Research, 192*(1), 1–17.
2. Emrouznejad, A., & Yang, G. (2018). A survey and analysis of the first 40 years of scholarly literature in DEA: 1978–2016. *Socio-Economic Planning Sciences, 61*(1), 1–5.
3. Fletcher, C. (2001). Performance appraisal and management: The developing research agenda. *Journal of Occupational & Organizational Psychology, 74*(4), 473.
4. Murphy, K. R., & Cleveland, J. N. (1995). *Understanding performance appraisal. Social, organizational and goal-based perspectives*. SAGE Publications.
5. Lohman, L. (2021). Evaluation of university teaching as sound performance appraisal. *Studies in Educational Evaluation, 70*, 101008.
6. Dasanayaka, C. H., Abeykoon, C., Ranaweera, R. A. A. S., & Koswatte, I. (2021). The impact of the performance appraisal process on job satisfaction of the academic staff in higher educational institutions. *Education Science, 11*, 623.
7. Dugelova, M., & Strenitzerova, M. (2015). The using of data envelopment analysis in human resource controlling. *Procedia Economics and Finance, 26*, 468–475.
8. Zbranek, P. (2013). Data envelopment analysis as a tool for evaluation of employees performance. *Review of Agricultural and Applied Economics, 16*(1), 12–21.
9. Shirouyehzad, H., Hosseinzadeh Lotfi, F., Aryanezhad, M. B., & Dabestani, R. (2011). A data envelopment analysis approach for measuring the efficiency of employees: A case study. *The South African Journal of Industrial Engineering, 23*(1), 191–201.

10. Osman, I. H., Berbary, L. N., Sidani, Y., Al-Ayoubi, B., & Emrouznejad, A. (2011). Data envelopment analysis model for the appraisal and relative performance evaluation of nurses at an intensive care unit. *Journal of Medical Systems, 35*, 1039–1062.
11. Oral, M., Oukil, A., Malouin, J.-L., & Kettani, O. (2014). The appreciative democratic voice of DEA: A case of faculty academic performance evaluation. *Socio-Economic Planning Sciences, 48*, 20–28.
12. Oukil, A. (2018, July 10–11). Building a consistent system for faculty appraisal using data envelopment analysis. In *Towards new practices in learning & teaching for higher education* (pp. 1–17). Algiers, Algeria. https://doi.org/10.13140/RG.2.2.29318.93768
13. Cooper, W. W., Seiford, L. M., & Tone, K. (2007). Data envelopment analysis. Springer.
14. Allen, R., Athanassopoulos, A., Dyson, R. G., & Thanassoulis, E. (1997). Weights restrictions and value judgements in data envelopment analysis: Evolution. development and future directions. *Annals of Operations Research, 73*, 13–34.
15. Koltai, T., & Tamás, A. (2022). Performance evaluation of teams in business simulation games with weight restricted data envelopment analysis models. *The International Journal of Management Education, 20*(3).
16. Hillier, F. S., & Lieberman, G. J. (1995). *Introduction to operation research*. McGraw-Hill.
17. Cooper, W. W., Ruiz, J. L., & Sirvent, I. (2007). Choosing weights from alternative optimal solutions of dual multiplier models in DEA. *European Journal of Operational Research, 180*(1), 443–458.

Regional Surgical Systems Response to COVID-19 Pandemic—Preliminary Insights Through Longitudinal DEA

Sean Shao Wei Lam, Yao Ge, Ashish Kumar, Ginny Zhenzhi Chen, Ahmadreza Pourghaderi, Ma Wai Wai Zaw, Pierce K. H. Chow, and Hiang Khoon Tan

Abstract *Goal:* The objectives of this benchmarking study are: (1) to evaluate the efficiency changes, and; (2) to elicit factors associated with efficiency improvements for surgical systems in South and Southeast Asia during the COVID-19 pandemic. *Methods*: We performed a two-stage benchmarking analysis for 9 hospital and health services (HHS) across 6 countries from Jan 2020 till Dec 2021. The first stage involves quantifying efficiency changes in response to COVID-19 via the Malmquist Index (MI). The second stage evaluates an explanatory model to elicit the factors associated with efficiency improvements. *Results*: Over the study period, the trend in MIs reveals that the countries started with low efficiency improvement during the initial phases of the pandemic. The situation improved with efficiency gains varying over the course of the pandemic due to the dynamically evolving policies, resurgence of new variants and introduction of vaccines. Measures related to safeguarding the wellbeing

S. S. W. Lam (✉) · Y. Ge · G. Z. Chen
Health Services Research Centre, Singapore Health Services Pte Ltd, Singapore, Singapore
e-mail: gmslasws@nus.edu.sg

S. S. W. Lam · A. Kumar · G. Z. Chen · M. W. W. Zaw · P. K. H. Chow · H. K. Tan
Duke-NUS Medical School, National University of Singapore, Singapore, Singapore

A. Pourghaderi
School of Public Health and Preventive Medicine, Monash University, Melbourne, Australia

M. W. W. Zaw
Department of Anaesthesiology, Singapore General Hospital, Singapore, Singapore

S. S. W. Lam · P. K. H. Chow · H. K. Tan
SingHeath Duke-NUS Global Health Institute, Singapore, Singapore

P. K. H. Chow · H. K. Tan
Division of Surgery and Surgical Oncology, National Cancer Centre Singapore and Singapore General Hospital, Singapore, Singapore

S. S. W. Lam · P. K. H. Chow · H. K. Tan
Surgery Academic Clinical Programme, Duke-NUS Medical School, Singapore, Singapore

S. S. W. Lam
Lee Kong Chian School of Business, Singapore Management University, Singapore, Singapore

A. Emrouznejad et al. (eds.), *Advances in the Theory and Applications of Performance Measurement and Management*, Lecture Notes in Operations Research,
https://doi.org/10.1007/978-3-031-61597-9_17

of healthcare workers and preserving the structural capacity of HHS during the pandemic are found to be significantly associated with improvements in efficiency. *Conclusion*: This benchmarking study allows us to evaluate the relative efficiencies of health systems in dealing with the pandemic and to identify the best practices associated with efficiency improvements. Our analysis framework provides HHS- and country-specific insight that will help to inform policy responses to future public health crises.

Keywords Surgical systems · COVID-19 · South and Southeast Asia · Efficiency · Benchmarking · Malmquist index

1 Introduction

Severe acute respiratory syndrome coronavirus 2(SARS-CoV-2) resulted in the global COVID-19 pandemic of unprecedented scale. As of the July 2023, there has been over 750 million cases and nearly 7 million deaths reported from this pandemic [1]. The pandemic necessitated lockdowns that had "far-reaching effects in different strata of life" [2]. On the frontlines, healthcare workers battled risks of infection and mortality. Policymakers had to make complex trade-offs under situation of extreme uncertainty and ambiguity. Immediate health system control measures have to be developed and implemented rapidly [3]. Public health measures such as social distancing, or "lockdown", have been implemented and scaled back at the national and regional level. Such measures aim to limit calibrate physical interactions amongst the population [3–5]. Personal protective equipment (PPE) measures have also been implemented in varying degrees across different countries and regions as evidence of SARS-CoV-2 virus might spread through aerosols from respiratory droplet emerged [4]. A variety of control measures have been continuously refined over the pandemic as we see the development of new vaccines and multiple resurgence of new variants.

Within health systems, hospitals and health services (HHS) have dynamically calibrated their own measures to ensure that surgical services can continue to be provided effectively [6]–7. The "unique and intimate relationship between patient and surgeon" does not have a substitute, and consequently surgical service providers faced difficult challenges during the pandemic [8]. A global review on the impact of the first wave of COVID-19 found that the reorganisation of surgical services to deal with the pandemic comprised six domains [6], and that "cancelling or postponing 'non-urgent' surgeries was almost universal"; such actions were initiated both by hospitals and by patients [9]. As the immediate effects of the pandemic recede, there is a need to draw lessons from the efforts to manage any repercussions. Quantitative benchmarking allows us to evaluate the relative efficiencies of health systems in dealing with the pandemic and to identify the best practices associated with efficiency improvements [10].

The innovative contribution of this benchmarking study is the focus on 6 South and Southeast Asian countries covering 9 HHS that are benchmarked longitudinally to elicit key lessons for the improvement of health system efficiency. In this context of the resource-limited surgical systems of South and Southeast Asia, our study of the efficiency of response to the COVID-19 pandemic addresses a problem domain that has a significant bearing on the well-being of a large and under-served population. The country representation spread across heterogeneous political and administrative structures, and public health systems and is part of a larger study with eleven different regional health system participation from India, Indonesia, Malaysia, Pakistan, Vietnam and Singapore, amongst others [11]. The insight from the study will help to inform policy responses to future public health crises.

2 Data and Methodology

2.1 *Study Design*

We performed a two-stage analysis to evaluate the government COVID-19 policy responses in south and southeast Asia. To capture the longitudinal shifts in relative efficiencies across time, we leveraged on the data envelopment analysis (DEA) model-based Malmquist Index (MI) [12] in the first stage. This stage involves quantifying efficiency changes of the 6 selected countries in response to COVID-19 via the DEA-based MI [12]. In the second stage, we used Tobit regression [13] to investigate the relationship between efficiency changes and HHS policies to identify the policies that are associated with an improvement in the national level efficiency. The explanatory model allows us to understand the HHS specific interventions associated with improvement in the relative efficiency determined in the first stage [14]. Figure 1 is a conceptual summary of our study design.

2.2 *Data Sources*

For the first stage of analysis, we extracted a panel dataset of 6 countries in South and Southeast Asia from January 2020 to December 2021 to evaluate the MI model. The model's inputs are: (i) GDP per capita (constant 2015 US$); (ii) population density (population/ km^2); (iii) International Health Regulations core capacity scores (IHRCCS) [15]; (iv) National Response Index. The outputs are (i) COVID-19 new cases; (ii) COVID-19 deaths.

GDP per capita, population density and population size are extracted from World Bank database [16, 17]. GDP per capita is often considered as an indicator of a country's standard of living and size of economy, which might be a potential factor influencing health system efficiency. Population density can have a significant impact

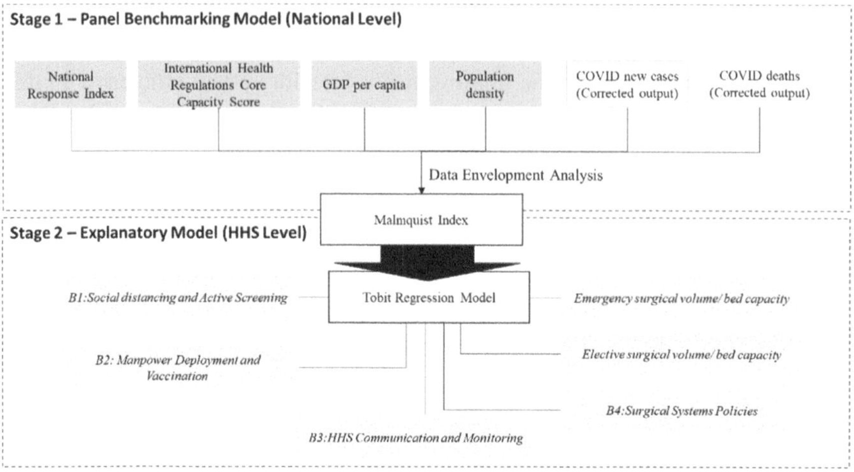

Fig. 1 Conceptual summary of study design

on the spread of COVID-19. The reciprocal of population density (Area per individual) was included as an input measure. In densely populated areas, the virus can spread more easily [18, 19]. The IHRCCS [15] is a self-assessment questionnaire that is completed by countries to evaluate their progress in implementing the core capacities in place to detect, assess, notify, and respond to public health emergencies of international concern, including pandemics like COVID-19. The National Response Index(NRI) is based on the Government Response Index in Oxford Covid-19 Government Response Tracker (OxCGRT) [20]. The NRI scores are updated regularly to reflect changes in government responses over time.

The output measures are COVID-19 new cases and deaths derived from the World Health Organization (WHO) COVID-19 Dashboard [21]. Both of these measures are important in evaluating a country's performance in controlling the pandemic. A high number of new cases is similar to the incidence rates which indicates that the virus is spreading rapidly within a community. Similarly, a high mortality rate indicates that the virus is causing severe illness and death within a community. Given that incidence and mortality are undesirable health results, they cannot be used directly in the DEA as output measures. To correct for this, the output variables are inverted (M minus the number of new cases and deaths, where M is a suitably large constant) [22, 23]. The inverted measures are normalized against the population size.

For the second stage, a hospital response index (HRI) was developed. The HRI took reference from the NRI derived from the OxCGRT [20] and validated by the survey team with senior surgical leaders in the partnering HHS. 9 partnering HHS institutions are asked to fill up a questionnaire across the study period to derive the HRI. Each monthly questionnaire contains a set of binary variables for 4 groups of factors, associated with flags that indicate whether it is a targeted policy measure (restricted to some groups or departments) versus a generic measure that is applied across the HHS. The HRI includes the following factors:

1. Social distancing policies (Cancel public events, restrictions on gatherings, visiting and internal movement, visitor controls) and active screening policies (perimeter and workforce screening)
2. HHS COVID manpower policies (Deployment of staff for COVID 19 activities, manpower vaccination rates)
3. HHS communication and monitoring policies (Routine instructions, testing policy, temperature monitoring)
4. Surgical systems policies (Splitting surgical teams, personal protective equipment requirements).

With the responses from the survey, scores ($I_{i,j,t}$) are then computed for factor j of country i in month t as $I_{i,j,t} = \frac{\max[(v_{i,j,t} - 0.5 f_{i,j,t}), 0]}{N_{j,t}}$ where $v_{i,j,t}$ is the binary response variable for the absence/presence of the policy for HHS i, factor j in month t. $f_{i,j,t}$ is the binary response variable for whether the policy is a targeted measure or applied generically across the HHS. $N_{j,t}$ is a normalization constant for $I_{i,j,t}$ to scale the scores to between 0 and 1. Apart from the HRI, data on monthly surgical volumes was sourced from the partnering HHS who participated in the study. Elective and emergency surgery volumes were collated from our panel. The surgery volume data were normalized against the number of beds in the health system (surgical volume per bed) as additional contextual variables in the model.

2.3 Model and Formulation

DEA as a non-parametric method, need not have a pre-specified production function [23]. DEA estimates the relative efficiency of each DMU with respect to the efficient frontier. The input-oriented DEA model with constant returns to scale is adopted for this study since the data consists of normalized numbers [24]. The Malmquist index (MI) is a measure of efficiency changes over time based on the efficiency scores derived from the non-parametric DEA model [25]. MI considers both technical progress (frontier-shift) and the catch-up effect. Frontier shift is evaluated as: $F_{i,t-1,t} = \sqrt{\frac{\delta^{t-1}\left[(x_i, y_i)^{t-1}\right]\delta^{t-1}\left[(x_i, y_i)^{t}\right]}{\delta^{t}\left[(x_i, y_i)^{t-1}\right]\delta^{t}\left[(x_i, y_i)^{t}\right]}}$; whereas the catch-up effect is evaluated as: $C_{i,t-1,t} = \frac{\delta^{t}\left[(x_i, y_i)^{t}\right]}{\delta^{t-1}\left[(x_i, y_i)^{t-1}\right]}$. $\delta^{t_1}\left[(x_i, y_i)^{t_2}\right]$ is computed based on the input-oriented radial DEA model for DMU i as $\min\left\{\theta | \theta x_i^{t_1} \geq X^{t_2}\lambda; y_i^{t_1} \leq Y^{t_2}\lambda; \lambda \geq 0\right\}$. These models are solved for the intra-temporal efficiency score, $\delta^{t}\left[(x_i, y_i)^{t}\right]$; and the intertemporal efficiency score, $\delta^{t}\left[(x_i, y_i)^{t-1}\right]$. The MI for DMU i is given by $MI_{i,t-1,t} = F_{i,t-1,t} C_{i,t-1,t}$ [24]. MI > 1 shows efficiency improvement whilst deterioration is seen when MI < 1. MI $= 1$ indicates no changes. The MI is evaluated with the DEA-SOLVER-LV8 [19].

With the panel MI estimates, Tobit regression was used to evaluate the policies that are associated with efficiency improvements across time [26]. As the MI estimates are based on the comparison of efficiencies in the current period (t) relative to the previous period, *t* − *1*, the proposed Tobit explanatory model is thus: $MI_{i,t-1,t} = \beta_0 + \sum_{j=1}^{4}\beta_j B_{j,i,t-1} + \alpha_1 E_{i,t-1} + \alpha_2 L_{i,t-1} + \varepsilon_{i,t-1}$, where $MI_{i,t-1,t}$ represents the MI of country *i* at time *t* relative to *t-1* and $B_{j,i,t-1}$ represents the level of COVID-19 policies of country *i* at time $t - 1$ (in month). $E_{i,t-1}$ and $L_{i,t-1}$ denote volumes of emergency and elective surgeries per bed in HHS *i* for month $t - 1$ respectively. β_j represents the effects of various COVID-19 related factors in response to COVID-19, whereas $\alpha_{j\in\{1,2\}}$ denotes the effects of these contextual factors related to the volume of surgeries in the preceding month. The model was evaluated with the VGAM [27] library on R 4.0.4 [28]. The COVID policies for country *i*, time *t* are:

$B_{1,i,t}$: Social distancing (Cancel public events, restrictions on gatherings, visiting and internal movement, visitor controls) and active screening policies (e.g. Perimeter and workforce screening)

$B_{2,i,t}$: HHS COVID manpower policies (Deployment of staff for COVID 19 activities, Manpower vaccination rates)

$B_{3,i,t}$: HHS communication and monitoring policies (Routine instructions, testing policy, temperature monitoring)

$B_{4,i,t}$: Surgical systems policies (Splitting surgical teams, personal protective equipment requirements)

3 Results and Discussion

Table 1 shows the descriptive statistics for the inputs, outputs across 24 months (January 2020–December 2021) and contextual variables across 18 months (January 2020–June 2021) for the 9 HHS in 6 different countries. The mean NRI across the study period is 0.604. The range of IHRCCS varies from 0.92 in Singapore to 0.52 in Pakistan. GDP per capita shows significant skewness, with median of $3,595 but a mean of $14,033. The highest GDP per capita is $67,176 while the lowest is only $1,410. Area/population (AP) is also highly dispersed. Consistent with geographical and socioeconomic heterogeneity, COVID outcomes vary considerably amongst these countries. At the end of December 2021, COVID-19 case fatality rates (confirmed deaths as a percentage of confirmed cases) ranged from 3.38% in Indonesia to 0.3% in Singapore against a world average of 1.92%; the percentage of fully vaccinated population varied from 29% in Pakistan to 83% in Singapore against a world average of 48%. Non-pharmaceutical Interventions (NPI) deployed in these countries also have varying degree of severity and effectiveness. NPIs are actions taken to flatten the disease transmission curve without the use of medicine and vaccines [29].

Figure 2 shows the trend of MI for 5 countries. The average MI started with a relatively low value of 0.95 in February 2020. The difference in MI values across countries were greater during the initial months; it started at a high of 0.26, but within three months it had narrowed to 0.04 and stayed within a relatively narrow

Table 1 Descriptive statistics of the inputs, outputs and contextual factors

Variable	Mean (SD)	Median (IQR)	Max	Min
Input variables				
GDP/capita (100,000U$)	0.140 (0.223)	0.036 (0.085)	0.672	0.001
Area/population (km^2/person)	0.004 (0.003)	0.003 (0.005)	0.010	0.000
IHRCCS	0.752 (0.130)	0.760 (0.170)	0.920	0.520
NRI	0.604 (0.168)	0.644 (0.171)	0.886	0.011
India	0.652 (0.190)	0.688 (0.122)	0.886	0.030
Indonesia	0.572 (0.171)	0.594 (0.135)	0.730	0.045
Malaysia	0.644 (0.174)	0.675 (0.117)	0.809	0.071
Pakistan	0.559 (0.171)	0.612 (0.113)	0.695	0.011
Singapore	0.603 (0.136)	0.633 (0.144)	0.742	0.190
Vietnam	0.596 (0.155)	0.633 (0.108)	0.774	0.022
Output variables				
Corrected new cases/population	0.380 (0.799)	0.071 (0.261)	1.834	0.001
India	0.006 (0.002)	0.006 (0.001)	0.007	0.001
Indonesia	0.036 (0.001)	0.036 (0.001)	0.037	0.032
Malaysia	0.296 (0.006)	0.298 (0.008)	0.301	0.279
Pakistan	0.043 (0.001)	0.043 (0.001)	0.044	0.043
Singapore	1.794 (0.037)	1.787 (0.075)	1.834	1.755
Vietnam	0.102 (0.002)	0.103 (0.001)	0.103	0.098
Corrected deaths/population	0.006 (0.785)	0.001 (0.004)	0.028	0.000
India	0.000 (0.000)	0.000 (0.000)	0.000	0.000
Indonesia	0.001 (0.000)	0.001 (0.000)	0.001	0.000
Malaysia	0.004 (0.001)	0.004 (0.000)	0.005	0.004
Pakistan	0.001 (0.000)	0.001 (0.000)	0.001	0.001
Singapore	0.027 (0.001)	0.027 (0.001)	0.028	0.026
Vietnam	0.002 (0.000)	0.002 (0.000)	0.002	0.001
Contextual variables				
Malmquist index	1.010 (0.306)	0.997 (0.013)	3.950	0.281
India	0.993 (0.033)	0.995 (0.017)	1.076	0.906
Indonesia	0.998 (0.020)	0.997 (0.011)	1.064	0.964
Malaysia	0.967 (0.093)	0.998 (0.008)	1.076	0.720
Pakistan	0.956 (0.175)	0.998 (0.013)	1.069	0.281
Singapore	1.155 (0.722)	0.998 (0.016)	3.950	0.749
Vietnam	0.989 (0.021)	0.997 (0.010)	1.006	0.929
Elective surgeries/inpatient bed	0.465 (0.324)	0.445 (0.568)	1.104	0.004
Emergency surgeries/inpatient bed	0.463 (0.333)	0.433 (0.354)	1.576	0.060

(continued)

Table 1 (continued)

Variable	Mean (SD)	Median (IQR)	Max	Min
HRI	0.483 (0.191)	0.549 (0.172)	0.750	0.000
$B_{1,i,t}$	0.565 (0.263)	0.621 (0.327)	0.875	0.000
$B_{2,i,t}$	0.212 (0.139)	0.180 (0.201)	0.589	0.000
$B_{3,i,t}$	0.647 (0.243)	0.694 (0.222)	0.944	0.000
$B_{4,i,t}$	0.527 (0.266)	0.625 (0.375)	1.000	0.000

range for most of the study period. The wide variations in policy responses in the initial period can be largely attributed to the lack of knowledge about the emerging infectious diseases [30]. Most countries struggled during the initial phase of the pandemic which was declared a pandemic by the WHO on 11th March 2020 [31]. A slew of NPIs were rapidly implemented and changes were made as more scientific evidence emerged [29].

Vaccination programs and the emergence of new variants of concern (VOC) subsequently resulted in changes to the policy measures, such as the re-introduction of "lock-downs" that were earlier relaxed [32]. These resulted in changes of the MI across time as NPI and vaccination rates varied due to the changing pandemic situations. Despite the uncertainties, Malaysia seemed to be the most stable in performance across the countries, having the lowest MI variations across time. Singapore was able to see efficiency gains after the initial 3 months of the pandemic (Feb–April 2020). This can be attributed to the refinement of NPIs, introduction of vaccines and other

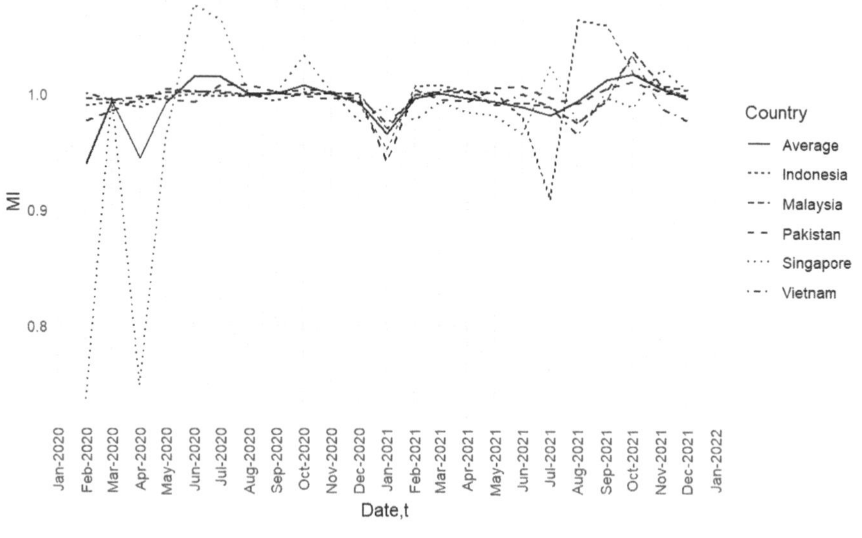

Fig. 2 Overall longitudinal trends in MI for across countries (excluding India)

measures, as more knowledge is gained about the pandemic [29, 30, 32]. For the next 6 months after the initial 3 months of the pandemic, a period of recovery in efficiency was observed across all DMUs.

Subsequently, lower values of MI were seen in the countries at the beginning of 2021, with India seeing the worst deterioration in performance over months 15–17 (see Table 2). These were likely to be associated with the effects of the second wave driven by VOCs, particularly from March 2021 (month 15) [33, 34]. Nonetheless, India recovered and stabilized from month 19 onwards. Countries with a lower base, or deteriorated to a lower base, may find it easier to improve relative to previous levels. These trends were reflected in the MI values.

Policy responses to COVID-19 was necessarily complex, with policies being taken in dynamic, highly uncertain and politically sensitive situations, resulting in diverse outcomes [35]. To understand the association of specific HHS policies on the efficiency, an explanatory model was evaluated with HHS specific contextual factors that include surgical system interventions. Table 3 shows the results of the Tobit regression model where the contextual factors describe NPIs measures related to social distancing, active screening, HHS manpower deployment, communication and monitoring and surgical systems policies. From the model, the effects of HHS manpower policies and elective surgical volumes per bed are significant. These results are in line with earlier findings. HHS COVID manpower policies are important to keep the health workforce safe and to maintain the effective functioning of HHS. A successful response to the significant influx of COVID-19 patients hinges on the effective deployment of the healthcare workforce during the pandemic [11, 36]. Similar to evidence reported in existing literature, higher vaccination rates have also been shown to bring more benefits for the HHS [37]. Apart from the manpower policies, HHS also implemented other surgical systems policies to preserve structural capacity [11]. Some of these policies were lifted as HHS returns to business as usual [7]. Amongst the reference set of countries, policies to preserve HHS capacities through proactive reduction of non-essential elective surgeries (α_2) seemed to have significant association on the MI.

The key insights gleaned from our study can be grouped into the longitudinal dynamics of efficiency and factors that can result in the improvement of HHS efficiency. Over the study period, the trends in MIs show that the countries were not prepared to face the crisis. This initial trough of Singapore over the initial wave and the rapid decrease of efficiency in India from month 15 due to VOCs are good learning lessons for future crises. Scenarios for emerging pandemics are often difficult to predict, regardless of whether the emergent pandemic are due to novel viruses or VOCs. The retrospective analysis of HHS policies will inform more realistic scenarios which can help to refine the playbook to improve HHS management policies during the next pandemic's initial phase. Measures related to safeguarding the wellbeing of healthcare workers and preserving the structural capacity of HHS during the pandemic are found to be significant. From a managerial viewpoint, a better appreciation is needed on averted health systems costs when Health Care Workers (HCWs) and health systems are better protected [15], and the view of expenditure on health systems as necessary investment to ensure their viability during emergent crisis [38].

Table 2 Trends in MI

	Month																						
DMU ↓	2	3	4	5	6	7	8	9	10	11	12	13	14	15	16	17	18	19	20	21	22	23	24
India	0.98	0.99	1.00	0.99	0.98	0.93	0.91	0.96	**1.08**	**1.07**	**1.03**	0.96	**1.02**	0.98	***0.72***	***0.28***	**3.95**	**1.14**	**1.05**	**1.04**	1.00	**1.02**	**1.00**
Indonesia	0.99	0.99	0.99	1.00	1.00	1.00	1.00	0.99	**1.00**	1.00	0.99	0.95	**1.01**	**1.01**	**1.00**	**1.00**	0.98	0.91	**1.06**	**1.06**	**1.02**	**1.01**	1.00
Malaysia	1.00	0.99	1.00	**1.00**	**1.00**	**1.00**	1.00	1.00	1.00	**1.00**	**1.00**	0.97	1.00	**1.00**	**1.00**	0.99	0.99	0.99	0.97	0.99	**1.04**	**1.01**	**1.00**
Pakistan	0.98	0.99	0.99	1.00	0.99	**1.01**	**1.01**	**1.00**	1.00	1.00	1.00	0.94	**1.00**	1.00	0.99	**1.01**	**1.01**	1.00	0.99	**1.01**	**1.01**	**1.00**	1.00
Singapore	*0.74*	1.00	*0.75*	0.96	**1.08**	**1.06**	1.00	**1.00**	**1.03**	**1.00**	0.98	0.99	0.98	0.99	0.98	0.98	0.97	**1.02**	0.98	1.00	0.99	**1.02**	**1.00**
Vietnam	**1.00**	0.99	0.99	**1.00**	**1.00**	**1.00**	1.00	**1.00**	**1.00**	**1.00**	1.00	0.97	1.00	**1.00**	**1.00**	0.99	1.00	0.99	0.96	**1.00**	**1.03**	0.99	0,98

Table 3 Tobit regression effects on MI

Effect	Description	Coefficient (SE)
β_1	Social distancing and active screening policies	0.295 (0.175)
β_2	HHS COVID manpower policies	0.759 (0.006)*
β_3	HHS communication and monitoring policies	−0.261 (0.189)
β_4	Surgical systems policies (excluding elective reduction)	−0.247 (0.129)
α_1	Emergency surgical volumes per bed capacity	−0.055 (0.572)
α_2	Elective surgical volumes per bed capacity	−0.207 (0.052)*

*Statistical significance at the $p < 0.10$ level

4 Conclusion

This study has three main limitations. Firstly, in terms of coverage, the work is limited to the reference set of HHS between whom a research collaboration agreement was set up. Though there is a large administrative effort required to establish such collaborations, similar analysis with larger reference sets would add to the body of knowledge on surgical systems efficiency under crises. Secondly, our national-level analysis is constrained by the reliability of data on COVID-19 new cases and deaths. We believe that working within this constraint is necessary in the interest of deriving insight that enhances preparedness for future crises. Thirdly, our analysis of efficiency is restricted to changes in efficiency, rather than efficiency levels per se. While this analysis of efficiency dynamics is insightful, it will also be valuable to assess overall impact on efficiency across the pandemic. The uncertainty inherent in pandemic and healthcare data further suggests that applying robust DEA [39]–40 will be promising.

Despite the limitations and potential future research directions, this study brings new insight from the South and Southeast Asian regional perspective on national and HHS efforts in ensuring a resilient health system through the pandemic. The use of the DEA approach has demonstrated practical value in this benchmarking study across HHS. Measures related to safeguarding the wellbeing of healthcare workers and preserving the structural capacity of HHS during the pandemic are found to be significantly associated with improvements in efficiency. The close collaborations with regional HHS partners continue to provide impetus for continuous learning and strengthening of regional collaboration for better global health.

Acknowledgements This study was funded by the SGH Health Development Fund (grant no. FRGR01PNDM20, Response and Outcome Analysis of Tertiary Surgical Unit in Asia). We thank all the participants from the 9 HHS across the region for taking time out of their busy schedules for this project. We are grateful for the support by the SingHealth International Collaboration Office for the coordination of the study.

References

1. WHO. (2023). 'WHO Coronavirus (COVID-19) Dashboard'. Retrieved July 30, 2023, from https://covid19.who.int/
2. Onyeaka, H., Anumudu, C. K., Al-Sharify, Z. T., Egele-Godswill, E., & Mbaegbu, P. (2021). COVID-19 pandemic: A review of the global lockdown and its far-reaching effects. *Science Progress, 104*(2), 003685042110198. https://doi.org/10.1177/00368504211019854
3. Chua, A. Q., et al. (2020). Health system resilience in managing the COVID-19 pandemic: Lessons from Singapore. *BMJ Global Health, 5*(9), e003317. https://doi.org/10.1136/bmjgh-2020-003317
4. Chu, D. K., et al. (2020). Physical distancing, face masks, and eye protection to prevent person-to-person transmission of SARS-CoV-2 and COVID-19: A systematic review and meta-analysis. *The Lancet, 395*(10242), 1973–1987. https://doi.org/10.1016/S0140-6736(20)31142-9
5. Thu, T. P. B., Ngoc, P. N. H., Hai, N. M., & Tuan, L. A. (2020). Effect of the social distancing measures on the spread of COVID-19 in 10 highly infected countries. *Science of The Total Environment, 742*, 140430. https://doi.org/10.1016/j.scitotenv.2020.140430
6. O'Rielly, C., et al. (2021). Surgery and COVID-19: A rapid scoping review of the impact of the first wave of COVID-19 on surgical services. *British Medical Journal Open, 11*(6), e043966. https://doi.org/10.1136/bmjopen-2020-043966
7. Abdullah, H. R., et al. (2022). Resuming elective surgery after COVID-19: A simulation modelling framework for guiding the phased opening of operating rooms. *International Journal of Medical Informatics, 158*, 104665. https://doi.org/10.1016/j.ijmedinf.2021.104665
8. Kibbe, M. R. (2020). Surgery and COVID-19. *JAMA, 324*(12), 1151. https://doi.org/10.1001/jama.2020.15191
9. Søreide, K., et al. (2020). Immediate and long-term impact of the COVID-19 pandemic on delivery of surgical services. *British Journal of Surgery, 107*(10), 1250–1261. https://doi.org/10.1002/bjs.11670
10. Ettorchi-Tardy, A., Levif, M., & Michel, P. (2012). Benchmarking: A method for continuous quality improvement in health. *Healthc Policy, 7*(4), e101-119.
11. Huang, W. N., Xin, X., Rao, V., Wong, T. H., Chow, P., & Tan, H. K. (2023). Battling against the great disruption to surgical care in a pandemic: Experiences of 11 South and Southeast Asian countries. *British Medical Journal Open, 13*(4), e060770. https://doi.org/10.1136/bmjopen-2022-060770
12. Caves, D. W., Christensen, L. R., & Diewert, W. E. (1982). Multilateral comparisons of output, input, and productivity using superlative index numbers. *The Economic Journal, 92*(365), 73–86. https://doi.org/10.2307/2232257
13. Baltagi, B. H. (2021) Econometric analysis of panel data. *In Springer texts in business and economics*. Springer International Publishing. https://doi.org/10.1007/978-3-030-53953-5
14. Banker, R. D., & Natarajan, R. (2008). Evaluating contextual variables affecting productivity using data envelopment analysis. *Operations Research, 56*(1), 48–58. https://doi.org/10.1287/opre.1070.0460
15. Bolas, T., Werner, K., Alkenbrack, S., Uribe, M. V., Wang, M., & Risko, N. (2023). The economic value of personal protective equipment for healthcare workers. *PLOS Glob Public Health, 3*(6), e0002043. https://doi.org/10.1371/journal.pgph.0002043
16. The World Bank. (2023). 'World Development Indicators (people per sq km of land area)'. Retrieved July 30, 2023, from https://data.worldbank.org/indicator/EN.POP.DNST
17. The World Bank. (2023). 'World Development Indicators, Population'. Retrieved July 30, 2023, from https://data.worldbank.org/indicator/SP.POP.TOTL
18. Sigler, T., et al. (2021). The socio-spatial determinants of COVID-19 diffusion: The impact of globalisation, settlement characteristics and population. *Global Health, 17*, 56. https://doi.org/10.1186/s12992-021-00707-2
19. Cooper, W. W., Seiford, L. M., Tone, K., & Cooper, W. W. (2006). *Introduction to data envelopment analysis and its uses: With DEA-solver software and references*. Springer.

20. Hale, T., et al. (2021). A global panel database of pandemic policies (Oxford COVID-19 Government Response Tracker). *Nature Human Behaviour, 5*(4), 529–538. https://doi.org/10.1038/s41562-021-01079-8
21. WHO. (2023). 'World Health Organization Coronavirus (COVID-19) Dashboard'. Retrieved July 27, 2023, from https://covid19.who.int/data
22. Cordero, J. M., Nuño-Solinís, R., Orueta, J. F., Polo, C., Del Río-Cámara, M., & Alonso-Morán, E. (2016). Evaluación de la eficiencia técnica de la atención primaria pública en el País Vasco, 2010–2013. *Gaceta Sanitaria, 30*(2), 104–109. https://doi.org/10.1016/j.gaceta.2015.11.002
23. González-de-Julián, S., Barrachina-Martínez, I., Vivas-Consuelo, D., Bonet-Pla, Á., & Usó-Talamantes, R. (2021). Data envelopment analysis applications on primary health care using exogenous variables and health outcomes. *Sustainability, 13*(3), 1337. https://doi.org/10.3390/su13031337
24. Cooper, W. W., Seiford, L. M., & Zhu, J. (2004). *Handbook on data envelopment analysis*. Kluwer Academic.
25. Färe, R., Grosskopf, S., Lindgren, B., & Roos, P. (1994). Productivity developments in Swedish hospitals: A Malmquist output index approach. In *Data envelopment analysis: Theory, methodology, and applications* (pp. 253–272). Springer Netherlands. https://doi.org/10.1007/978-94-011-0637-5_13
26. Falavigna, G., Ippoliti, R., & Ramello, G. B. (2018). DEA-based Malmquist productivity indexes for understanding courts reform. *Socio-Economic Planning Sciences, 62*, 31–43. https://doi.org/10.1016/j.seps.2017.07.001
27. Yee, T. W. (2015). Vector generalized linear and additive models: With an implementation in R. In *Springer series in statistics*. Springer New York. https://doi.org/10.1007/978-1-4939-2818-7
28. Core Team, R. (2022). R: A language and environment for statistical computing. R foundation for statistical computing. Vienna, Austria. https://www.R-project.org/
29. Zhang, W., et al. (2023). Non-pharmaceutical interventions for COVID-19 reduced the incidence of infectious diseases: A controlled interrupted time-series study. *Infectious Diseases of Poverty, 12*(1), 15. https://doi.org/10.1186/s40249-023-01066-3
30. Agyapon-Ntra, K., & McSharry, P. E. (2023). A global analysis of the effectiveness of policy responses to COVID-19. *Science and Reports, 13*(1), 5629. https://doi.org/10.1038/s41598-023-31709-2
31. Cucinotta, D., & Vanelli, M. (2020). WHO declares COVID-19 a pandemic. *Acta Bio Medica Atenei Parmensis, 91*(1), 157–160. https://doi.org/10.23750/abm.v91i1.9397
32. Summan, A., & Nandi, A. (2022). Timing of non-pharmaceutical interventions to mitigate COVID-19 transmission and their effects on mobility: A cross-country analysis. *The European Journal of Health Economics, 23*(1), 105–117. https://doi.org/10.1007/s10198-021-01355-4
33. Samarasekera, U. (2021). India grapples with second wave of COVID-19. *The Lancet Microbe, 2*(6), e238. https://doi.org/10.1016/S2666-5247(21)00123-3
34. Song, H., Fan, G., Liu, Y., Wang, X., & He, D. (2021). The second wave of COVID-19 in South and Southeast Asia and the effects of vaccination. *Frontiers in Medicine, 8*, 773110. https://doi.org/10.3389/fmed.2021.773110
35. Amul, G. G., Ang, M., Kraybill, D., Ong, S. E., & Yoong, J. (2022). Responses to COVID-19 in Southeast Asia: Diverse paths and ongoing challenges. *Asian Economic Policy Review, 17*(1), 90–110. https://doi.org/10.1111/aepr.12362
36. Sun, S., Xie, Z., Yu, K., Jiang, B., Zheng, S., & Pan, X. (2021). COVID-19 and healthcare system in China: Challenges and progression for a sustainable future. *Global Health, 17*(1), 14. https://doi.org/10.1186/s12992-021-00665-9
37. Nurchis, M. C., et al. (2022). COVID-19 vaccination campaign among the health workers of Fondazione Policlinico Universitario Agostino Gemelli IRCCS: A cost-benefit analysis. *IJERPH, 19*(13), 7848. https://doi.org/10.3390/ijerph19137848
38. Luce, B. R., Mauskopf, J., Sloan, F. A., Ostermann, J., & Paramore, L. C. (2006). The return on investment in health care: From 1980 to 2000. *Value in Health, 9*(3), 146–156. https://doi.org/10.1111/j.1524-4733.2006.00095.x

39. Peykani, P., Mohammadi, E., Saen, R. F., Sadjadi, S. J., & Rostamy-Malkhalifeh, M. (2020). Data envelopment analysis and robust optimization: A review. *Expert Systems, 37*(4), e12534. https://doi.org/10.1111/exsy.12534
40. Toloo, M., Mensah, E. K., & Salahi, M. (2022). Robust optimization and its duality in data envelopment analysis. *Omega, 108*, 102583. https://doi.org/10.1016/j.omega.2021.102583

An Analysis of the Malmquist-DEA Productivity Index Applied to the Segment of Electric Power Transmission in Brazil

Joice Figueiredo, Solange Kileber, Leonardo Campos Filho, Rafael Cancella, Erika Nogueira, Mayara Silva, Amaro Pereira, Ubaldo Rios, Ligianne Carvalho da Silva Dâmaso, Andrea Damico, and Bianca Morais

Abstract The Malmquist productivity index is used in the segment of Electric Power Transmission in Brazil to adjust the Allowed Annual Revenue (AAR) by reducing the costs of operation and maintenance of companies. The regulator, the National

---

The authors thank the National Electric Power Agency (ANEEL) in the Research & Development project "RD-06904-0001/2021—An Analysis of Tariff Revision and Transmission Auctions". [R&D Leadership: ISA CTEEP].

J. Figueiredo (✉) · S. Kileber · L. C. Filho
SIGLASUL, Rio de Janeiro, Brazil
e-mail: jfigueiredo@siglasul.com.br

S. Kileber
e-mail: skileber@siglasul.com.br

L. C. Filho
e-mail: lcampos@siglasul.com.br

R. Cancella · E. Nogueira · M. Silva · A. Pereira
Planejamento Energético, COPPE, Universidade federal do Rio de Janeiro, Rio de Janeiro, Brazil
e-mail: rafaelmorais@ppe.ufrj.br

E. Nogueira
e-mail: erika@ppe.ufrj.br

M. Silva
e-mail: mayara.silva@ppe.ufrj.br

A. Pereira
e-mail: amaro@ppe.ufrj.br

U. Rios · L. C. da Silva Dâmaso · A. Damico · B. Morais
ISA CTEEP, São Paulo, Brazil
e-mail: urios@isacteep.com.br

L. C. da Silva Dâmaso
e-mail: ldamaso@isacteep.com.br

A. Emrouznejad et al. (eds.), *Advances in the Theory and Applications of Performance Measurement and Management*, Lecture Notes in Operations Research,
https://doi.org/10.1007/978-3-031-61597-9_18

Electric Power Agency (in Portuguese, Agência Nacional de Energia Eletrica—ANEEL), intends to capture productivity gains arising from technological evolution to reduce tariffs. The methodology and parameters to be used in the calculation of productivity gains are defined quinquennially by ANEEL after public discussion. The last discussion occurred in 2021 and the productivity index was established as zero in 2022. In the current regulatory model, the Data Envelopment Analysis (DEA) is adopted to calculate the Malmquist distance functions, which are obtained using data from 2013 to 2019. Productivity is segregated between: (i) technological change; and (ii) efficiency change (segregated between scale change and pure efficiency change). This methodology also considers the estimation of confidence intervals for technological gains, which are obtained using bootstrap. In this article, we propose to update the information used in Malmquist to verify whether there have been changes in productivity. We also aim to make a diagnosis of the capture of technical and scale change in transmission auction concessions in Brazil. Since these concessions have the regulatory revenue determined by the lowest bid offered in the auction, it is important to understand whether the capture productivity gains are consistent considering the Malmquist interpretation.

Keywords Malmquist productivity index · Electric power transmission · Data envelopment analysis · Brazil

1 Introduction

Economic regulation seeks mechanisms to make monopoly environments, as the electric energy transmission segment, more similar to competitive ones. In this sense, its main purposes are to (i) generate the necessary incentives so that the regulated company can be efficient and provide a quality service; and (ii) transfer at least part of such productivity gains to the consumer.

In order to achieve these purposes, we highlight regulation based on incentives and the *franchise bidding*. In regulation based on incentives, prices/revenues are revised periodically according to the regulated company's efficient costs. The temporary price determination encourages the company to reduce costs, which is called "market competition". In *franchise bidding,* "market competition" is established, and the company acquires the right to provide the service for a specified period after winning an auction in which prices are established throughout the entire contract and generally adjusted only for inflation [1].

In this context, in the regulatory models are distinguished by the moment of capture of productivity gains. Incentive regulation tends to share efficiency and

A. Damico
e-mail: agonzaga@isacteep.com.br

B. Morais
e-mail: bsmorais@isacteep.com.br

productivity gains with consumers in periodic reviews or tariff adjustments. On the other hand, the *franchise bidding* captures such gains at the time of the auction, since the competitive environment induces participants to offer a lower price level.

In Brazil, the electric energy transmission segment is organized through concession contracts signed after public auctions. However, the discussion about capturing productivity gains is controversial. For concession contracts up to 2006, the regulator understood that the winning proposal would already deduct part of the potential productivity gains, called Business Efficiency Gain (*Ganho de Eficiência Empresarial*—GEE), from bid revenue. The consequence is tariffs reduction, in line with the *franchise bidding*. For this reason, such contracts had no periodic review clause on the Allowed Annual Revenue (*Receita Anual Permitida*—RAP) offered at auction.

However, later, the Brazilian Federal Court of Auditors (*Tribunal de Contas da União*—TCU) determined that the regulator should institute periodic review processes of the RAP, corroborating the ideas of regulation by incentives. Among TCU's requirements, there is the adequacy of public service concession contracts for electricity transmission in order to provide the mechanisms for appropriation of productivity gains by consumers.

Given the lack of consensus on the need for appropriation of these gains, the objective of this work is to verify whether the current methodology used to determine the GEE of the transmission concessionaires bid is adequate for the regulatory model used in the transmission segment in Brazil. Specifically, we expect to update and carry out a critical analysis of the methodology used to determine the GEE of the bid transmission concessionaires.

In order to achieve the purpose of this work, the methodology used involves: (i) an initial diagnosis about how the regulator determine productivity gains (Malmquist index, Data Envelopment Analysis and *bootstrap*) in Brazil, (ii) a survey of the most current database for the parameters used to calculate these gains, and (iii) GEE estimates for the electric energy transmission segment based on the information gathered.

In addition to this introduction, the article is structured in other four sections. In the second section, materials and methods, we detail the models used to calculate business efficiency gains and describe the analyzed scenarios. In the third section we present the results and discussion. In the fourth section we summarize the main conclusions and messages of the work and in the fifth section the bibliographical references are shown.

2 Materials and Methods

The calculation of productivity gains in electric energy transmission segment is performed by ANEEL using a retrospective method, with the breakdown of Malmquist index and using efficiencies calculated through the Data Envelopment Analysis (DEA) model. In addition, confidence intervals are built for the breakdown

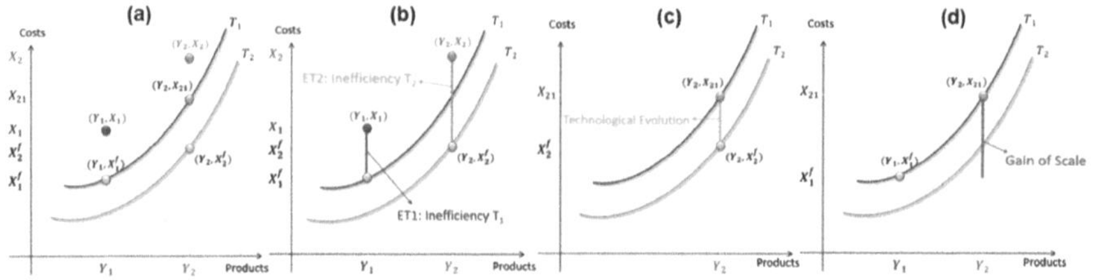

Fig. 1 Breakdown of productivity gains. *Source* Adapted from the Regulatory Impact Analysis Report No. 01/2021-SRM/ANEEL [2]

of the indexes obtained. Data from transmission concessionaires are organized in terms of operational groupings.

In Fig. 1 the levels of efficient inputs and outputs in a hypothetic company are presented. (Y_1, X_1) e (Y_2, X_2) are output and input levels at times T_1 e T_2. (Y_1, X_1^f) and (Y_2, X_2^f) are equivalent to the levels of efficient outputs and inputs in. A company's productivity gains can be broke down into three sources: (i) change in production efficiency, basically the distance between (Y_1, X_1) and (Y_2, X_2), and its efficient levels, (Y_1, X_1^f) and (Y_2, X_2^f), , getting closer or moving away from the cost frontier (Fig. 1b); (ii) technological change, denoted by the difference between the efficient costs of period T1 if producing Y2 and the efficient costs of period T2 for the same production level Y2,with displacement of the cost frontier (Fig. 1c); and (iii) scale gains, denoted by the difference between efficient costs as a function of increased production levels, from Y_1 to Y_2, given the technology in T_1 (Fig. 1d).

In the current methodology, ANEEL focuses only on productivity gains arising from technological evolution (ii) and scale gains (iii), called GEE (from Portuguese *Ganhos de Eficiência Empresarial*). It is understood that efficiency gains (i) are already internalized in the costs resulting from the bid's competitive process itself, which selects the most efficient companies.

The National Electric Energy Agency (*Agência Nacional de Energia Elétrica*—ANEEL) uses Malmquist index method [3] (Eq. 1) to calculate the productivity, technological and scale gains of the transmitters.

$$
\begin{aligned}
PTF_{t-1,t} &= (PTF_{t-1} \times PTF_t)^{1/2} \\
&= \left[\frac{D_0^{t-1}(X^t, Y^t)}{D_0^{t-1}(X^{t-1}, Y^{t-1})} \times \frac{D_0^t(X^t, Y^t)}{D_0^t(X^{t-1}, Y^{t-1})}\right]^{1/2} \qquad (1)
\end{aligned}
$$

where PTF is the Total productivity of the factors; D_0: distance function; $\frac{D_0^{t-1}(x^t, y^t)}{D_0^{t-1}(x^{t-1}, y^{t-1})}$: productivity gain between periods t and $t-1$, in reference to the instant $t-1$; $\frac{D_0^t(x^t, y^t)}{D_0^t(x^{t-1}, y^{t-1})}$: productivity gain between periods t and $t-1$, in reference to the instant t.

The Malmquist index can be broken down into three components (Eq. 2) [4]: (i) pure efficiency gains; (ii) technological evolution; and (iii) pure scale gains. In (i) and (ii) there are variable scale returns, whereas in scale gains (iii) the returns are constant in the numerator of each analyzed relationship and variable in the denominator.

$$\begin{aligned} M_0\left(x^t, y^t, x^{t-1}, y^{t-1}\right) &= (i)x(ii)x(iii) \\ &= \left(\frac{D^t_{oV}\left(X^t, Y^t\right)}{D^{t-1}_{oV}\left(X^{t-1}, Y^{t-1}\right)}\right) \\ &\times \left[\left(\frac{D^{t-1}_{0V}\left(X^t, Y^t\right)}{D^t_{0V}\left(X^t, Y^t\right)}\right) \times \left(\frac{D^{t-1}_{0V}\left(X^{t-1}, Y^{t-1}\right)}{D^t_{0V}\left(X^{t-1}, Y^{t-1}\right)}\right)\right]^{1/2} \\ &\times \left[\frac{\left(\frac{D^{t-1}_0\left(X^t, Y^t\right)}{D^{t-1}_{0V}\left(X^t, Y^t\right)}\right) \times \left(\frac{D^t_0\left(X^t, Y^t\right)}{D^t_{0V}\left(X^t, Y^t\right)}\right)}{\left(\frac{D^{t-1}_0\left(X^{t-1}, Y^{t-1}\right)}{D^{t-1}_{0V}\left(X^{t-1}, Y^{t-1}\right)}\right) \times \left(\frac{D^t_0\left(X^{t-1}, Y^{t-1}\right)}{D^t_{0V}\left(X^{t-1}, Y^{t-1}\right)}\right)}\right]^{1/2} \end{aligned} \quad (2)$$

ANEEL uses the non-parametric DEA model to estimate the Malmquist distance functions (Eq. 3 and constraints (4)). In DEA, these distances are estimated using a mathematical optimization model that aims to compare the performance of each decision unit (DMU) in relation to an efficiency frontier, maximizing the relationship between outputs and inputs for each DMU. Efficiency is calculated in terms of how well a given DMU uses its inputs to produce outputs, compared to the other DMUs.

$$\max_{u,v,\varphi} h_0 = \sum\nolimits_{j=1}^{m} v_j \times y_j^0 + \varphi, s.a. \quad (3)$$

Subject to

$$\sum\nolimits_{i=1}^{n} u_i \times x_i^0 \le 1, \sum\nolimits_{j=1}^{m} v_j \times y_j^k - \sum\nolimits_{i=1}^{n} u_i \times x_i^k \le 0 \ (k = 1,2, \ldots, K) \text{ and } u_i, v_j \ge 0 \quad (4)$$

where h_0 is the efficiency of the company under analysis; v_j is the "weight" assigned to product "j"; y_j^k is the vector of products of company "k"; φ is the "scale factor"; u_i is the "weight" assigned to input "i"; x_i^k is the input vector of company "k"; n is the total inputs; m is the total number of products.

The Shephard distances are used [5–7] in the formulation to calculate Malmquist, which is the inverse of the Farrell technical efficiencies estimated in DEA model (input-oriented model). Thus, in the analysis of results, a Malmquist index higher than the unit is interpreted as an increase in productivity (progress), while an index lower than the unit shows a reduction in productivity (regression).

Next, ANEEL calculates confidence intervals for DEA efficiencies and, consequently, for Malmquist index and its components. In order to do so, the regulator adopts the bootstrap approach [8, 9], simplified in the step-by-step shown in Fig. 2 [9].

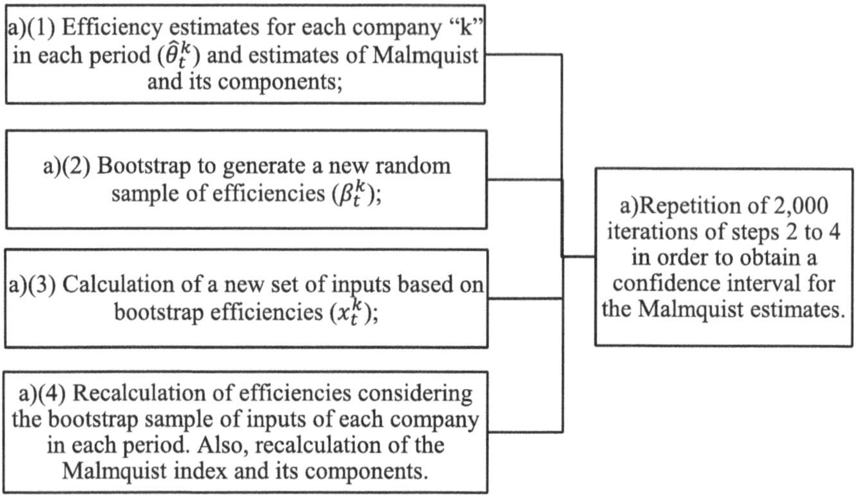

Fig. 2 Step by step in 4 steps to measure the confidence intervals for DEA efficiencies and Malmquist index

Based on the estimated efficiencies using DEA model ($\hat{\theta}_t^k$) for the "k" companies analyzed in step 1, a new random sample of efficiencies is estimated for each year (β_t^k), to expand the base of observations and derives estimates of parameters of the population that originated the initial sample. To avoid peaks in the empirical distribution to be constructed, it is necessary to use the smoothed bootstrap method [10] shown in step 2. Based on the smoothed efficiencies obtained in the previous step, a new set of inputs is calculated (step 3). Finally, the efficiencies are recalculated with the bootstrap sample of companies and Malmquist index and its components (step 4).

Steps 2–4 are repeated 2,000 times in order to define an empirical distribution function and their confidence intervals. The repetitions generate two thousand estimates of Malmquist and its components for each company in the sample. To compute an index representative of the sector, the median of the companies' estimates is calculated, resulting in 2,000 medians. The lower and upper limits are defined using the 5% and 95% percentiles of these medians, respectively.

2.1 Regulatory Operating Costs and Malmquist Index

In order to review the revenue referring to operating costs, ANEEL: (i) identifies the portion of Allowed Annual Revenue that corresponds to regulatory operating costs (COM from Portuguese *custos operacionais e de manutenção*) (Eq. 5); and (ii) applies the productivity gain arising from technological change and scale gain over this cost (Eq. 6).

$$\mathrm{COM}(t) = \theta \times I \tag{5}$$

$$COM'(t) = COM\,(t) \times (1 - \rho)^n \tag{6}$$

where *COM*: regulatory operating cost resulting from the review; θ: cost percentage; I: regulatory capex; ρ: percentage of annual productivity gains; and n: number of years considered (equal to 5 years). *COM'*: regulatory operating cost deducted from productivity gains

The productivity gains are deducted from the regulatory operating costs in order to share the gains with the consumers. Although Eqs. 5 and 6 are not directly applied in this work, they demonstrate how the regulatory agency uses the technological change and scale gains calculated by Eqs. 1–4.

2.2 Data Base and Scenarios

In this study, a database was built for the power transmission segment in Brazil, with information on inputs and assets for the years 2017–2021. The inputs considered are operational costs: Personnel, Materials, Third-Party Services, Insurance, Taxes, Others and Labor Convictions (PMSO) accounts, in Brazilian reais of 2022 [11]. The assets are: total extension of networks with voltage between 69 and 750 kV; power transformers and converters with voltage between 10.5 and 750 kV; and modular switching units, with voltage between 4.16 and 750 kV [12, 13].

The database is composed of 19 *holdings* and another 52 concessionaires that do not share costs, totaling 71 DMUs. However, one of them has zero PMSOs and was excluded, with the initial database consisting of 70 DMUs. Besides, in order to obtain more robust data, two procedures are carried out in the database, as follows:

- *Procedure 01*: (i) exclusion of specific GEE estimates in the reference biennium for companies that, after grouping into *holdings*, do not have all the products considered in the analysis; and (ii) exclusion of *outliers* identified in DEA estimates in the reference biennium by the Jackknife method [14].
- *Procedure 02*: application of a filter to remove companies that have at least one of the following characteristics: (i) are undergoing court and out-of-court reorganization processes; (ii) their assets entered into commercial operation at the end of the period under analysis; (iii) concessions where most assets are operated in direct current; and (iv) failed to submit all requested information to the regulator.

Thus, two scenarios for GEE simulations were built originating from procedures based on 70 DMUs, namely:

- *Scenario 01*: application of procedure 01 results in 40 concessionaires for analysis;
- *Scenario 02*: application of procedures 01 and 02 results in 32 concessionaires to be analyzed.

Table 1 Descriptive statistics

Scenario	Statistic	PMSO	Ext. of network	Power transf.	Modular swit. units
Scenario 01	Mean	194.081,97	3.356,58	302,30	11.912,76
	SD	365.396,32	4.918,30	578,23	21.672,40
	Median	42.550,52	948,40	47,50	4.280,00
	Minimum	0,00	0,00	0,00	0,00
	Maximum	1.877.744,14	19.982,05	2.598,00	113.236,50
Scenario 02	Mean	228.373,54	3.608,44	363,23	12.996,55
	SD	412.695,40	5.427,10	632,08	23.995,11
	Median	32.295,99	720,76	59,00	2.550,00
	Minimum	0,00	0,00	0,00	0,00
	Maximum	1.760.965,17	19.982,05	2.598,00	113.236,50

Descriptive statistics for the databases referring to scenarios 01 and 02 are presented in Table 1. In general, Scenario 02 presents relative higher mean and standard deviation for all variables when compared to Scenario 01.

3 Results and Discussion

In this section we present and discuss the results of the DEA model update obtained by the breakdown of Malmquist index applied to the power transmission segment in Brazil, based on data from 2017 to 2021. In the current methodology, this information is from 2015 to 2019. In the two scenarios described above, the evolution of all components of Malmquist index, composed of technological advances and gains in scale, and their confidence intervals will be analyzed.

Based on the samples defined for each scenario, DEA model was estimated to obtain the distance functions to be used in the calculation of Malmquist index.

Next, the total productivity index and its components (technological evolution, pure scale gains and pure efficiency gains) were calculated according to the breakdown proposed by Ray and Desli [4], for each company in the sample and for each biennium in each scenario. In Figs. 3 and 4 the medians of the results obtained in each biennium are presented. The percentage shown equals the productivity index (1/Shephard distance) minus the unit.

The results obtained in these scenarios indicate that: (i) Malmquist index shows a productivity gain in almost all biennia, except in the 2018/2019 period for scenario 02; (ii) scale gains were equal to zero in all analyzed biennia, reflecting the limited impact on operating costs due to the inclusion of reinforcements (additional assets in relation to the bid purpose) in the transmission system; (iii) efficiency gains are predominantly negative; and (iv) there is an increasing trend of technological evolution in the first three biennia, with a reversal in the 2020/2021 period.

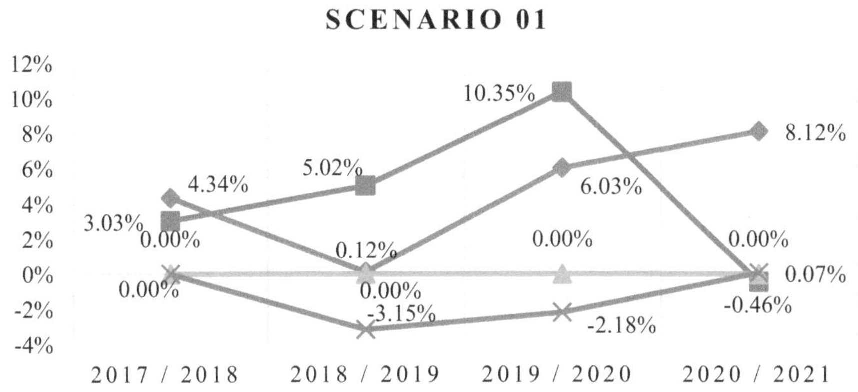

Fig. 3 Breakdown of Malmquist index scenario 01

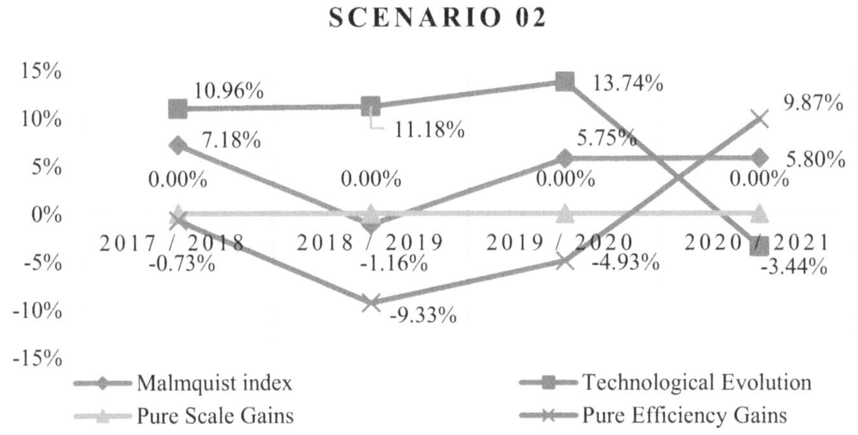

Fig. 4 Breakdown of Malmquist index scenario 02

The analysis of Figs. 3 and 4 also allows us to find that, despite the path of Malmquist index and its components having a similar trend in scenarios 01 and 02, the magnitude of the values found in these two analyzes is significantly different, especially for efficiency gains and technological evolution. This evidence reflects the high sensitivity of DEA model, used to estimate Malmquist, against the database used.

In order to calculate the GEE, in particular, only technological evolution and scale gains are found, and these values are obtained through the median of the estimated indexes for all transmission companies assessed throughout the period of analysis for each scenario. Table 2 shows the punctual result obtained for each GEE component in scenarios 01 and 02.

Table 2 Punctual GEE results

Index	Scenario 01 (%)	Scenario 02 (%)
Technological evolution	4.91	8.72
Scale gain	0.00	0.00

The figures in Table 2 indicate a possible technological evolution and lack of scale gains in both scenarios assessed. However, in terms of technological evolution, when assessing the estimated indexes for each company, per biennium, significantly volatile results can be seen. For example, as seen in Table 3, in scenario 02, transmission company TP Sul recorded a progress of 50.9% in the biennium 2018/2019, followed by a technological return of 32.8% in the biennium 2019/2020 and a progress of 30.7% in the biennium 2020/2021. These results are questionable and contradictory, since this is a stable segment in terms of technology, and it is not expected that the power sector productivity presents such sudden changes due to technological changes, even more so with annual changes between progress and return of technology.

Since the sample variation proved to be significantly relevant to the results obtained, in particular for technological evolution, the analysis of punctual estimates must consider a source of uncertainty [9]. Because of that, confidence intervals were estimated for GEE components as shown in Table 4. In calculating *bootstrap*, a significance level of 5% was considered.

At Table 4, regarding technological evolution and efficiency gains, for both scenarios, there is no sample evidence that the median of Malmquist index is different from zero, considering the level of significance analyzed, since the confidence intervals include this value. Thus, it is not possible to objectively conclude that the analyzed period showed a gain or loss in productivity due to technological evolution and pure efficiency gains in the power transmission segment in Brazil.

On the other hand, for scale gains, the upper and lower bands, for all significance levels, were equal to the median value of 0.00%. Therefore, there is evidence that in

Table 3 GEE results for TP Sul in scenario 02

	2017/2018	2018/2019	2019/2020	2020/2021
TP Sul	–	1,509	0,672	1,307

Table 4 95% confidence interval

	Scenario 01		Scenario 02	
	Lower limit	Upper limit	Lower limit	Upper limit
Malmquist index	0.041	0.046	0.032	0.047
Technological evolution	−0.033	0.157	−0.038	0.193
Efficiency gain	−0.097	0.076	−0.123	0.084
Scale gain	0	0	0	0

the analyzed period there were no scale gains to be applied to the transmission sector in Brazil. The ranges of the general productivity index were positive in all scenarios.

That said, the analysis of Table 4 indicates that, although there are productivity gains in the transmission segment in Brazil, it is not possible to say what the source of these gains is, whether due to technological evolution or efficiency gains—considering scale gains equal to zero. Furthermore, there is an instability that may be related to the volatility of input data. One of the reasons for the high variability found is the time mismatch between the entry of a project in the asset base and the accounting of associated costs. This is because operational costs are incurred in the construction phases, when the asset has not yet entered commercial operation.

In practice, for new DMUs, this effect related to the time mismatch was purged as proposed by ANEEL in Subsidy Intake Nº. 14/2022 [15]. Therefore, annual records in which the input is positive and assets equal to zero were excluded from the sample. However, for DMUs that already have assets in the database, it is not possible to identify which part of the PMSO increase is related to the construction phase. Thus, it gives rise artificial productivity gains/decreases. Holding other factors constant, the PMSO increase not accompanied by the commercial operation of an asset in the same year decreases the productivity. In the following year, when the asset enters the base, there is an artificial productivity gain. These losses and gains do not necessarily have the same effects for the same DMU. And, when this happens for several companies at the same time, it tends to further aggravate the situation.

In addition, Bragança and Camacho [1] state that technological advances are a very common type of productivity gain in telecommunications, which is a technology-intensive sector, but not very significant in sectors such as transport, power, and water and sewerage, where technological changes are smoother. Bragança and Camacho [1]. In Brazil, for example, in airports and highways sectors, where bidding is also done, productivity gains are not applied. Thus, these authors claim that it seems unnecessary to capture productivity gain resulting from technological evolution in the transmission segment in Brazil, corroborating with the conclusions of this article. Furthermore, the authors point out that drastic errors in the use and/or measurement of this variable can, at one extreme, affect the economic feasibility of companies/ investments and, at the other, penalize the consumer with excessive tariffs.

Additionally, it is understood that in practice, companies do not differentiate whether the productivity gains expected at the time of the auction come from efficiency gains or technological evolution. Which means that, regardless of where it comes from, productivity gains are already expected in the competitive auction environment and should not be subject to analysis in tariff reviews.

Therefore, there is no reason to capture productivity gains from operating costs, given that efficiency gains and possible technological evolution are already captured at the time of the auction. One can only discuss the application of scale gains in the event of introducing reinforcements in the purpose of the bidding, however, this impact proved to be null.

A limitation of this article is that it does not assess other methods, such as the parametric model based on Stochastic Frontier Analysis (SFA). Additionally, the results of DEA model are data sensitive: the inclusion or exclusion of any unit from the analysis causes significant changes in the result, as the model measures the relative efficiency compared to the other units. Since there is instability in the analyzed database, the current database may not be enough for assessment with this methodology. This instability is caused by the time mismatch between the entry into commercial operation of assets and the accounting of associated costs, as previously mentioned. One option would be to use a more homogeneous database, with updates between operating costs and assets.

In this article only one database update was performed using the current methodology, but it is possible to use other methodologies. In other words, if one continues to calculate GEE for bid transmission companies, future work can be performed to improve the regulation, for example: (i) estimating other models of breakdown of productivity gains, (ii) estimating the Malmquist distance functions from methods other than DEA, as is the case of SFA model, (iii) estimating Malmquist index with defined gains from scale as defined by the breakdown formulas, (iv) estimating DEA model with the inclusion of weight restrictions, (v) testing the incorporation of other asset variables, such as total reactive power and substation equipment, (vi) assessing the model consistency results only with bidding companies, (vii) excluding the operational scale filter to assess the impact of keeping small companies in the sample, and (viii) analyzing the impact of a longer time period on the productivity component.

4 Conclusions

In this article, we sought to update the calculation of the productivity gain to be applied to bid electric power transmission concessionaires in Brazil, according to the methodology determined by ANEEL. In addition, the adequacy of inserting the Malmquist index in the regulatory model used for these companies was discussed, that is, the *franchise bidding* regulation.

Initially, the results obtained from the breakdown of Malmquist index indicate that there are productivity gains in the Brazilian transmission segment, even in the case considering confidence intervals built for this index. However, it was not possible to confirm the origin of these gains, as the estimated ranges for efficiency gains and technological evolution were not conclusive. In addition, it was found that scale gains are null, even in the case of the entry of reinforcements in the system.

It was concluded that, as this is a competitive environment, the offers made in the transmission auctions already incorporate possible productivity gains, so there should not be a reduction of revenues of bid transmission companies because of productivity gains during the concession contract period. It was understood that, in practice, companies do not differentiate whether the variation in productivity foreseen at the time of the auction is due to changes in production efficiency or in technology. Thus, regardless of where they come from, productivity gains are already foreseen and should not be analyzed in tariff reviews.

Another important finding concerns the characteristics of data in the transmission segment in Brazil. For a new asset, there is a time mismatch between the entry date of the inputs and products in the database since costs may be recognized while the asset is still in the construction phase. For new DMUs, it is possible to eliminate this effect by excluding from the sample the annual records in which PMSO is positive and the assets equal to zero. However, for DMUs that already have assets in the database, it is not possible. Thus, it gives rise to artificial productivity gains/decreases. These losses and gains do not necessarily have the same effects for the same DMU. And, when this happens for several companies at the same time, it tends to further aggravate the situation.

Finally, it is understood that there is still much to be discussed in the development of an assertive regulation that favors the efficiency of bidding processes in the transmission segment in Brazil. For example, if GEE calculation is kept, future work can be performed to improve the regulation, with improvements in the assumptions and variables used in DEA or the use of other methodologies that are more adequate to the database of transmission companies.

References

1. Bragança, G. G. F., de Camacho, F. T. (2012). *Uma Nota Sobre o Repasse de Ganhos de Produtividade Em Setores de Infraestrutura No Brasil (Fator X).*
2. Aneel. (2021). Revisão Metodológica—Fator X e Ganhos de Eficiência Empresarial Relatório de Análise de Impacto Regulatório N° 1/2021-SRM/ANEEL Anexo Da Nota Técnica N° 20/2021-SRM/ANEEL Processo N° 48500.001216/2021-11. Brasília, Brasil
3. Malmquist, S. (1953). Index numbers and indifference surfaces. *Trabajos de Estadística, 4*, 209–242. https://doi.org/10.1007/BF03006863
4. Ray, C. S., Desli, E. (1997). Productivity growth, technical progress, and efficiency change in industrialized countries: Comment. *American Economic Review, 87*.
5. Shephard, R. W. (1953). *Theory of cost and production functions*. Princeton.
6. Coelli, T., Rao, P., & Battese, G. E. (1998). *An introduction to efficiency and productivity analysis*. MA, Springer, US.
7. Lekprichakul, T., Bank, A. D., & Chakravorty, U. (2003). Efficiency and technical change in the Philippine rice sector: A Malmquist total factor productivity analysis. *American Journal of Agricultural Economics, 85*, 943–963.
8. Figueiredo, A. T. (2007). Mensuração e Análise Da Evolução Da Produtividade Total Dos Fatores Agregada No Brasil: Aplicação Da Abordagem de Bootstrap Ao Índice de Malmquist., Universidade federal do Rio Grande do Sul.
9. Simar, L., & Wilson, P. W. (1999). Estimating and bootstrapping malmquist indices. *European Journal of Operational Research, 115*, 459–471. https://doi.org/10.1016/S0377-2217(97)00450-5
10. Efron, B. (1979). Bootstrap methods: Another look at the Jackknife. *Annals of Statistics, 7*, 1–26.
11. ANEEL Balancete Mensal Padronizado-Transmissão. Retrieved January 20, 2023, from http://informacoesbmp.aneel.gov.br/ConsultarBMPAberto.aspx
12. ONS SINDAT—Sistema de Informações Geográficas Cadastrais Do SIN. http://sindat.ons.org.br/SINDAT/Home/ControleSistema

13. ANEEL Sistema de Gestão Da Transmissão—SIGET Retrieved January 20, 2023, from https://dadosabertos.aneel.gov.br/dataset/sistema-de-gestao-da-transmissao-siget
14. Miller, R. G. (1974). The Jackknife-a review. *Biometrika, 61*, 1–15.
15. ANEEL. (2022). Tomada de Subsídios 014/2022. Brasília, Brasil.
16. Coelli, T. J., Rao, P., Donnell, C. J., Battese, G. E. (2005). *An introduction to efficiency and productivity analysis*. Springer S.

Primary Healthcare Efficiency Assessment Using Panel Data

Silvia González-de-Julián, Isabel Barrachina-Martínez, and David Vivas-Consuelo

Abstract The objective of this study is to analyze the efficiency of Primary Healthcare Centers (PHCs) within a health district of the Valencian Community (Spain). The methodology employs Data Envelopment Analysis (DEA) with input orientation and variable returns to scale, utilizing panel data spanning a 5-year period (2015–2019). Inputs considered are number of physicians, nurses, and pharmacy costs, while outputs include number of consultations, hospital emergencies, referrals, avoidable hospitalizations, avoidable mortality, prescription efficiency, and the exogenous variable of morbidity. Three models are developed focusing on healthcare activity, outcomes, and both. Results reveal differences in efficiency scores depending on output variables, with certain PHCs consistently efficient or closer to the efficient frontier, while others remain inefficient. The study indicates a general decrease in efficiency levels over the assessed period, particularly notable when only activity outputs are included. Efficiency analysis using DEA proves valuable as a managerial tool for identifying inefficiencies in PHCs, albeit with the need to align assessment objectives with desired outcomes to ensure accurate results.

Keywords Efficiency · Primary healthcare centers · Healthcare outcomes

1 Introduction

Healthcare is one of the fundamental pillars of the welfare state, together with education and the social protection system (which includes the pension system, sickness and disability benefits, unemployment and social services), and is one of the public services with the highest volume of expenditure. In recent years there has been an increase in this expenditure, driven by the progressive aging of the population, a

S. González-de-Julián (✉) · I. Barrachina-Martínez · D. Vivas-Consuelo
Research Center for Health Economics and Management, Universitat Politècnica de València, Camino de Vera s/n, 46022 Valencia, Spain
e-mail: silgonde@upv.edu.es

A. Emrouznejad et al. (eds.), *Advances in the Theory and Applications of Performance Measurement and Management*, Lecture Notes in Operations Research,
https://doi.org/10.1007/978-3-031-61597-9_19

growing demand for services from patients and the incorporation of more advanced, but also more expensive, technology.

Within the health services, primary healthcare services represent the gateway to the health system and play an agency role, so that their proper functioning determines that performance of specialized healthcare and hospitalization, which has a direct impact on the health of the population [1, 2]. The sustainability of the public health system requires the efficiency of primary healthcare services.

In this study, real world data from the Primary Healthcare Centers (PHC) in a health district of the Valencian Community (Spain) are used. The objective is to develop models to measure the efficiency of the PHCs and to evaluate how the variables introduced in the models influence the efficiency scores, to design models that incorporate healthcare quality variables and healthcare outcomes as outputs, and to assess the evolution of the efficiency of the PHCs.

2 Materials and Methods

2.1 Data and Variables

The Valencian Community (east of Spain) health map is divided into 24 health districts. Each of them contains a number of PHCs that may differ from one health district to another. The Valencia Clínico—La Malvarrosa Health District is one of the biggest health districts of the Valencian Community. It is divided into 18 PHCs serving an approximate population of 320,000 inhabitants.

The PHCs of the analyzed district are very different in terms of size and characteristics of the population they serve, their location (urban and rural or semi-rural), etc., which implies a great heterogeneity between PHCs and provides a greater wealth of data for the analysis by being able to establish comparisons between the different areas.

The databases of the Regional Ministry of Health (*Conselleria de Sanitat Universal i Salut Pública*) and the Valencia Clínico—La Malvarrosa Health District have been integrated to obtain for each PHC: characteristics of the covered population, healthcare activity and human resources of the PHCs. The information sources used were: the Population Information System (SIP), the Hospital Minimum Data Set (MDS), the database of emergencies and referrals from the Hospital Information System (HIS), the mortality data from the Mortality Register from the General Direction for Public Health and Addictions, and the electronic outpatient clinical records (ABUCASIS), which include the Ambulatory Information System (SIA) and the Pharmacy Prescriptions Manager (GAIA).

The data obtained were anonymised and the study has strictly complied with the current personal data protection regulations, specifically Regulation 2016/679 of the European Parliament and of the Council of 27 April 2016 on the protection of natural persons with regard to the processing of personal data and on the free movement of such data, as well as the Organic Law 3/2018, of December 5 concerning protection of personal data and guarantee of digital rights.

2.2 Methodology

Data envelopment analysis with panel data over a 5 year-period (2015–2019) is used to assess the efficiency, in which each of the 18 PHCs constitutes a Decision-Making Unit (DMU). Inputs included are rates per 10,000 inhabitants of: physicians, nurses and pharmacy costs; as outputs the rates per 10,000 inhabitants of: consultations, hospital emergencies, referrals, avoidable hospitalizations, avoidable mortality and prescription efficiency.

The case-mix (multimorbidity index) is used as exogenous variable, as it does not directly participate in the production process and cannot be controlled, but do affect it, given that the activity and healthcare results of the PHCs are conditioned by the characteristics of the population they serve.

It is also important to include outputs that measure the quality of the services provided. Most of the studies carried out only include measures or indicators of activity, such as the number of consultations or the number of emergencies, since it is not easy to have indicators that measure healthcare outcomes (avoidable hospitalizations or avoidable mortality). But this should be the ultimate goal. These data are not easy to obtain and usually have to be constructed from other data.

The DEA model used is the basic radial model, with variable returns to scale and input orientation of the 18 PHCs of the Valencia Clínico—La Malvarrosa Health District for the years 2015–2019. A database has been generated that includes all the variables for each of the PHC and each of the years, which is known as panel data.

The DEA methodology has a restriction as to the maximum number of variables that can be entered as inputs and outputs, which is set to a third of the DMUs evaluated. For this particular health district, with 18 PHCs, will set the limit to enter 6 variables between inputs and outputs. In order to avoid this dimensionality problem, a panel data is used, converting the sample of 18 units into a panel of 90 observations (18 units for 5 years). This approach suggests grouping all cross-sectional sections forming a single intertemporal set that uses and treats separately all observations or DMUs of all periods. It considers each unit evaluated, in this case each PHC, as if it was a different unit in each of the reference periods. Thus, the use of longitudinal data makes it possible to compare a unit in a specific period with its own performance in multiple years or periods, as well as with the performance of the rest of the units, to discriminate between efficient units, to provide greater robustness to the data, and to reduce the problem of studies with small samples [3, 4].

The methodology with input orientation and variable returns to scale is used. For each production unit, an input series x and an output series y are considered. This analysis is expressed in its input orientation according to the following expression:

$$\min \theta - \varepsilon\left(\sum_{i=1}^{m} S_i^- + \sum_{r=1}^{s} S_r^+\right)$$

$$s.t. \sum_{j=1}^{n} \lambda_j x_{ij} + s_i^- = \theta x_{io} i = 1,2, \ldots, m; \tag{1}$$

$$\sum_{j=1}^{n} \lambda_j y_{rj} - s_r^+ = Y_{r0} r = 1,2, \ldots, s;$$

$$\lambda_j \geq 0 j = 1,2, \ldots, n;$$

$$\sum\nolimits_{j=1}^{n} \lambda_j = 1$$

Variable returns to scale (VRS) are used in our analysis. Unlike the constant returns to scale (CRS) method, the VRS does not assume that all PHCs have the same scale of production and, therefore, considers the different dimensions that each of them may have.

It is worth mentioning that the rate of physicians and nurses is considered a non-discretionary input. These are variables that cannot be reduced or increased by management. They are exogenously fixed, since the number of professionals in the centers is determined by the Regional Ministry of Health, and the PHCs, and their managers, have limited capacity for maneuver. It is important to take them into account in the analysis, since not doing so may lead to the estimation of biased scores. The approach developed by Banker and Morey (1986) is used, knowing that it has some limitations [5, 6].

The rate of emergencies, avoidable hospitalizations and avoidable mortality are considered undesirable outputs. It is very important to take this into account when designing the models and carry out a prior transformation so that they are imputed in the appropriate way during the analysis. The treatment of undesirable outputs is a complex issue, and different alternative approaches can be found in the literature [7]. In our study, the simplest approach has been chosen, in which the original values are replaced by ones modified by subtracting a sufficiently high fixed amount (multiplying the result by −1 to make it positive) [8–11]. By doing this, the traditional DEA model can be used, although it must necessarily be applied with variable returns to scale.

With regard to exogenous or non-controllable variables, these are variables that are exogenously fixed and cannot be managed by the PHCs, but which have a direct effect on their activity and results. The treatment of these variables is also complex because there are multiple methodological options, all of which have advantages and disadvantages. We have opted for the one developed in the book by Cooper et al. [12].

One of the factors that influence the analysis is that the DMUs analyzed are of different sizes, and more robust results are obtained when the DMUs are of similar sizes. The analysis of the data collected from this district manifests the heterogeneity of the PHCs in terms of covered population, the level of activity and use of resources.

To mitigate this inconvenience, and so that the results are not affected, the original data are transformed into rates per covered population in each PHC, specifically, into rates per 10,000 inhabitants. These rates constitute the variables used in the models.

The efficiency scores of each PHCs have been analysed according to three models with different approaches. All the models incorporate the same inputs and different combinations of outputs related to: healthcare activity (first model), healthcare outcomes (second model) and healthcare activity + outcomes (third model, in which all outputs are included). Each model is analysed with and without the exogenous variable.

3 Results

Several models are considered, in order to subsequently select the one that allows for the clearest differentiation of PHC efficiency and that considers the results on the health of the population, rather than healthcare activity.

Our selection of variables was based on previous analysis, on the review of variables used in other studies, but also considering the limitations of the existing data collection process in the District and with the aim of obtaining a model that is useful for better management of health resources.

Specifically, we considered three inputs and up to six outputs to describe the production technology of the PHCs. In addition, we include variables that reflect the characteristics of the covered population. As inputs, we selected the number of physicians and nurses, and the pharmacy costs (prescriptions by family physicians and pediatricians). As outputs, we used variables that measure the activity of the centers (number of consultations, number of emergencies and number of referrals) and variables that measure healthcare results or outcomes (avoidable hospitalizations, avoidable mortality and an indicator of prescription efficiency). As for non-controllable or exogenous variables, the case-mix is considered.

As already indicated, rates per 10,000 population are used to try to mitigate the effect that different PHC sizes may have on the results of the DEA.

Thus, three possible models with different specifications have been evaluated, which are detailed in Table 1.

In all of them, the rates of pharmacy costs and physicians and nurses have been considered as inputs. The variables related to personnel are considered non-discretionary inputs, given the rigidity of the Spanish Public Health System and the limited capacity of PHC manage the number of physicians or nurses available.

As for the outputs, a first model has been designed to evaluate the activity of the centers and uses the rate of consultations, emergencies and referrals as variables. The rate of emergencies is treated as an undesirable output.

The second model attempts to evaluate the population healthcare outcomes and uses avoidable hospitalizations, avoidable mortality and prescription efficiency as

Table 1 Models specifications

Role	Model 1	Model 2	Model 3
Input	Pharmacy costs[a]	Pharmacy costs[a]	Pharmacy costs[a]
	Physicians[a]	Physicians[a]	Physicians[a]
	Nurses[a]	Nurses[a]	Nurses[a]
Output	Consultations[a]	Avoidable hospitalizations[a]	Consultations[a]
	Emergencies[a]	Avoidable mortality[a]	Emergencies[a]
	Referrals[a]	Prescription efficiency (%)	Referrals[a]
			Avoidable hospitalizations[a]
			Avoidable mortality[a]
			Prescription efficiency (%)
Exogenous	Case-mix	Case-mix	Case-mix

[a]Rates per 10,000 inhabitants

output variables. Hospitalizations and avoidable mortality are treated as undesirable outputs.

In the third and last model, all the output variables are included, so that both the pure activity of the centers and the healthcare outcomes are evaluated.

In addition, each of these models has been evaluated without considering the exogenous variable, or incorporating the case-mix as an uncontrollable input.

Table 2 presents the main descriptive statistics for the entire sample, that is, for the 90 observations available. The high values of the standard deviation and the maximum and minimum values of each of the variables reveal the existence of significant heterogeneity among PHCs, with very diverse sizes and large variations in both their resource endowment and their results, despite using rates per 10,000 inhabitants.

Figure 1 shows the average values of each variable for each year of the period studied. Firstly, we can see that the evolution of the inputs, i.e., the number of physicians and nurses, together with the pharmacy costs (represented with a different scale on the right axis of the graph). It can be seen how these three inputs show an increasing trend. In relation to personnel is because there were hirings in 2018 (more evident in the case of physicians) and the pharmacy costs trend is due to the increase in the number of prescriptions and, especially, the incorporation of more expensive drugs.

As for the exogenous variable (or non-controllable input), no major variations are observed over these five years. These are the characteristics of the population served, in terms of disease burden, and it is usual for there to be no significant changes in such short periods of time.

Regarding outputs, the variables measuring the activity of the centers are presented first. An upward trend is observed in the rate of emergencies and referrals (represented by the scale on the right axis of the graph), while the rate of consultations (left axis) shows a clear decrease, especially since 2018, due to a change in

Table 2 Descriptive statistics for total sample observations

Role	Variable	Average	Standard deviation	Maximum	Minimum
Input	Pharmacy costs[a]	2,068,435	327,727	2,863,437	1,445,219
	Physicians[a]	7.3	0.9	9.7	5.8
	Nurses[a]	5.4	0.8	7.3	3.4
Output	Consultations[a]	65,110	9,360	81,610	43,468
	Emergencies[a]	3,890	495	5,143	2,548
	Referrals[a]	3,801	608	5,549	2,184
	Avoidable hospitalizations[a]	21.4	6.0	43.4	8.7
	Avoidable mortality[a]	16.1	5.2	31.2	5.6
	Prescription efficiency (%)	59.3	14.4	87.7	25.8
Exogenous	Case-mix	47.3	5.9	59.1	37.6

[a]Rates per 10,000 inhabitants

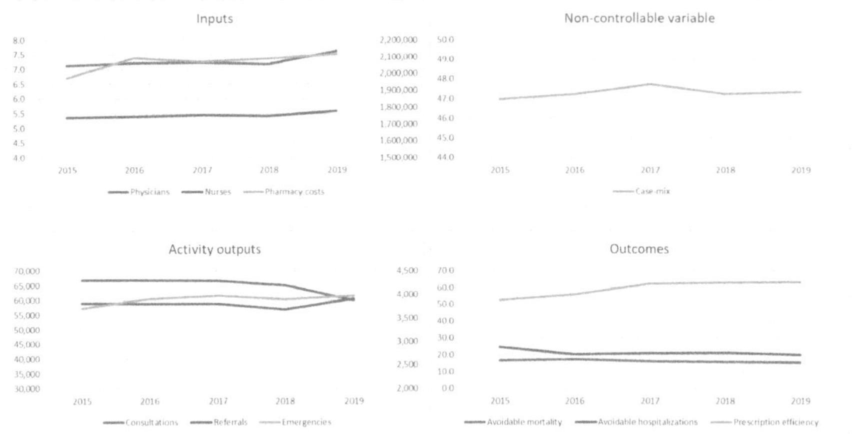

Fig. 1 Evolution of average values over the period

the way of recording some of the tasks usually performed by the nursing staff, such as extractions, injectables, cures, etc., and which are not strictly considered consultations.

Finally, the evolution of the variables measuring healthcare outcomes is shown. No major variations were observed in the rates of hospitalizations and avoidable mortality, which remained at similar values over the five years. In terms of prescription efficiency, there is a clear upward trend, indicating that an effort is being made by the centers to prescribe the most appropriate drugs at all times, for example, antibiotics or anti-inflammatory drugs only when they are strictly necessary, or those drugs recommended in the clinical guidelines for certain pathologies.

The following Table 3 summarizes the main descriptive statistics (average, standard deviation, maximum and minimum) of all the units evaluated in a dynamic context, i.e. 18 PHCs over a period of 5 years (2015–2019), making a total of 90 observations, and for the different models analyzed: activity, healthcare outcomes and including activity and outcomes jointly, with input orientation, and excluding non-controllable variable and incorporating the case-mix.

The average values of the efficiency scores estimated with the 3 alternative models (activity, healthcare outcomes and activity + healthcare outcomes) differ considerably, being significantly lower in model 2, where the lowest average value (0.7719) is obtained when the non-controllable variable is not included. Furthermore, it is in model 2 where the greatest differences between the units evaluated are also observed, as can be seen by analysing the standard deviation, with the lowest score being obtained when the case-mix is not included in the analysis (0.5047).

When variables that measure activity (number of consultations, referrals and emergencies) are taken into account, the scores obtained are higher, and there is less dispersion among the units evaluated. This occurs in models 1 and 3, being much more evident in the latter model. It is in model 3, with case-mix, where the average score is highest (0.9862) and in model 3, with case-mix, the highest minimum value is obtained (0.8492).

All this indicates that some of the units analyzed perform better when the activity of the centers is evaluated (model 1), while others perform better when the healthcare outcome variables are used (model 2). When all the output variables are used together (model 3), some indicators are compensated by others and the overall scores improve.

One of the objectives of this work is to analyze the efficiency of the PHCs in a dynamic context, so, once the overall results have been analyzed, Fig. 2 shows the temporal evolution of the average efficiency scores estimated with each of the 3 models over the period 2015–2019.

Table 3 Descriptive statistics of efficiency scores estimated with the alternative models

Model	Average scores	Standard deviation	Maximum	Minimum
1. Activity				
Without exogenous variable	0.8917	0.0996	1	0.6253
Including case-mix	0.9579	0.0530	1	0.7780
2. Outcomes				
Without exogenous variable	0.7719	0.1536	1	0.5047
Including case-mix	0.9400	0.0807	1	0.7160
3. Activity + outcomes				
Without exogenous variable	0.9230	0.1020	1	0.6289
Including case-mix	0.9862	0.0326	1	0.8492

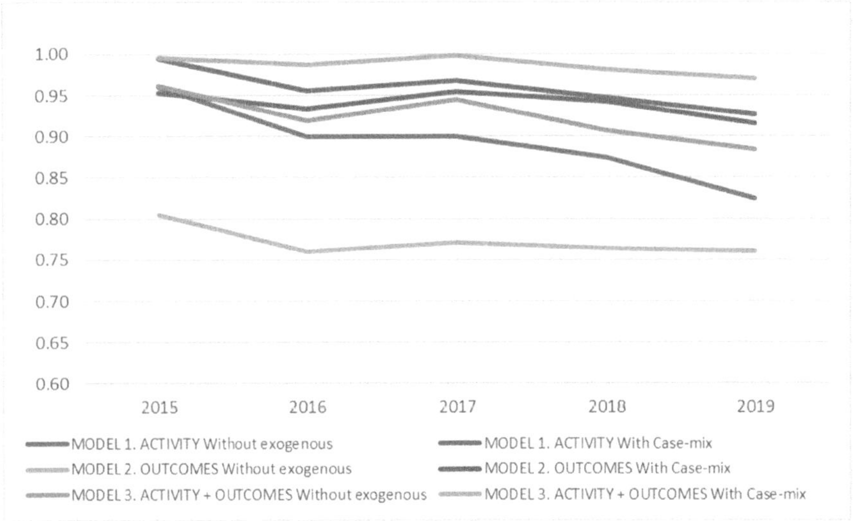

Fig. 2 Evolution of efficiency scores estimated with the alternative models

It can be seen how the scores follow a clearly decreasing trend over time, especially when activity measures are used (models 1 and 3). It is in 2015 when the evaluated units obtain the highest scores and then decrease, especially in 2016 and 2019. This trend is slightly attenuated when incorporating the non-controllable variable in the models. In model 2 the trend is not as pronounced, but a slight decrease is still observed. It should be borne in mind that the variables used in this model involve few cases per year for each PHC (around 20 avoidable admissions and 15 cases of avoidable mortality per 10,000 inhabitants), so that small variations in each of the years can significantly affect the results. On the other hand, the decreasing evolution of the scores is influenced by an increasing trend in the inputs or resources used (greater personnel and greater pharmacy costs) and a decreasing trend in the outputs or results, especially in the number of consultations. All this translates, therefore, into an increase in the level of inputs and a reduction in the level of outputs, which could jointly explain the decrease in the efficiency scores for those years.

The scores of the different models are presented individually for each center in Table 4. The table shows the average score for the 5 years analyzed for each PHC. Those PHCs that obtain a score of 1 (maximum value) obtain the highest score in the 5 years and are therefore considered to be fully efficient.

Significant differences are observed between the results obtained with each of the models, and when the exogenous variable is incorporated into the analysis, the scores improve in general.

In the first model, which measures activity, there is no PHC that is efficient in all cases. PHC13 and PHC16 are fully efficient in all 5 years when using the case-mix. In model 2, when healthcare outcomes are evaluated, it is observed that the scores obtained are lower, in general. It can be seen how PHC17, which was already among

Table 4 Efficiency scores for each PHC in the period 2015–2019

DMU	Model 1. Activity		Model 2. Outcomes		Model 3. Activity + outcomes	
	Without exogenous variable	Including case-mix	Without exogenous variable	Including case-mix	Without exogenous variable	Including case-mix
PHC1	0.9724	0.9850	0.9338	0.9435	0.9899	0.9956
PHC2	0.8326	0.8589	0.6461	0.7565	0.9191	0.9697
PHC3	0.8306	0.9315	0.7369	0.9329	0.8690	0.9451
PHC4	0.8156	0.9896	0.6481	0.9581	0.8191	1
PHC5	0.8860	0.9105	0.7448	0.8734	0.9640	0.9726
PHC6	0.9832	0.9874	0.8249	0.9014	0.9832	0.9900
PHC7	0.9854	0.9882	0.9721	0.9728	0.9880	0.9896
PHC8	0.8945	0.9091	0.6471	0.8348	0.9801	0.9801
PHC9	0.9422	0.9509	0.9940	0.9943	1	1
PHC10	0.7944	0.9559	0.7408	0.9771	0.8087	0.9883
PHC11	0.8204	0.9239	0.7436	0.8814	0.8234	0.9460
PHC12	0.8642	0.9915	0.8668	0.9899	0.9153	0.9938
PHC13	0.9770	1	0.9431	1	0.9916	1
PHC14	0.9586	0.9926	0.5746	0.9670	0.9813	1
PHC15	0.8589	0.9633	0.7383	0.9522	0.8882	0.9811
PHC16	0.9563	1	0.5463	0.9842	1	1
PHC17	0.9861	0.9866	1	1	1	1
PHC18	0.6923	0.9181	0.5926	1	0.6934	1

the most efficient units in model 1, is fully efficient in all years, with and without exogenous variable. It has a very low rate of hospitalizations and avoidable mortality, the best in the Department, and the prescription efficiency indicator is above average. All this leads it to obtain the best scores.

The case of PHC18 is noteworthy, as it has the highest rates of avoidable hospitalizations and avoidable mortality in the District, being considerably higher than those obtained by the other PHCs. It can be seen how it has one of the worst scores in the model without the exogenous variable, but once the burden of disease is taken into account, it becomes efficient.

Finally, in model 3, which jointly evaluates activity and healthcare outcomes, the scores of all the PHCs improve. In this model, practically all the PHCs obtain very high scores close to 1, in all cases they are above 0.8 (except PHC18 in the variant without case-mix) and few differences are observed between units. This indicates that the PHCs are performing in a fairly homogeneous manner when doing a more global analysis of their performance. PHCs that score higher in model 1 somewhat compensate for having not so good results on the health indicators. While PHCs that are more efficient in health outcomes compensate with these indicators for not scoring as high in activity.

DEA methodology identifies efficient PHCs as a whole. The results show that there are differences in the efficiency scores estimated depending on the variables introduced as output, although it is observed that certain PHCs are always efficient or remain closer to the efficient frontier, while others are always inefficient. Considering healthcare activity outputs generates changes in the scores with an increase in the number of efficient PHCs. It is possible to detect that PHCs experienced, in general, a clear decrease in their efficiency levels throughout the period assessed. This decrease is more pronounced when only activity outputs are included.

4 Discussion

This study analyzes the efficiency of the 18 PHCs of the Valencia Clínico—La Malvarrosa Health District over a 5-year period (2015–2019). This is the first study of efficiency evaluation in primary healthcare carried out in the Valencian Community. The data envelopment analysis technique, a methodology that has been widely used in previous studies in the health sector [13–18], is used to estimate the efficiency scores of PHCs with panel data [3, 19] and the efficiency results obtained with three models with different specifications are compared.

DEA is a useful methodology in the evaluation of the efficiency of PHCs and provides valuable information for managers. What is of interest is to compare PHCs with best practices, and to determine, for PHCs that fall below this frontier, possible improvements, i.e., the resources that should be reduced or the results that should be improved.

The results suggest that there has been a decrease in the efficiency levels of PHCs over the period under study, especially when evaluated using activity variables. In the case of using variables that measure healthcare outcomes, the worsening is not as evident, but it is possible to identify more differences between the PHCs and it is observed that there is still ample room for improvement for most of them.

From the point of view of healthcare activity, the estimated efficiency scores achieved by the PHCs are higher. However, when incorporating quality variables or healthcare outcomes (hospitalizations and avoidable mortality), more differences are detected between the PHCs (especially in model 2). The importance of an adequate selection of the variables to be used is evident [13].

There are still relatively few studies evaluating efficiency in primary healthcare. In most of them, the analysis is carried out from the point of view of activity, given the impossibility on many occasions of accessing indicators that allow the quality of the healthcare provided by the centers to be evaluated. Not taking quality indicators into account may end up rewarding in some way those centers that have a higher activity than others, simply because they are operating with lower quality standards [20].

The effect of incorporating morbidity into the analysis has a significant influence on the scores obtained by the PHCs, which is why its incorporation into the analysis is fundamental, as shown by other authors [7, 21].

The efficiency found with this methodology only allows comparison between the set of PHCs considered. In this case, no major differences are observed in the score obtained between the components of the group, especially with models 1 and 3, which implies that they all work in a similar way.

5 Conclusions

Data envelopment analysis proves to be a valuable methodology for assessing the efficiency of PHCs and it is useful as a managerial tool in terms of resource allocation. It allows the analysis of the inefficiencies of the PHCs, although it is necessary to identify the objectives of the PHCs, as the variables included in the models and the perspective of the analyses influence the results. It is essential to include healthcare quality variables and healthcare outcomes in the analysis, as well as to take into account the characteristics of the population served in each PHC.

References

1. Starfield, B., Shi, L., & Macinko, J. (2005). Contribution of primary care to health systems and health. *Milbank Quarterly, 83*, 457–502. https://doi.org/10.1111/j.1468-0009.2005.00409.x
2. Caminal, J., Starfield, B., Sánchez, E., Casanova, C., & Morales, M. (2004). The role of primary care in preventing ambulatory care sensitive conditions. *European Journal of Public Health, 14*, 246–251. https://doi.org/10.1093/eurpub/14.3.246
3. Cordero, J. M., García-García, A., Lau-Cortés, E., & Polo, C. (2023). Assessing Panamanian hospitals' performance with alternative frontier methods. *International Transactions in Operational Research, 30*, 394–420. https://doi.org/10.1111/itor.13013
4. O'Neill, L., Rauner, M., Heidenberger, K., & Kraus, M. (2008). A cross-national comparison and taxonomy of DEA-based hospital efficiency studies. *Socio-Economic Planning Sciences, 42*, 158–189. https://doi.org/10.1016/J.SEPS.2007.03.001
5. Banker, R. D., & Morey, R. C. (1986). Efficiency analysis for exogenously fixed inputs and outputs, *34*, 513–521, https://doi.org/10.1287/OPRE.34.4.513
6. Ruggiero, J. (2007). Non-discretionary inputs. In *Modeling data irregularities and structural complexities in data envelopment analysis* (pp. 85–101). Springer US.
7. Cordero, J. M., Alonso-Morán, E., Nuño-Solinis, R., Orueta, J. F., & Arce, R. S. (2015). Efficiency assessment of primary care providers: A conditional nonparametric approach. *European Journal of Operational Research, 240*, 235–244. https://doi.org/10.1016/j.ejor.2014.06.040
8. Pastor, J. T. (1996). Translation invariance in data envelopment analysis: A generalization. *Annals of Operations Research, 66*, 93–102.
9. Seiford, L. M., & Zhu, J. (2002). Modeling undesirable factors in efficiency evaluation. *European Journal of Operational Research, 142*, 16–20. https://doi.org/10.1016/S0377-2217(01)00293-4
10. Färe, R., & Grosskopf, S. (2004). Modeling undesirable factors in efficiency evaluation: Comment. *European Journal of Operational Research, 157*, 242–245. https://doi.org/10.1016/S0377-2217(03)00191-7

11. Seiford, L. M., & Zhu, J. (2005). A response to comments on modeling undesirable factors in efficiency evaluation. *European Journal of Operational Research, 161*, 579–581. https://doi.org/10.1016/j.ejor.2003.09.018
12. Cooper, W. W., Seiford, L. M., & Tone, K. (2007). *Data envelopment analysis: A comprehensive text with models, applications, references and DEA-solver software: Second edition*, ISBN 0387452818.
13. Zakowska, I., & Godycki-Cwirko, M. (2020). Data envelopment analysis applications in primary health care: A systematic review. *Family Practice, 37*, 147–153. https://doi.org/10.1093/fampra/cmz057
14. Amado, C. A., da Santos, E. F., dos, S. P. (2009). Challenges for performance assessment and improvement in primary health care: The case of the Portuguese health centres. *Health Policy, 91*, 43–56, https://doi.org/10.1016/J.HEALTHPOL.2008.11.008
15. Caballer-Tarazona, M., Moya-Clemente, I., Vivas-Consuelo, D., & Barrachina-Martínez, I. (2010). A model to measure the efficiency of hospital performance. *Mathematical and Computer Modelling, 52*, 1095–1102. https://doi.org/10.1016/j.mcm.2010.03.006
16. Pérez-Romero, C., Ortega-Díaz, M. I., Ocaña-Riola, R., & Martín-Martín, J. J. (2017). Análisis de la eficiencia técnica en los hospitales del Sistema Nacional de Salud español. *Gaceta Sanitaria, 31*, 108–115. https://doi.org/10.1016/j.gaceta.2016.10.007
17. Kohl, S., Schoenfelder, J., Fügener, A., & Brunner, J. O. (2018). The use of data envelopment analysis (DEA) in healthcare with a focus on hospitals. *Health Care Management Science, 22*, 245–286. https://doi.org/10.1007/S10729-018-9436-8
18. Hollingsworth, B., Dawson, P. J., & Maniadakis, N. (1999). Efficiency measurement of health care: A review of non-parametric methods and applications. *Health Care Management Science, 2*, 161–172. https://doi.org/10.1023/A:1019087828488
19. Siciliani, L. (2006). Estimating technical efficiency in the hospital sector with panel data: A comparison of parametric and non-parametric techniques. *Applied Health Economics and Health Policy, 5*, 99–116. https://doi.org/10.2165/00148365-200605020-00004
20. Pelone, F., Kringos, D. S., Romaniello, A., Archibugi, M., Salsiri, C., & Ricciardi, W. (2015). Primary care efficiency measurement using data envelopment analysis: A systematic review. *Journal of Medical Systems, 39*. https://doi.org/10.1007/s10916-014-0156-4
21. Cordero-Ferrera, J. M., Crespo-Cebada, E., & Murillo-Zamorano, L. R. (2011). Measuring technical efficiency in primary health care: The effect of exogenous variables on results. *Journal of Medical Systems, 35*, 545–554. https://doi.org/10.1007/s10916-009-9390-6

Maximizing Efficiency and Mobility in Education Funding: A Three-Stage Centralized Data Envelopment Analysis Model

Bo Hsiao and Kok Fong See

Abstract Financial resource planning is essential for resource allocation and overall university performance enhancement. Allocating financial resources across universities by the central government can be a complex and time-consuming task, particularly when the Ministry of Education is responsible for coordinating and monitoring a group of universities in China. In addition, education funding from the central government is always limited, and therefore, such funding must be allocated efficiently. However, the importance of the mobility of these financial resources across universities is often ignored. To address this issue, this chapter proposes a three-stage centralized data envelopment analysis method to analyze university efficiency from the financial resource perspective, focusing on the transfer-out and transfer-in of different universities. This approach aims to maximize the efficiency of individual universities while also considering overall adjustments. The results can provide an essential reference for the overall planning of financial resources policies, timely adjustment of the organization's operating conditions, and enhancement of the organization's ability to adapt to changes in the financial resources environment to improve operational performance and maintain sustainable management.

Keyword Centralized DEA · Education funding · Resource allocation · China

B. Hsiao
Department of Information Management, National Yunlin University of Science and Technology, Douliu, Taiwan
e-mail: bhsiao@yuntech.edu.tw

K. F. See (✉)
Economics Program, School of Distance Education, Universiti Sains Malaysia, Pulau Pinang, Malaysia
e-mail: kfsee@usm.my

A. Emrouznejad et al. (eds.), *Advances in the Theory and Applications of Performance Measurement and Management*, Lecture Notes in Operations Research,
https://doi.org/10.1007/978-3-031-61597-9_20

1 Introduction

Higher education nowadays is a competitive industry, and each university is compared from several dimensions, including research, employability, and reputation. To compete with other universities worldwide, the government plays an important role in supporting the development of higher education in the country. The government allocates funding to support teaching and research activities, especially for public universities. For instance, public universities administered by the Ministry of Education (MOE) in China will receive educational allocations from the government for teaching activities. However, educational funding is limited, and as such, it should be allocated efficiently to enhance university performance.

University performance has attracted the interest of many researchers and policymakers. Several performance assessments have been conducted in the past from different perspectives, namely, technical efficiency and productivity measures. The former focuses on how effectively inputs are converted into outputs, while the latter measures the ratio of output over input. To carry out such analyses, two popular methods, data envelopment analysis (DEA) and stochastic frontier analysis, have been utilized in the past. However, these analyses mainly focus on how resources are used to produce outputs, except for centralized DEA, which addresses resource allocation within the same system.

Despite the increasing trend in the total number of higher education efficiency studies, the investigations of resource allocation in higher education remain limited. For instance, Contreras and Lozano [1] examined the Spanish public university system using the lexicographic goal programming approach. The results of the study are made using central decision maker preferences for each input and output aggregate goal and relative importance weights, and specific technical efficient targets for each university are identified. While Contreras and Lozano [1] applied a centralized DEA model to deal with the problem of allocating additional resources at 9 selected universities in Spain.

Xiong et al. [2] is the only study on Chinese HEIs that considered fixed resource allocation over a multiple-year period. The study is primarily focused on how to allocate research resources more effectively during different time periods. A sample of 61 public universities in China is selected to examine the reallocated government research funding across these MOE-administered universities. To complement their research, our current study aims to examine whether education funding has been allocated more effectively by the government. Given that Chinese public universities heavily depend on funding from the central government, studying educational resource allocation is essential for understanding how efficiently these educational funds are allocated.

The majority contribution of this study is to examine the efficiency of the financial funding allocation mechanism implemented by the MOE of China for selected universities under their administration. This study provides significant insights into resource mobility and movement direction, aiming to ensure the transfer of resources between universities based on their production and service requirements.

The rest of the chapter is structured as follows: Sect. 2 presents a literature review on resource allocation using the centralized DEA model, and Sect. 3 outlines the methodology of the centralized DEA model. In Sect. 4, we present the results and discussion on the allocation of government educational resources across public universities. Finally, Sect. 5 provides a summary of the conclusions.

2 A Short Literature on the Centralized DEA Model

Data envelopment analysis was originally proposed by Charnes et al. [3] under constant returns to scale and later extended by Banker et al. [4] to accommodate variable returns to scale. The DEA model offers several advantages, such as not requiring a prior functional form and handling multiple inputs and outputs, which have been highlighted in the literature. In the early 2000s, some researchers began applying the DEA model from a centralized perspective to allocate resources. Lozano and Villa [5] conducted an earlier study in which they introduced the centralized DEA approach to project the sampled DMUs and adjust input and output measures. This work on centralized DEA has since been extended to allocate resources more effectively across various industries. Over the last decade, more than 50 research papers have focused on the centralized DEA model. These studies can be categorized into two main groups: first, modifications of the existing centralized DEA model, and second, applications of the centralized DEA model using real-world data.

The centralized DEA model has been extended by various scholars over the past decade. For example, Hosseinzadeh Lotfi et al. [6] and Yu et al. [7] considered the Russell efficiency measure, Lozano [8] utilized the slack-based concept, and Tohidi et al. applied the directional distance function in the centralized DEA model. Furthermore, the centralized DEA model has been integrated with other DEA models. For instance, Fang [9] introduced the centralized DEA model under the generalized DEA model, and Yu and Chen [10] applied the centralized DEA concept in the network DEA model. However, most of these studies, except for Hsiao and Chen [11], have primarily focused on adjusting input resources without analyzing the movement of input resources from one DMU to another. Hsiao and Chen [11] stand as the only study that considers internal resource mobility, which is essential for effective resource allocation to enhance overall performance. Such a model could offer valuable insights into how to reallocate input resources more efficiently among DMUs.

In the latter, the application of the centralized DEA model in empirical research studies is popular, particularly in transportation studies for allocating input resources. For example, Chang et al. [12], Yu and Chen [10], and Chen et al. [13] employed the centralized DEA model for shipping; Yu et al. [7] and Ripoll-Zarraga and Lozano [14] for airports; and Hsiao and Chen [11] and Hsiao et al. [15] for railways. However, there have been fewer applications of the centralized DEA model in other industries, such as Ray [16] for banking; Lopez-Torres and Prior [17] for public schools; and Contreras and Lozan [1, 18] and Xiong et al. [2] for universities. Notably, in higher education studies, most of these works mainly focus on increasing or decreasing internal input resources, but analyzing the mobility of internal resources may be more practical for examining existing funding allocation practices and assisting the government in allocating internal resources more effectively between universities to enhance overall performance.

3 Centralized DEA Framework

The centralized DEA model, as proposed in Hsiao and Chen [11], has been adapted in the current study to suit the context within the higher education industry. The notations used in the following centralized DEA model are summarized in Table 1.

The study utilizes a three-stage centralized DEA framework, which is presented in the following details.

Stage I: The maximum overall efficiency of universities can be estimated by solving the following centralized DEA model, as shown in Model (1):

$$\text{Max}\ \theta \tag{1}$$

$$\sum_{r=1}^{N}\sum_{j=1}^{N}\lambda_{jr}x_{ij} = \sum_{r=1}^{N}x_{ir},\ i = 1,\ldots,n_i \tag{1.1}$$

$$\sum_{j=1}^{N}\lambda_{jr}x_{fj} \leq x_{fr};\ f = 1,\ldots,n_f;\ r = 1,\ldots,N \tag{1.2}$$

$$\sum_{r=1}^{N}\sum_{j=1}^{N}\lambda_{jr}y_{oj} \geq \theta\sum_{r=1}^{N}y_{or};\ o = 1,\ldots,n_o \tag{1.3}$$

$$\sum_{j=1}^{N}\lambda_{jr}y_{oj} \geq y_{or};\ o = 1,\ldots,n_o;\ r = 1,\ldots,N \tag{1.4}$$

$$\sum_{j=1}^{N}\lambda_{jr} = 1;\ r = 1,\ldots,N \tag{1.5}$$

$$\forall\lambda_{jr} \geq 0;\ \forall j, r = 1,\ldots,N \tag{1.6}$$

Equation (1.1) indicates that the total amount of each adjustable input after adjustment is equal to the total amount before the adjustment. Equation (1.2) shows that the amount of each nonadjustable input for the r-th university after adjustment is less than or equal to the amount before the adjustment. This inequality constraint follows the approach for handling nondiscretionary inputs proposed by Banker and Morey

Table 1 Descriptions of variables

Variables/ notations	Descriptions/items
N	Total number of universities
n_i	Total number of adjustable inputs (e.g. funding)
n_f	Total number of nonadjustable inputs
n_o	Total number of outputs
$j, r\ (j, r = 1, \ldots, N)$	Indexes for universities
$i\ (i = 1, \ldots, n_i)$	Indexes for adjustable inputs
$f\ (f = 1, \ldots, n_f)$	Indexes for nonadjustable inputs
$o\ (o = 1, \ldots, n_o)$	Indexes for outputs
θ	Increase proportion of outputs in the first stage
θ^*	Optimal efficiency score obtained from the first stage
x_{ij}	i-th adjustable input of the j-th university
x_{fj}	f-th nonadjustable input of the j-th university
y_{oj}	o-th output of the j-th university
λ_{jr}	Intensity variable for projecting r-th university
s_{ir}^-	Negative slack variable for i-th adjustable input of the r-th university
s_{ir}^+	Positive slack variable for i-th adjustable input of the r-th university
s_{ir}^{-*}	Optimal number of negative slack variable for i-th adjustable input of the r-th university
s_{ir}^{+*}	Optimal number of positive slack variable for i-th adjustable input of the r-th university
$T_{r \to j}^i$	Number of i-th slack variable for i-th adjustable input that can be transferred from r-th university n to j-th university
$T_{j \to r}^i$	Number of i-th slack variable for i-th adjustable input that can be transferred from j-th university to r-th university

[19]. Equation (1.3) implies that the total amount of each output should be increased as much as possible after adjustment. Equation (1.4) shows that the amount of each output for the r-th university after adjustment is more than or equal to the amount before the adjustment. Equation (1.5) indicates that this model operates under the assumption of variable returns to scale, given that universities vary in size. Equation (1.6) illustrates that any intensity variable should be greater than or equal to zero.

Stage II: Given the optimal total output production of the selected MOE-administered universities obtained in the first stage, the maximum value of the negative slack variables and the minimum value of the positive slack variables are determined by solving the following centralized DEA model (Model 2):

$$\text{Max} \sum_{r=1}^{N}\left(\sum_{i=1}^{n_i}\frac{(s_{ir}^{-}-s_{ir}^{+})}{x_{ir}}\right) \tag{2}$$

$$\sum_{j=1}^{N}\sum_{r=1}^{N}\lambda_{jr}x_{ij}=\sum_{r=1}^{N}(x_{ir}+s_{ir}^{+}-s_{ir}^{-});\ i=1,\ldots,n_i \tag{2.1}$$

$$\sum_{j=1}^{N}\lambda_{jr}x_{ij}\le x_{ir}+s_{ir}^{+}-s_{ir}^{-};\ i=1,\ldots,n_i;\ r=1,\ldots,N \tag{2.2}$$

$$\sum_{r=1}^{N}s_{ir}^{+}=\sum_{r=1}^{N}s_{ir}^{-};\ i=1,\ldots,n_i \tag{2.3}$$

$$\sum_{j=1}^{N}\lambda_{jr}x_{fj}\le x_{fr};f=1,\ldots,n_f;\ r=1,\ldots,N \tag{2.4}$$

$$c_a^b\sum_{r=1}^{N}\sum_{j=1}^{N}\lambda_{jr}y_{oj}=\theta^*\sum_{r=1}^{N}y_{or};\ o=1,\ldots,n_o \tag{2.5}$$

$$\sum_{j=1}^{N}\lambda_{jr}y_{oj}\ge y_{or};\ o=1,\ldots,n_o;\ r=1,\ldots,N \tag{2.6}$$

$$\sum_{j=1}^{N}\lambda_{jr}=1;\ r=1,\ldots,N \tag{2.7}$$

$$\forall\lambda_{jr}\ge 0;\ \forall j,r=1,\ldots,N \tag{2.8}$$

Equation (2.1) indicates that the total amount of each adjustable input after adjustment is equal to the total amount before adjustment, accounting for positive slack and subtracting negative slack. Equation (2.2) shows that the amount of each adjustable input for the r-th university after adjustment is equal to the amount before adjustment, considering positive slack and subtracting negative slack. Equation (2.3) indicates that the total amount of adjustable inputs remains unchanged. Equation (2.5) shows that the overall output expansion equals the optimal efficiency score. Equations (2.4) and (2.6)–(2.8) mirror Eqs. (1.2) and (1.4)–(1.6).

Stage III: Given the optimal amounts of positive and negative input slacks obtained in the first case of the second stage, the maximum amount of move-out input slacks and the minimum amount of move-in input slacks without movement costs are determined by solving the following centralized DEA model, as shown in Model (3):

$$\text{Max} \sum_{r=1}^{N} \left(\sum_{i=1}^{n_i} \sum_{j=1, j \neq r}^{N} \frac{(T^i_{r \to j} - T^i_{j \to r})}{x_{ir}} \right) \tag{3}$$

$$\sum_{r=1}^{N} \sum_{j=1}^{N} \lambda_{jr} x_{ij} = \sum_{r=1}^{N} (x_{ir} + \sum_{j=1, j \neq r}^{N} T^i_{j->r} - \sum_{j=1, j \neq r}^{N} T^i_{r->j}), i = 1, \ldots, n_i; \tag{3.1}$$

$$\sum_{j=1}^{N} \lambda_{jr} x_{ij} \leq x_{ir} + \sum_{j=1, j \neq r}^{N} T^i_{j->r} - \sum_{j=1, j \neq r}^{N} T^i_{r->j}; r = 1, \ldots, N, i = 1, \ldots, n_i \tag{3.2}$$

$$\sum_{r=1}^{N} \sum_{j=1, j \neq r}^{N} T^i_{j->r} = \sum_{r=1}^{N} \sum_{j=1, j \neq r}^{N} T^i_{r->j}; i = 1, \ldots, n_i \tag{3.3}$$

$$\sum_{j=1, j \neq r}^{N} T^i_{j->r} = s^{+*}_{ir}; r = 1, \ldots, N, i = 1, \ldots, n_i \tag{3.4}$$

$$\sum_{j=1, j \neq r}^{N} T^i_{r->j} = s^{-*}_{ir}; r = 1, \ldots, N, i = 1, \ldots, n_i \tag{3.5}$$

$$\sum_{r=1}^{N} \sum_{j=1, j \neq r}^{N} T^i_{r->j} \times \sum_{r=1}^{N} \sum_{j=1, j \neq r}^{N} T^i_{j->r} = 0; i = 1, \ldots, n_i \tag{3.6}$$

$$\sum_{j=1}^{N} \lambda_{jr} x_{fj} \leq x_{fr}, f = 1, \ldots, n_f, r = 1, \ldots, N \tag{3.7}$$

$$\sum_{r=1}^{N} \sum_{j=1, j \neq r}^{N} \lambda_{jr} y_{oj} = \theta^*_r \sum_{r=1}^{N} y_{or}; o = 1, \ldots, n_o \tag{3.8}$$

$$\sum_{j=1}^{N} \lambda_{jr} y_{oj} \geq y_{or}; o = 1, \ldots, n_o, r = 1, \ldots, N \tag{3.9}$$

$$\sum_{j=1}^{N} \lambda_{jr} = 1, r = 1, \ldots, N \tag{3.10}$$

$$\forall \lambda_{jr} \geq 0, T^i_{r->j}, T^i_{j->r} \text{ are restricted to integer values.} \tag{3.11}$$

Equation (3.1) states that the total amount of each adjustable input after adjustment should be equal to the total amount before adjustment, increased by the total amount of move-out slacks, and decreased by the total amount of move-in slacks. Equation (3.2) indicates that the amount of each adjustable input for the r-th university after adjustment should be less than or equal to the amount before adjustment, plus the total amount moving from other universities to the r-th university, and minus the total amount moving from the r-th university to other universities. Equation (3.3) suggests that the total amount of each adjustable input remains unchanged. Equation (3.4) illustrates that the total amount of each move-in slack for the r-th university is equal to the optimal amount of each positive slack for the r-th university obtained from the second stage. Equation (3.5) illustrates that the total amount of each move-out slack for the r-th university is equal to the optimal amount of each negative slack for the r-th university obtained from the second stage. Equation (3.6) indicates that the move-out and move-in slack variables for each adjustable input will not exist simultaneously. Equations (3.7)–(3.11) are similar to Eqs. (2.4) and (2.6)–(2.8). In another scenario, the consideration of resource mobility, accounting for movement costs (e.g., regional considerations, policy requirements), can be found in Appendix.

4 Results and Discussion

The study is conducted using a sample of 57 Chinese universities in 2019 that fall under the administration of the MOE. The study used seven input measures and six output measures. The inputs included the number of current students at both the undergraduate and postgraduate levels, the total number of university employees (professor, nonprofessor and administrative staff), the building floor area, and the education funding allocated by the MOE. On the other hand, the outputs considered were the number of graduate students at both levels, the income from technology transfer, the research funding received, and the total number of article publications in international and local journals. The data for the study were extracted from the "2019 Statistical Report on MOE's Administered Higher Education Institutions." The selection of inputs and outputs was based on previous research conducted by Fu et al. [20] and Fu and See [21].

4.1 Descriptive Statistics

Table 2 presents a summary of the statistics pertaining to the selected inputs and output measures used in the study. On average, 21,680 current students were enrolled in undergraduate programs, and 23,422 were enrolled in postgraduate programs. The universities employed an average of 755 professors, 1,738 nonprofessor staff, and 1,993 administrative and other personnel. To gauge the capital resources, the

Table 2 Descriptive statistics

	Mean	Stdev	Min	Max
Current undergraduate students	21,680.33	8426.711	7536	43,257
Current postgraduate students	23,421.56	11,610.52	5909	51,207
Professors	755	388.7482	192	1728
Non-professor	1738.211	534.4536	705	3035
Total admin and others	1993.474	1682.463	497	10,478
Building floor area (m^2)	1,561,775	688,204.4	307,645	3,193,649
MOE education funding (RMB'000)	1,876,989	861,783.8	776,190	4,769,810
Undergraduate graduates	5037.491	2068.029	1244	10,468
Master and Ph.D. graduates	4957.544	2375.021	1463	11,215
Technology transfer income received (RMB'000)	27,065.16	60,740.16	87	311,100
Research grant funding received (RMB'000)	1,076,006	994,391.8	87,897	5,452,151
International journal publications	3520.456	2529.287	404	9625
Local journal publications	3191.474	2227.321	449	10,324

building floor area was used as a proxy, with an average of 1,561,775m^2. Additionally, the average education funding allocated by the MOE in China for 2019 was RMB 1,876.99 million. Considering the input resources, the selected universities produced a total of 5,037 undergraduate and 4,958 postgraduate graduates. Moving on to research outputs, the average income from technology transfer and research grant funding received by the selected universities amounted to RMB 27.07 million and RMB 1,076 million, respectively. Furthermore, 3,520 and 3,191 articles were published in international and local journals, respectively.

4.2 Results of the Centralized DEA Model

The objective of the study is to examine the education funding allocation by the MOE of China for the selected 57 universities. To conduct this study, education funding from the MOE is considered a mobility resource given the set of input and output measures at each university. In other words, the current study primarily focuses on education funding while keeping other input resources and outputs unchanged. It is important to note that this concept of mobility resource allocation is introduced for the first time in higher education efficiency studies.

Table 3 presents the movement of education funding between the selected universities in China. The third column of Table 3 suggests the total movement of education funding for each university, with a positive value indicating extra education funding provided by the university and a negative value representing excess education funding experienced by the university, which needs to be reallocated to other universities.

As highlighted in Sect. 1, the study utilizes the proposed framework presented in Hsiao and Chen [11] to analyze the education funding allocation by the MOE of China. Based on the results of the centralized DEA model, we observe that the current education allocation method adopted by the Ministry of Education of China is fairly efficient in allocating funding to these universities. Out of the 57 universities in China, only 8 require a reallocation of education funding. The remaining universities do not require any resource allocation, as the funding given by the Ministry of Education of China is sufficient to support them according to the given set of outputs.

For instance, Beijing Jiaotong University requires approximately RMB 377.54 million in extra education funding from the government. This additional funding can be obtained by transferring the excess education funding of RMB 195.84 million from Xi'an Jiaotong University and RMB 181.70 million from Lanzhou University. Similarly, Tongji University requires extra education funding of RMB 175.29 million, which could be reallocated from Xiamen University and Xi'an Jiaotong University.

In contrast, East China University of Science and Technology has experienced excess education funding, amounting to RMB 54.07 million, which should be reallocated to another university that requires more education funding. In this case, the surplus funding of this university can be moved to Chongqing University. The same applies to Xiamen University, Xi'an Jiaotong University, and Lanzhou University; their excess education funding can be reallocated to other universities in need.

One of the advantages of this proposed framework is that each university from the sample that requires extra education funding from the MOE of China can do so without additional actual funding allocation. Such reallocation of education funding is mainly based on the current output measures given a set of other input resources. The mobility of education funding is relatively important since education funding is limited, and these funds should be allocated more efficiently to reduce the waste of resources.

The suggested mobility in education funding can improve the efficiency of the sampled universities by approximately 1.02%. This implies that the total of six outputs of overall universities, including the number of graduates, technology transfer, grant funding received, and research outputs, can be increased by 1.02% using the existing funding provided by the MOE of China.

Table 3 Mobility education funding within universities administered by the MOE of China

DMU#	University name	Total mobility education funding (RMB'000)	Education funding (In) (RMB'000)	Education funding (out) (RMB'000)
1	Peking University	0	0	0
2	Renmin University of China	0	0	0
3	Tsinghua University	0	0	0
4	Beijing Jiaotong University	377,544	52(195,841), 57(181,703)	0
5	The University of Science and Technology Beijing	0	0	0
6	Beijing University of Chemical Technology	0	0	0
7	Beijing University of Posts and Telecommunications	0	0	0
8	China Agricultural University	0	0	0
9	Beijing Forestry University	0	0	0
10	Beijing University of Chinese Medicine	0	0	0
11	Beijing Normal University	0	0	0
12	China University of Petroleum	0	0	0
13	North China Electric Power University	0	0	0
14	Nankai University	0	0	0
15	Tianjin University	0	0	0
16	Dalian University of Technology	0	0	0
17	Northeastern University	0	0	0
18	Jilin University	0	0	0
19	Northeast Forestry University	0	0	0
20	Fudan University	0	0	0
21	Tongji University	175,293	35(157,340), 52(17,953)	0

(continued)

Table 3 (continued)

DMU#	University name	Total mobility education funding (RMB'000)	Education funding (In) (RMB'000)	Education funding (out) (RMB'000)
22	Shanghai Jiao Tong University	0	0	0
23	East China University of Science and Technology	−54,066	0	47(54,066)
24	Donghua University	0	0	0
25	East China Normal University	83,316	35(83,316)	0
26	Nanjing University	0	0	0
27	Southeast University	0	0	0
28	China University of Mining and Technology	0	0	0
29	Hohai University	0	0	0
30	Jiangnan University	0	0	0
31	Nanjing Agricultural University	0	0	0
32	China Pharmaceutical University	0	0	0
33	Zhejiang University	0	0	0
34	Hefei University of Technology	0	0	0
35	Xiamen University	−267,669	0	21(157,340), 25(83,316), 47(27,013)
36	Shandong University	0	0	0
37	Wuhan University	0	0	0
38	Huazhong University of Science and Technology	0	0	0
39	China University of Geosciences	0	0	0
40	Wuhan University of Technology	0	0	0
40	Wuhan University of Technology	0	0	0
41	Huazhong Agricultural University	0	0	0
42	Central China Normal University	0	0	0

(continued)

Table 3 (continued)

DMU#	University name	Total mobility education funding (RMB'000)	Education funding (In) (RMB'000)	Education funding (out) (RMB'000)
43	Hunan University	0	0	0
44	Central South University	0	0	0
45	National Sun Yat-sen University	0	0	0
46	South China University of Technology	0	0	0
47	Chongqing University	81,080	23(54,067), 35(27,013)	0
48	Southwest University	0	0	0
49	Sichuan University	0	0	0
50	Southwest Jiaotong University	0	0	0
51	University of Electronic Science and Technology of China	0	0	0
52	Xi'an Jiaotong University	−213,795	0	4(195,841), 21(17,954)
53	Xidian University	0	0	0
54	Chang'an University	0	0	0
55	Northwest A&F University	0	0	0
56	Shaanxi Normal University	0	0	0
57	Lanzhou University	−181,703	0	4(181,703)

Note The number presented in the third column indicates the total mobility of financial resources. The figures in the fourth and fifth columns represent the movement of financial resources in and out of each respective DMU number, with parentheses indicating the total movement of financial resources

5 Conclusions

The study demonstrates the reallocation of education funding for Chinese universities under the administration of the MOE. This is the first study to consider mobility resources in the context of higher education efficiency. The primary contribution of this research is to provide comprehensive guidelines for allocating funding more efficiently based on input and output measures within the higher education industry. Moreover, understanding the movement of resources between universities offers valuable insights that can assist the government in managing scarce resources more effectively.

Overall, the education funding mechanism implemented by the MOE of China is fairly good when compared to the results of the centralized DEA model. The study revealed that only 8 out of 57 Chinese universities in 2019 needed some adjustment in their education allocation to enhance overall outputs by 1.02%. Nevertheless, there is room for improvement if the government considers these results for a better future in the higher education industry. Universities facing a shortage of education funding could obtain more resources by transferring funds between universities without additional actual costs.

The study applied data from annual statistical reports on Chinese universities, but there are other variables that could be considered in the future. For example, reallocating or sharing faculty members and facilities among universities within the system might be beneficial. However, the practicality of such resource allocation depends on the feasibility of mobility resources in Chinese higher education. Furthermore, it is important to incorporate relevant constraints related to transfers into the model, particularly concerning policy or business decisions. This inclusion is crucial for understanding how our analysis can influence governmental decisions in higher education. To obtain the robust efficiency score, considering the integration of robust optimization, as proposed by Toloo et al. [22, 23], our suggested centralized DEA framework emerges as a promising avenue for future research.

Appendix

Given the optimal amount of positive and negative input slacks obtained in the second phase, the maximum amount of move-out input slacks and the minimum amount of move-in input slacks with movement costs (i.e., region considerations or policy requirements) are determined by solving the following centralized DEA model, as shown in Model (A1):

$$\text{Max} \sum_{j=1}^{N} \left(\sum_{i=1}^{n_i} \sum_{j=1, j \neq r}^{N} \frac{(T^i_{r \to j} C^i_{r \to j}(d(r,j)) - T^i_{j \to r} C^i_{j \to r}(d(j,r)))}{x_{ir}} \right) \tag{A1}$$

Subject to

$$\sum_{r=1}^{N} \sum_{j=1}^{N} \lambda_{jr} x_{ij} = \sum_{r=1}^{N} (x_{ir} + \sum_{j=1, j \neq r}^{N} T^i_{j->r} C^i_{j->r}(d(j,r)) - \sum_{j=1, j \neq r}^{N} T^i_{r->j} C^i_{r->j}(d(r,j))), \tag{A1.1}$$

$i = 1, \ldots, n_i;$

$$\sum_{j=1}^{N} \lambda_{jr} x_{ij} \le x_{ir} + \sum_{j=1, j \ne r}^{N} T^{i}_{j->r} C^{i}_{j->r}(d(j,r)) - \sum_{j=1, j \ne r}^{N} T^{i}_{r->j} C^{i}_{r->j}(d(r,j)); r = 1, \ldots, N, i = 1, \ldots, n_i \tag{A1.2}$$

$$\sum_{r=1}^{N} \sum_{j=1, j \ne r}^{N} T^{i}_{j->r} C^{i}_{j->r} d(j,r) = \sum_{r=1}^{N} \sum_{j=1, j \ne r}^{N} T^{i}_{r->j} C^{i}_{r->j} d(r,j); i = 1, \ldots, n_i \tag{A1.3}$$

$$\sum_{j=1, j \ne r}^{N} T^{i}_{j->r} C^{i}_{j->r} d(j,r) = s^{+*}_{ir}; r = 1, \ldots, N, i = 1, \ldots, n_i \tag{A1.4}$$

$$\sum_{j=1, j \ne r}^{N} T^{i}_{r->j} C^{i}_{r->j} d(r,j) = s^{-*}_{ir}; r = 1, \ldots, N, i = 1, \ldots, n_i \tag{A1.5}$$

$$\sum_{r=1}^{N} \sum_{j=1, j \ne r}^{N} T^{i}_{r->j} C^{i}_{r->j} d(r,j) \times \sum_{r=1}^{N} \sum_{j=1, j \ne r}^{N} T^{i}_{j->r} C^{i}_{j->r} d(j,r) = 0; i = 1, \ldots, n_i \tag{A1.6}$$

$$\sum_{j=1}^{N} \lambda_{jr} x_{fj} \le x_{fr}, f = 1, \ldots, n_f, r = 1, \ldots, N \tag{A1.7}$$

$$\sum_{r=1}^{N} \sum_{j=1}^{N} \lambda_{jr} y_{oj} = \theta^{*} \sum_{r=1}^{N} y_{or}; o = 1, \ldots, n_o \tag{A1.8}$$

$$\sum_{j=1}^{\mathrm{N}} \lambda_{jr} y_{oj} \ge y_{or}; o = 1, \ldots, n_o, r = 1, \ldots, N \tag{A1.9}$$

$$\sum_{j=1}^{N} \lambda_{jr} = 1, r = 1, \ldots, N \tag{A1.10}$$

$$\forall \lambda_{jr} \ge 0, T^{h}_{r->j}, T^{i}_{j->r}, T^{m}_{r->j}, T^{m}_{j->r} \text{ are restricted to integer values.} \tag{A1.11}$$

where $C^{p}_{r->j}(d(r,j)) = \begin{cases} 1 \; d(r,j) \le D^{p}_{r,j} \\ 0 \; d(r,j) > D^{p}_{r,j} \end{cases}$, $p \in \{i, m\}$, denoted as the indicator of resource p moving from the r-th university to the j-th university. If cost $d(r,j) \le$ budget limitation $D^{p}_{r,j}$, then $C^{p}_{r->j}(d(r,j))$ is 1; otherwise 0.

References

1. Contreras, I., & Lozano, S. (2020). Allocating additional resources to public universities. A DEA bargaining approach. *Socio-Economic Planning Sciences 71*, 100752.
2. Xiong, X., Yang, G. L., Zhou, D. G., & Wang, Z. L. (2022). How to allocate multi-period research resources? Centralized resource allocation for public universities in China using a parallel DEA-based approach. *Socio-Economic Planning Sciences, 82*, 101317.
3. Charnes, A. C., Cooper, W. W., & Rhodes, E. I. (1978). Measuring the efficiency of decision making units. *European Journal of Operational Research, 2*, 429–444.
4. Banker, R. D., Charnes, A., & Cooper, W. W. (1984). Some models for estimating technical and scale inefficiencies in data envelopment analysis. *Management Science, 20*(9), 1087–1092.
5. Lozano, S., & Villa, G. (2004). Centralized resource allocation using data envelopment analysis. *Journal of Productivity Analysis, 22*, 143–161.
6. Hosseinzadeh Lotfi, F., Noora, A. A., Jahanshahloo, G. R., Gerami, J., & Mozaffari, M. R. (2010). Centralized resource allocation for enhanced Russell models. *Journal of Computational and Applied Mathematics, 235*(1), 1–10.
7. Yu, M. M., Chern, C. C., & Hsiao, B. (2013). Human resource rightsizing using centralized data envelopment analysis: Evidence from Taiwan's airports. *Omega, 41*(1), 119–130.
8. Lozano, S. (2014). Company-wide production planning using a multiple technology DEA approach. *The Journal of the Operational Research Society, 65*(5), 723–734.
9. Fang, L. (2013). A generalized DEA model for centralized resource allocation. *European Journal of Operational Research, 228*, 405–412.
10. Yu, M. M., & Chen, L. H. (2016). Centralized resource allocation with emission resistance in a two-stage production system: Evidence from a Taiwan's container shipping company. *Transportation Research Part A, 94*, 650–671.
11. Hsiao, B., & Chen, L. H. (2022). Internal resource mobility of a railway administration: A three-phase centralized data envelopment analysis. *Research in Transportation Business & Management, 45*, 100855.
12. Chang, S. M., Wang, J. S., Yu, M. M., Shang, K. C., Lin, S. H., & Hsiao, B. (2015). An application of centralized data envelopment analysis in resource allocation in container terminal operations. *Maritime Policy & Management, 42*(8), 776–788.
13. Chen, M. C., Yu, M. M., & Ho, Y. T. (2018). Using network centralized data envelopment analysis for shipping line resource allocation. *International Journal of Environmental Science and Technology, 15*(8), 1777–1792.
14. Ripoll-Zarraga, A. E., & Lozano, S. (2020). A centralised DEA approach to resource reallocation in Spanish airports. *Annals of Operations Research, 288*, 701–732.
15. Hsiao, B., Shu, L. C., Yu, M. M., Shen, L. K., & Wang, D. J. (2017). Performance evaluation of the Taiwan railway administration. *Annals of Operations Research, 259*(1–2), 119–156.
16. Ray, S. (2016). Cost efficiency in an Indian bank branch network: A centralized resource allocation model. *Omega, 65*, 69–81.
17. López-Torres, L., & Prior, D. (2016). Centralized allocation of human resources. An application to public schools. *Computers & Operations Research, 73*, 104–114.
18. Contreras, I., & Lozano, S. (2022). Size efficiency, splits and merger gains, and centralized resource reallocation of Spanish public universities. *Socio-Economic Planning Sciences, 81*, 101190.
19. Banker, R. D., & Morey, R. C. (1996). Estimating production frontier shifts: An application of DEA to technology assessment. *Annals of Operations Research, 66*, 181–196.
20. Fu, T. T., Sung, A. D., See, K. F., & Chou, K. W. (2019). Do optimal scale and efficiency matter in Taiwan's higher education reform? A stochastic cost frontier approach. *Socio-Economic Planning Sciences, 67*, 111–119.
21. Fu, T. T., & See, K. F. (2022). An integrated analysis of quality and productivity growth in China's and Taiwan's higher education institutions. *Economic Analysis and Policy, 74*, 234–249.

22. Toloo, M., & Mensah, E. K. (2019). Robust optimization with nonnegative decision variables: A DEA approach. *Computers & Industrial Engineering, 127*, 313–325.
23. Toloo, M., Mensah, E. K., & Salahi, M. (2022). Robust optimization and its duality in data envelopment analysis. *Omega, 108*, 102583.

Impact of China's Agricultural Water Policy Reforms on Water Usage Efficiency and Total Factor Productivity Change

Wasi Ul Hassan Shah and Nan Zhu

Abstract The Chinese government implemented policy reforms in the agricultural sector in 2012 to enhance the efficient utilization of limited agricultural water resources. This study utilized the SBM-DEA (Data Envelopment Analysis with Slack-Based Measure) and Malmquist productivity index to evaluate the Agricultural Water Usage Efficiency (AWUE) and total factor productivity growth across 31 Chinese provinces from 2000 to 2020. Additionally, Meta-frontier analysis was applied to estimate regional agricultural technological heterogeneity. The findings indicate an average AWUE score of 0.7032, suggesting a potential growth of 29.68%. Before the policy reforms (2000–2012) were implemented, the average AWUE score was 0.6743, which increased to 0.7502 after the policy reforms were introduced. The average Malmquist Productivity Index (MI) score for the entire period was 1.0749, representing an average growth in productivity of 7.49%. Technological change (TC) was the primary driver of this productivity growth, with an average TC value of 1.0655, surpassing the Efficiency Change (EC) value of 1.0101. However, there was a decline of 5.03% in MI after the policy implementation. The Technology Gap Ratio (TGR) value for the Sichuan Basin and surrounding regions was notably higher at 0.983, indicating a superior production technology in China's agriculture sector compared to other regions. The Mann–Whitney U and Kruskal Wallis test evaluate the significant statistical differences concerning AWUE, MI, and TGR among regions and periods.

Keywords AWUE · Technological gaps · Productivity growth · Chinese agricultural sector

---

W. U. H. Shah (✉)
School of Management, Zhejiang Shuren University, Hangzhou 310015, China
e-mail: wasi450@yahoo.com; wasiulhassanshah@zjsru.edu.cn

N. Zhu
The Western Business School, Southwestern University of Finance and Economics, Chengdu, China
e-mail: zhunan@swufe.edu.cn

A. Emrouznejad et al. (eds.), *Advances in the Theory and Applications of Performance Measurement and Management*, Lecture Notes in Operations Research,
https://doi.org/10.1007/978-3-031-61597-9_21

1 Introduction

Water scarcity threatens economic growth, human health, and environmental sustainability worldwide. The UN has prioritized water usage efficiency. SDG 6 promotes water cooperation, advocacy, governance, research, water footprint assessments, international agreements, and capacity building. Efficient water use protects ecosystems and guarantees clean water access. UN urged Individuals, communities, organizations, and industries to work together to solve water scarcity and secure a sustainable future [1]. Agriculture accounts for approximately 70% of the world's freshwater withdrawals, making water efficiency a top priority. Adopting water-efficient practices ensures a sustainable equilibrium between water resources and agricultural production, thereby promoting sustainable food production to meet the requirements of a growing population without depleting water resources [2]. It conserves water for essential uses, protects ecosystems, provides economic benefits, strengthens climate resilience, reduces water stress, and ensures long-term sustainability. Farmers play a vital role in securing food, preserving the environment, and promoting global sustainable development by adopting water-saving technologies and practices [3].

China is presently confronted with severe water scarcity problems that pose significant threats to its social, economic, and environmental stability. Rapid industrialization, urbanization, and population growth have increased the nation's water demand, while pollution and ineffective water management exacerbate the issue. The critical challenges stem from the unequal geographical distribution of water resources, with the northern regions experiencing severe water shortages compared to the south, which is relatively water-rich [4]. This disparity makes it difficult to satisfy the water needs of essential sectors such as agriculture, industry, and urban centers. Additionally, excessive groundwater extraction to support agricultural and industrial requirements has led to the depletion of aquifers, resulting in falling water tables in many regions, thereby threatening the long-term sustainability of water resources [5]. The problem of water pollution persists, propelled primarily by industrial discharges, agricultural runoff, and untreated sewage, which has a devastating effect on public health, aquatic ecosystems, and agricultural productivity [6].

Climate change has altered precipitation patterns, increasing the frequency of extreme weather events such as droughts and flooding. These climatic changes place additional pressure on water resources, resulting in water shortages and damage to vital infrastructure [7]. Agriculture is China's largest water consumer, and inefficient irrigation practices and water-intensive crops substantially contribute to water stress in agricultural regions. Moreover, rapid urbanization and population development in cities exert enormous pressure on water supply systems, heightening the demand for and strain on existing water resources. In addition, China's heavy reliance on water-intensive processes for energy production, such as hydropower and coal-fired power plants, exacerbates the competition for water across sectors, thereby exacerbating the problem of water scarcity [8].

As significant rivers such as the Yangtze and the Mekong traverse multiple countries, addressing transboundary water issues is also challenging. Regional cooperation and coordination are essential for effectively administering shared water resources and resolving potential conflicts [9]. In response to these urgent issues, the Chinese government has implemented several measures, including water conservation campaigns, substantial investments in water infrastructure, water pricing reforms, and initiatives promoting water-efficient technologies. In China's quest to guarantee water security and effectively address the multifaceted effects of water scarcity on the nation's well-being and development, achieving sustainable water management remains an ongoing and complicated challenge [10].

China's agricultural sector is the largest user of fresh water, with approximately over 60% of total freshwater resources. The Chinese government employs several strategies to address water scarcity in agriculture, including the promotion of water efficient irrigation techniques such as drip irrigation and sprinkler systems, the introduction of water pricing reforms to incentivize water conservation, and the encouragement of modern agricultural practices utilizing cutting-edge technologies [11]. In addition, they emphasize crop selection and rotation, invest in the development of water infrastructure, implement water rights systems, and support agriculture demonstration programs that conserve water [12]. In addition, the government provides awareness-raising education and training and promotes integrated water management to address water scarcity issues comprehensively. These initiatives seek to improve water use efficiency, reduce water stress, and implement sustainable agricultural practices for national water security [13]. However, regional heterogeneity in production technology poses a significant barrier to water utilization efficiency in China's agricultural sector. Due to the country's diverse topography and varying agroclimatic conditions, various agricultural practices and technologies are utilized in different regions. Thus, some regions may utilize water more efficiently than others, resulting in disparities in water management and agricultural output [14, 15].

In 2012, the Chinese central government introduced the 'National Agricultural Water Saving Outline (2012–2020),' a comprehensive plan to conserve agricultural water [16]. The plan set water-saving objectives for the agricultural sector, which included the development of water-saving plans, improvements in policies and regulations, stricter controls on total agricultural water usage and irrigation quotas, and the promotion of agricultural water pricing reforms to enhance water usage efficiency and productivity growth [17]. The proposal emphasized the widespread implementation of high-efficiency water-saving irrigation systems and advanced technology adoption in all provinces, especially in large and medium-scale irrigation districts. It also encouraged farmer-based water cooperation organizations to support government-led irrigation system operation and maintenance and improve water charge collection efficiency [18]. Numerous research studies have evaluated the efficiency of agricultural water utilization and total factor productivity change in China [15, 19–22]. However, the level of success in achieving agricultural water utilization efficiency, productivity growth, and homogeneous technological utilization across different regions of China remains unexplored and worth further investigation. To address these research objectives, the current study employs DEA SBM (Data Envelopment Analysis with Slack-Based Measure) and the Malmquist productivity index in its initial phase to evaluate agricultural water usage efficiency and total factor productivity change across 31 Chinese provinces from 2001 to 2020. The analysis focuses on efficiency and productivity changes in China's agriculture sector during this period and assesses the primary determinant (efficiency change or technology change) of total factor productivity change. Additionally, to examine the impact of the 2012 policy reforms, the study divides the 20 years into two segments: pre-policy reforms (2001–2011) and post-policy reforms (2012–2020), enabling a comparison of the mean efficiency and productivity scores between the two periods. Furthermore, Meta-frontier analysis is employed to evaluate regional production technology heterogeneity in China's agriculture sector. The Mann–Whitney test is applied to identify significant statistical differences between the two time periods (pre- and post-policy reform) concerning average agricultural water usage efficiency and total factor productivity change. Additionally, the Kruskal Wallis test is utilized to detect statistical differences in agricultural water usage efficiency, total factor productivity, and technology gap ratio among different agricultural regions in China. The remainder of the study is structured as follows: Sect. 2 presents the methodologies used, while Sect. 3 explains the Inputs-Outputs selection and data sources. Section 4 illustrates the results and discussion, and finally, Sect. 5 outlines the conclusion and policy implications of the study.

2 Methodology

Data Envelopment Analysis (DEA) is a commonly used linear programming technique employed to assess the relative efficiency of decision-making units (DMUs) within various industries [23, 23]. The foundational constant return to scale model (CSR) was introduced by Charnes et al. [25], and later, Banker et al. [25] modified it to the variable return to scale (VSR) model. Building upon DEA methodology, Tone [26] further contributed to the field by proposing the Slack-based Measure (SBM) model.

2.1 DEA-SBM Model

The Slacks-Based Measure (SBM) presents an alternative approach to the conventional radial methods used in Data Envelopment Analysis (DEA) to evaluate efficiency. One of its key advantages lies in its ability to directly assess the excess input and insufficient output. The SBM method incorporates the notion of slacks, which represents the disparities between inputs and outputs at the production frontier, into the efficiency assessment process. This method operates on the following guiding principles:

In this study, referred to as "Provinces," there are n Decision Making Units (DMUs), each consisting of m input indicators and s output indicators. The DMU indexed as B_j where j ranges from 1 to n, is characterized by its $m \times 1$ input indicators $\left[x_{ij}\right]$, with i ranging from 1 to m, and its $s \times 1$ output indicators $\left[y_{rj}\right]$, with r ranging from 1 to s. The efficiency of the j_0-th DMU, denoted as h_{j0}, represents its relative value. The output-focused Slacks-Based Measure Data Envelopment Analysis (SBM-DEA) model with variable returns to scale operates as follows:

$$\text{Min } h_{j0} = \theta$$

$$\text{s.t}\begin{cases} \sum_{j=1}^{n} \lambda_j x_{ij} \leq \theta x_{ij0}, \; i = 1, \ldots, m & (1) \\ \sum_{j=1}^{n} \lambda_j y_{rj} \geq y_{ij}, \; r = 1, \ldots, s & (2) \\ \sum_{j=1}^{n} \lambda_j = 1, \lambda_j \geq 0, j = 1, \ldots, n & (3) \end{cases}$$

The efficiency value for the j-th position is denoted as θ, where λ_j is a non-negative vector. A Decision-Making Unit (DMU) is considered efficient only when θ equals 1; otherwise, it is deemed inefficient.

2.2 DEA-Meta Frontier Model

The Meta-frontier Model provides greater precision in estimating DMU (Decision Making Unit) efficiency when comparing various groups. For an objective evaluation of DMUs within the same group, where all entities have access to the same technology, it is therefore recommended to conduct comparisons within that group. The Technological Gap Ratio (TGR) is a valuable tool for measuring the technological advancement disparity between distinct groups. This TGR can be presented to a specific audience to illustrate the disparities between the groups' technological development [27].

$$TGR = \frac{MAWUE}{GAWUEi} \tag{4}$$

The GAWUEi denotes the Agricultural Water Use Efficiency (AWUE) of Decision-Making Units (DMUs) within a specific group. At the same time, MWUE represents the Meta-AWUE of DMUs from the entire population, encompassing all groups. The Technology Gap Ratio (TGR) quantifies the discrepancy ratio between these two types of DMUs by evaluating the meta-frontier technology against the frontier technology of a specific group [28]. A TGR value 1 indicates no technological disparity between the group and the Meta frontier.

2.3 DEA-Malmquist Productivity Index

Malmquist productivity indices allow a Decision-Making Unit (DMU) to track efficiency improvements over time. However, the effectiveness of this index depends on the assumption that the production function accurately represents the current technological state. DEA models are used to determine the position of this efficiency limit. The characteristics of a particular DMU (DMU_0) are defined by the output variation between two consecutive periods, "t" and "t + 1" [29].

$$M_0 = \frac{D_0^{t+1}\left(x_0^{t+1}, y_0^{t+1}\right)}{D_0^t\left(x_0^t y_0^t\right)}\left[\frac{D_0^t\left(x_0^{t+1}, y_0^{t+1}\right)}{D_0^t\left(x_0^t, y_0^t\right)}\frac{D_0^{t+1}\left(x_0^{t+1}, y_0^{t+1}\right)}{D_0^{t+1}\left(x_0^t y_0^t\right)}\right]^{1/2} \tag{5}$$

- Where: $D_0^t\left(x_0^t, y_0^t\right)$ indicate the TE estimation of DMU_0 for period t,
- $D_0^{t+1}\left(x_0^{t+1}, y_0^{t+1}\right)$ indicate the TE measurement of DMU_0 for period t + 1.
- $D_0^t\left(x_0^{t+1}, y_0^{t+1}\right)$ shows the change in TE from t to t + 1.
- The TE of a specific DMU (DMU_0) at time t + 1, denoted as $D_0^{t+1}\left(x_0^t, y_0^t\right)$, is determined by using its data from period t and comparing it with the data from period t + 1.

2.4 Mann Whitney U and Kruskal–Wallis Tests

Wilcoxon introduced the Mann–Whitney U test in 1945; later, Mann and Whitney modified it in 1947 [30]. This widely used non-parametric test compares two groups with different hypotheses. The test assesses demographic suitability by comparing the distributions of the populations, specifically by comparing the medians of both groups. Both H_0 and H_1 involve comparing the means of two unrelated groups. In situations with more than two groups, the Kruskal–Wallis test can be employed to assess statistical significance [31]. The current study identified a significant statistical difference between pre-and post-reform policy implementation concerning AWUE and MI mean scores. The hypotheses under consideration are as follows:

H_{01}: The average AWUE score is the same before and after policy reforms.

H_{02}: The average (MI) is the same before and after policy reforms.

The Kruskal–Wallis test is utilized to identify the significant statistical differences among the nine Chinese regions concerning the average Agricultural Water Use Efficiency (AWUE) and TGR. The hypothesis for this analysis is as follows:

H_{03}: The AWUE is the same in Nine agricultural regions.

H_{04}: The TGR is the same in Nine agricultural regions.

Table 1 Variables used to gauge the AWUE and TFP

Inputs	Outputs
The labour involved in agriculture: 10 000 persons	Agriculture output: million Yuan
Agricultural capital: 100 million RMB Water consumed in agriculture: 100 million cu.m	
Agricultural land: 10,000 hectares	

3 Inputs-Outputs Selection and Data Sources

Several research studies have utilized different combinations of inputs and outputs to evaluate water usage efficiency and total factor productivity change [32–35]. Table 1 presents the input and output variables based on previous research findings and data availability. The dataset used in this study was sourced from China's statistical books covering the years 2000 to 2020, and it includes data from 31 provinces.

4 Results and Discussion

This section presents the outcomes of three analytical methods: DEA-SBM, Meta-frontier analysis, and the Malmquist productivity index.

4.1 Agricultural Water Usage Efficiency in China (2000–2020)

The study used the DEA-SBM with inputs including labor, capital, water consumption and land utilized to produce gross agriculture output and gauge the China's agricultural water usage efficiency irom 2000 to 2020. The average AWUE score is 0.7032, indicating a protentional of 29.68%. It illustrates that Chinese provinces could improve their agricultural water usage efficiency by efficiently utilizing resources to produce agricultural output. This target could be achieved by reducing the inputs of water, capital, land and labour or enhancing the output. To check the policy reforms impact on AWUE, the study divides the period into two chunks and found that after the implementation of policy reforms, the AWUE increase gradually over the study period. Figure 1 shows that the average AWUE score before the policy reforms (2000–2012) was 0.6743, while after the reforms policy implementation, the average AWUE score increased to 0.7502. It indicates that policy reforms are essential in enhancing AWUE over the study period.

Table 2 presents the AWUE results of 31 provinces. Results indicate that Beijing, Chongqing, Henan, Shaanxi, Tianjin and Tibet are the top performers, with an AWUE of 1 for the study period. Shanghai, Inner Mongolia, Guangdong, Hainan, Jilin,

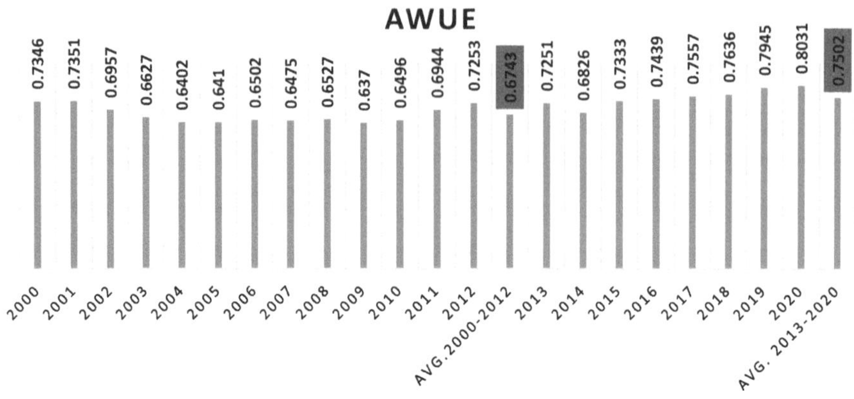

Fig. 1 Agricultural water usage efficiency in China (2000–2020)

Table 2 Average AWUE in 31 Chinese provinces (2000–2020)

Province	AWUE	Province	AWUE
Anhui	0.7413	Jiangxi	0.5675
Beijing	1	Jilin	0.8271
Chongqing	1	Liaoning	0.5419
Fujian	0.6841	Ningxia	0.2001
Gansu	0.3297	Qinghai	0.355
Guangdong	0.8627	Shaanxi	1
Guangxi	0.5111	Shandong	0.4304
Guizhou	0.6624	Shanghai	0.998
Hainan	0.8573	Shanxi	0.7022
Hebei	0.8182	Sichuan	0.8267
Heilongjiang	0.532	Tianjin	1
Henan	1	Tibet	1
Hubei	0.6829	Xinjiang	0.0908
Hunan	0.6232	Yunnan	0.5265
Inner Mongolia	0.9506	Zhejiang	0.7846
Jiangsu	0.7054		

Sichuan, Hebei, Zhejiang, Anhui, Jiangsu and Shanxi scored between 0.70 and 0.99. The average AWUE of Fujian, Hubei, Guizhou, Hunan, Jiangxi, Liaoning, Heilongjiang, Yunnan, Guangxi and Shandong ranges between 0.40 to 0.69. Finally, Qinghai, Gansu, Ningxia and Xinjiang were the least performer for the study period, with AWUE of less than 0.36.

Figure 2 presents each agricultural region's performance and found that Sichuan Basin, the surrounding Loess Plateau and Huang-Huai-Hai Plain are the top three performing regions with an average range of AWUE from 0.84 to 0.91. At the same

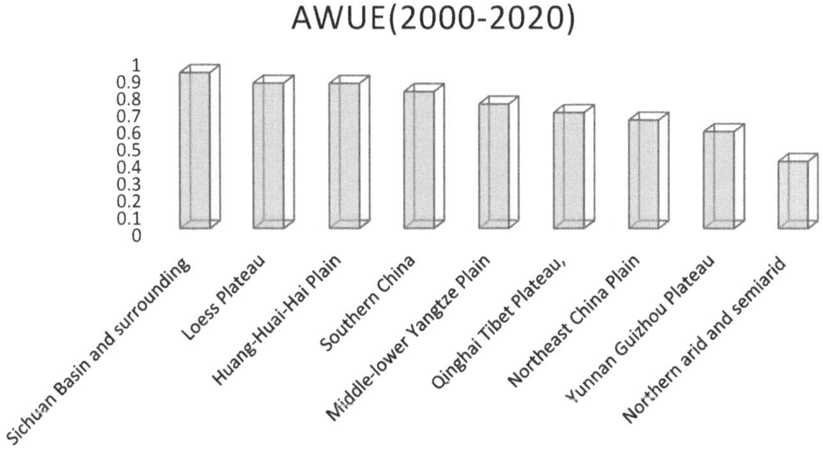

Fig. 2 AWUE in different agricultural regions of China

time, Northern arid and semiarid is the least performer with AWUE score of less than 0.40. Southern China, Middle-lower Yangtze Plain, Qinghai Tibet Plateau, Northeast China Plain, and Yunnan Guizhou Plateau performed moderately, and their scores are between 0.55 to 0.81.

4.2 Production Technology Heterogeneity

Table 3 presents the results of the Meta-frontier analysis. Results revealed that the production technology in 9 different agricultural regions is heterogeneous. The GF shows the group frontier efficiency scores within a specific region, while MF shows the AWUE in the meta frontier, including all the provinces. TGR indicate the technology gap ratio between a meta frontier and a group frontier. A value of 1 indicates no gap while less than 1 indicates the gap between the production technology of a particular province from its group provinces and national level. Elaborating on the regional heterogeneity in production technology, the results illustrate that the TGR value of the Sichuan Basin and surrounding regions is higher among all, with an average value of 0.983, which is closer to 1 indicating superior production technology in the agriculture sector of China. Huang-Huai-Hai Plain and Loess Plateau are ranked second and third in agricultural production technology of China. The Northern arid and semiarid regions have the most inferior agricultural production technology. Beijing, Henan, Tianjin, Shaanxi, Shanghai, Tibet and Chongqing are the provinces with the highest TGR value of 1 which shows that these provinces keep superior agricultural production technology among all 31 Chinese provinces. Jiangxi, Qinghai and Ningxia are the lowest scores for average TGR. Numerous research studies are

Table 3 Meta-frontier, Group Frontier and technology gap ratio for nine regions and 31 provinces of China

Region	Province	GF	MF	TGR
Huang-Huai-Hai Plain	Beijing	1	1	1
	Hebei	1	0.8182	0.8182
	Henan	1	1	1
	Shandong	0.5858	0.4304	0.7347
	Tianjin	1	1	1
Avg. Huang-Huai-Hai Plain		**0.9172**	**0.8497**	**0.9106**
Loess Plateau	Shaanxi	1	1	1
	Shanxi	1	0.7022	0.7022
Avg. Loess Plateau		**1**	**0.8511**	**0.8511**
Middle-lower Yangtze Plain	Anhui	0.9934	0.7413	0.7462
	Hubei	0.9716	0.6829	0.7029
	Hunan	0.952	0.6232	0.6546
	Jiangsu	0.8488	0.7054	0.8311
	Jiangxi	0.9265	0.5675	0.6125
	Shanghai	1	1	1
	Zhejiang	0.9889	0.7846	0.7934
Avg. Middle-lower Yangtze Plain		**0.9545**	**0.7293**	**0.763**
Northeast China Plain	Heilongjiang	0.8484	0.532	0.6271
	Jilin	1	0.8271	0.8271
	Liaoning	0.741	0.5419	0.7313
Avg. Northeast China Plain		**0.8631**	**0.6337**	**0.7285**
Northern arid and semiarid region	Gansu	0.4816	0.3297	0.6846
	Inner Mongolia	0.9519	0.9506	0.9986
	Ningxia	1	0.2001	0.2001
	Xinjiang	0.1232	0.0908	0.737
Avg. Northern arid and semiarid region		**0.6392**	**0.3928**	**0.6551**
Qinghai Tibet Plateau	Qinghai	1	0.355	0.355
	Tibet	1	1	1
Avg. Qinghai Tibet Plateau		1	0.6775	0.6775
Sichuan Basin and surrounding regions	Chongqing	1	1	1
	Sichuan	0.8558	0.8267	0.966
Avg. Sichuan Basin and surrounding regions		**0.9279**	**0.9134**	**0.983**
Southern China	Fujian	1	0.6841	0.6841
	Guangdong	1	0.8627	0.8627
	Hainan	0.9266	0.8573	0.9252
Avg. Southern China		**0.9755**	**0.8014**	**0.824**

(continued)

Table 3 (continued)

Region	Province	GF	MF	TGR
Yunnan Guizhou Plateau	Guangxi	0.5693	0.5111	0.8978
	Guizhou	0.9082	0.6624	0.7294
	Yunnan	0.5709	0.5265	0.9222
Avg. Yunnan Guizhou Plateau		**0.6828**	**0.5667**	**0.8498**

aligned with study results and indicate the technological differences among different provinces and regions in China.

4.3 *Total Factor Productivity Change*

Table 4 and Fig. 3 show the productivity change in China's agricultural sector. The average MI (2000–2020) score is 1.0749. It illustrates that, on average, there is a growth of 7.49% in productivity change. Technological change is the primary determinant of productivity growth as average TC = 1.0655 > EC = 1.0101. The average MI for 2000–2012 is 1.0964, and the MI value for 2013–2020 is 1.0429, indicating a decline of 5.03% after the policy reforms. It shows that the average productivity change after the policy reforms declined and did find an adverse effect on growth.

4.4 *Statistically Significant Differences Among Different Periods and Regions*

From the results of Sects. 4.1, 4.2 and 4.3, we found that AWUE, TGR, and MI values differ for different periods and regions; however, whether they are statistically different is a significant concern. Therefore, Sect. 4.4 describes the Kruskal Wallis and Mann–Whitney U test results to strengthen the study results. Table 5 and Fig. 4 show that the sig value of hypothesis 1 is less than 0.05, indicating a significant statistical difference among the AWUE of before and after policy reforms. However, the sig value of hypothesis 2 is higher than 0.05, indicating that MI for two time periods is not significantly different. Similarly, hypothesis 3 and 4 sig values are less than 0.05. It illustrates that AWUE and TGR significantly differ in China's nine agricultural regions.

Table 4 TFPC in the agriculture sector of China

Years	MI	EC	TC
	1.3318	1.0691	1.2918
	1.0189	0.962	1.0496
	1.0583	0.9531	1.1147
	1.1638	0.9624	1.2076
	1.0449	1.006	1.0376
	1.0157	1.0123	1.0014
	1.1344	1.0017	1.1313
	1.1288	1.0061	1.1207
	0.988	0.9785	1.011
	1.1114	1.0259	1.0847
	1.1534	1.079	1.0704
	1.0079	1.0232	0.9837
Avg. 2000–2012	**1.0964**	**1.0067**	**1.092**
	0.9967	0.9974	0.9971
	0.9095	0.9392	0.9808
	1.2098	1.0873	1.1106
	1.0139	1.0166	0.9957
	0.9669	1.0221	0.9454
	1.0453	1.0101	1.0331
	1.0789	1.0372	1.0384
	1.1198	1.0135	1.1042
Avg. 2013–2020	**1.0426**	**1.0154**	**1.0257**
Avg. 2000–2020	**1.0749**	**1.0101**	**1.0655**

5 Conclusion and Policy Implications

To gauge the impact of agricultural water policy reforms (2012) on AWUE and total factor productivity change, this study used the SBM-DEA and Malmquist productivity index for 31 provinces of China from 2000–2020. Further production technology heterogeneity in 9 different agricultural regions is evaluated through DEA meta-frontier analysis. Results indicate that the average AWUE score is 0.7032 indicating an improvement potential of 29.68%. It illustrates that Chinese provinces could improve their agricultural water usage efficiency by efficiently utilizing resources to produce agricultural output. This target could be achieved by either reducing the inputs of water, capital, land and labor or enhancing the output. The average AWUE score before the policy reforms (2000–2012) was 0.6743; after the reforms policy implementation, the average AWUE score increased to 0.7502. It indicates that policy reforms are essential in enhancing AWUE over the study period. The average MI for 2000–2012 is 1.0964, and the MI value for 2013–2020 is 1.0429, indicating a

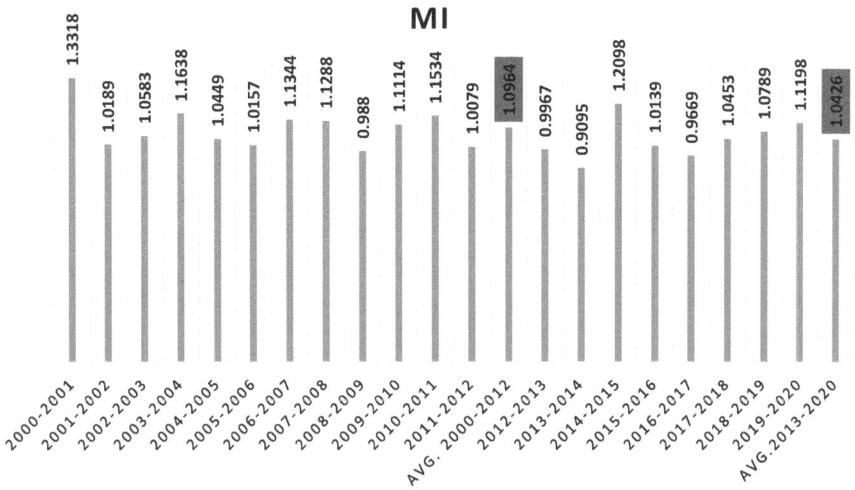

Fig. 3 TFPC in agriculture sector of China (2000–2020)

Table 5 Kruskal Wallis test and Mann–Whitney U results

Hypothesis test summary				
	Null hypothesis	Test	Sig.	Decision
1	The average AWUE score is the same before and after policy reforms	Mann–Whitney U test	0.001	Reject the null hypothesis
2	The average MI score is the same before and after policy reforms		0.238	Reject the null hypothesis
3	The average AWUE is the same in Nine Chinese regions	Kruskal–Wallis test	0.003	Reject the null hypothesis
4	The average TGR is the same in Nine Chinese regions		0.002	Reject the null hypothesis

decline of 5.03% after the policy reforms. Technology change is the primary determinant in total factor productivity growth, as TC is higher than EC during the study period. The TGR value of the Sichuan Basin and surrounding regions is higher among all, with an average value of 0.983, which is closer to 1, indicating superior production technology in the agriculture sector of China. Beijing, Henan, Tianjin, Shaanxi, Shanghai, Tibet and Chongqing are the provinces with the highest TGR

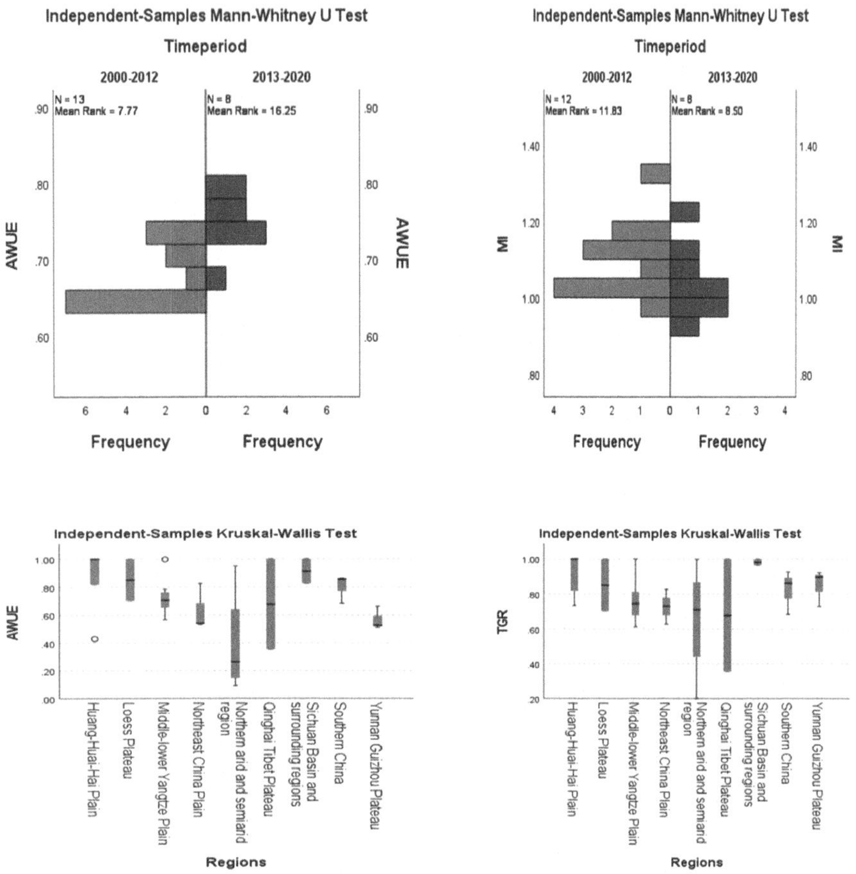

Fig. 4 Statistically significant difference among different time periods and rgions

value of 1. The results of the Mann–Whitney U test strengthen this argument that the average AWUE is statistically different before and after the policy reforms and shows growth. In contrast, the average MI is not significantly different for the two periods. Kruskal Wallis test proved that AWUE and TGR significantly differ for nine different agricultural regions of China.

Based on the above conclusions, this study recommends following policy implications for China's agricultural sector. Although the AWUE witnessed growth after Agricultural water policy reforms, it still has the potential to increase its efficiency. Efficient utilization of inputs, including agricultural water, labor, land and capital, to produce maximum agricultural output is a key to increase efficiency. The agricultural planning, operational strategies and appropriate utilization of resources of efficient provinces and regions are benchmarks for inefficient or less efficient Chinese provinces. Water-saving technology and best practices should be supported financially and technically. As results illustrate that a massive technological heterogeneity in China's agricultural sector exists; therefore, this study recommends that the agricultural regions and provinces with less developed provinces either invest in research and development to increase production technology advancement or acquire relatively advanced technology from other regions with advanced agricultural technologies. Policy measures should be implemented to promote and incentivize water-efficient agricultural practices such as drip irrigation, rainwater harvesting, and precision agriculture techniques. Technology drives total factor productivity growth. Policies should promote modern agricultural technology that maximizes resource use and output. TFP growth and technological supremacy vary by area. To promote technology transfer and best practices, encourage information exchange and collaboration between regions with greater and lower TFP. Make sure agricultural organizations can implement and monitor water policy. Institutional capability increases agricultural water policy enforcement and cooperation. Given the growing relevance of water resources management in climate change, policies should increase agricultural resilience. It may include water storage, climate-smart agriculture, and extreme weather tactics.

References

1. Sampedro, R. (2021). The sustainable development goals (SDG). *Carreteras*. https://doi.org/10.1201/9781003080220-8.
2. Shen, J., Zhao, Y., & Song, J. (2022). Analysis of the regional differences in agricultural water poverty in China: Based on a new agricultural water poverty index. *Agricultural Water Management*. https://doi.org/10.1016/j.agwat.2022.107745.
3. Saleh, E. M. A. (2020). Can irrigation water saving options cope with water scarcity in Egypt? *International Journal of Agricultural and Environmental Research (IJAAER)*.
4. Han, X., Zhao, Y., Gao, X., Jiang, S., Lin, L., & An, T. (2021). Virtual water output intensifies the water scarcity in Northwest China: Current situation, problem analysis and countermeasures. *Science of the Total Environment*. https://doi.org/10.1016/j.scitotenv.2020.144276.
5. Li, W., Hai, X., Han, L., Mao, J., & Tian, M. (2020). Does urbanization intensify regional water scarcity? Evidence and implications from a megaregion of China. *Journal of Cleaner Production*. https://doi.org/10.1016/j.jclepro.2019.118592.
6. Wang, Q., & Yang, Z. (2016). Industrial water pollution, water environment treatment, and health risks in China. *Environmental Pollution*. https://doi.org/10.1016/j.envpol.2016.07.011.
7. Kundzewicz, Z. W., Krysanova, V., Benestad, R. E., Hov, Piniewski, M., & Otto, I. M. (2018). Uncertainty in climate change impacts on water resources. *Environmental Science and Policy*. https://doi.org/10.1016/j.envsci.2017.10.008.
8. Zhang, X. N., Guo, Q. P., Shen, X. X., Yu, S. W., Qiu, G. Y. (2015). Water quality, agriculture and food safety in China: Current situation, trends, interdependencies, and management. *Journal of Integrative Agriculture*.
9. Guan, X., Zhang, J., Bao, Z., Liu, C., Jin, J., & Wang, G. (2021). Past variations and future projection of runoff in typical basins in 10 Water Zones, China. *Science of the Total Environment*. https://doi.org/10.1016/j.scitotenv.2021.149277.
10. Liu, J., Zang, C., Tian, S., Liu, J., Yang, H., Jia, S., You, L., Liu, B., & Zhang, M. (2013). Water conservancy projects in China: Achievements, challenges and way forward. *Global Environmental Change*. https://doi.org/10.1016/j.gloenvcha.2013.02.002.

11. Udimal, T. B., Jincai, Z., Ayamba, E. C., & Mensah Owusu, S. (2017). China's water situation; the supply of water and the pattern of its usage. *International Journal of Sustainable Built Environment*. https://doi.org/10.1016/j.ijsbe.2017.10.001.
12. Zhao, J., Yang, Y., Zhang, K., Jeong, J., Zeng, Z., Zang, H. (2020). Does crop rotation yield more in China? A meta-analysis. *Field Crops Research*.
13. Chang, I. S., Zhao, M., Chen, Y., Guo, X., Zhu, Y., Wu, J., & Yuan, T. (2020). Evaluation on the integrated water resources management in China's major cities–based on city blueprint® approach. *Journal of Cleaner Production*. https://doi.org/10.1016/j.jclepro.2020.121410.
14. Song, M., Wang, R., & Zeng, X. (2018). Water resources utilization efficiency and influence factors under environmental restrictions. *Journal of Cleaner Production*. https://doi.org/10.1016/j.jclepro.2018.02.259.
15. Zhang, F., Xiao, Y., Gao, L., Ma, D., Su, R., & Yang, Q. (2022). How agricultural water use efficiency varies in China—a spatial-temporal analysis considering unexpected outputs. *Agricultural Water Management*. https://doi.org/10.1016/j.agwat.2021.107297.
16. Zhang, H., Zhou, Q., & Zhang, C. (2021). Evaluation of agricultural water-saving effects in the context of water rights trading: An empirical study from China's water rights pilots. *Journal of Cleaner Production*. https://doi.org/10.1016/j.jclepro.2021.127725.
17. Xu, H., & Yang, R. (2022). Does agricultural water conservation policy necessarily reduce agricultural water extraction? Evidence from China. *Agricultural Water Management, 274*, 107987. https://doi.org/10.1016/j.agwat.2022.107987.
18. Qian, X. Y., & Liang, Q. M. (2021). Sustainability evaluation of the provincial water-energy-food nexus in China: Evolutions, obstacles, and response strategies. *Sustainable Cities and Society*. https://doi.org/10.1016/j.scs.2021.103332.
19. Zhang, M., Qin, J., Tan, H., Mao, H., Tu, X., & Jian, J. (2023). Education level of farmers, market-oriented reforms, and the utilization efficiency of agricultural water resources in China. *Economic Change and Restructuring*. https://doi.org/10.1007/s10644-022-09474-5.
20. Liu, P., Fang, Z., Lv, C., & Ruan, A. (2020). China's agricultural water-use efficiency and its influencing factors under the constraint of pollution emission. *International Journal of Design and Nature and Ecodynamics*. https://doi.org/10.18280/ijdne.150416.
21. Cao, X., Zeng, W., Wu, M., Guo, X., & Wang, W. (2020). Hybrid analytical framework for regional agricultural water resource utilization and efficiency evaluation. *Agricultural Water Management*. https://doi.org/10.1016/j.agwat.2020.106027.
22. Wang, F., Yu, C., Xiong, L., Chang, Y. (2019). How can agricultural water use efficiency be promoted in China? A spatial-temporal analysis. *Resources, Conservation and Recycling*. https://doi.org/10.1016/j.resconrec.2019.03.017.
23. Emrouznejad, A., & Yang, G. L. (2018). A survey and analysis of the first 40 years of scholarly literature in DEA: 1978–2016. *Socio-Economic Planning Sciences*. https://doi.org/10.1016/j.seps.2017.01.008.
24. Khoshroo, A., Izadikhah, M., & Emrouznejad, A. (2022). Total factor energy productivity considering undesirable pollutant outputs: A new double frontier based Malmquist productivity index. *Energy*. https://doi.org/10.1016/j.energy.2022.124819.
25. Charnes, A., Cooper, W. W., & Rhodes, E. (1978). Measuring the efficiency of decision making units. *European Journal of Operational Research*. https://doi.org/10.1016/0377-2217(78)90138-8.
26. Tone, K. (2001). Slacks-based measure of efficiency in data envelopment analysis. *European Journal of Operational Research*. https://doi.org/10.1016/S0377-2217(99)00407-5.
27. Sun, J., Wang, Z., & Li, G. (2018). Measuring emission-reduction and energy-conservation efficiency of Chinese cities considering management and technology heterogeneity. *Journal of Cleaner Production*. https://doi.org/10.1016/j.jclepro.2017.12.042.
28. Chiu, C. R., Lu, K. H., Tsang, S. S., & Chen, Y. F. (2013). Decomposition of meta-frontier inefficiency in the two-stage network directional distance function with quasi-fixed inputs. *International Transactions in Operational Research, 20*, 595–611. https://doi.org/10.1111/itor.12008.

29. Färe, R., Grosskopf, S., Lindgren, B., & Roos, P. (1992). Productivity changes in Swedish Pharamacies 1980–1989: A non-parametric Malmquist approach. *Journal of Productivity Analysis*. https://doi.org/10.1007/BF00158770.
30. Mann, H. B., & Whitney, D. R. (1947). On a test of whether one of two random variables is stochastically larger than the other. *The Annals of Mathematical Statistics*. https://doi.org/10.1214/aoms/1177730491.
31. Theodorsson-Norheim, E. (1986). Kruskal-Wallis test: BASIC computer program to perform nonparametric one-way analysis of variance and multiple comparisons on ranks of several independent samples. *Computer Methods and Programs in Biomedicine*. https://doi.org/10.1016/0169-2607(86)90081-7.
32. Wang, G., Chen, J., Wu, F., & Li, Z. (2015). An integrated analysis of agricultural water-use efficiency: A case study in the Heihe River Basin in Northwest China. *Physics and Chemistry of the Earth*. https://doi.org/10.1016/j.pce.2015.10.009.
33. Xu, H., Yang, R., & Song, J. (2021). Agricultural water use efficiency and rebound effect: A study for China. *International Journal of Environmental Research and Public Health*. https://doi.org/10.3390/ijerph18137151.
34. Chen, X., Zhang, Y., Liu, L., & Wang, L. (2020). Water use efficiency in Shaanxi and Shanxi Province analyzed using the super efficiency DEA model. *Journal of Irrigation and Drainage*. https://doi.org/10.13522/j.cnki.ggps.2020029.
35. Yasmeen, R., Tao, R., Shah, W. U. H., Padda, I. U. H., Tang, C. (2020). The nexuses between carbon emissions, agriculture production efficiency, research and development, and government effectiveness: Evidence from major agriculture-producing countries. *Environmental Science and Pollution Research*, 19431. https://doi.org/10.1007/s11356-022-19431-4.

Using DEA to Identify Hispanic Serving Institutions for Best Practices in Increasing STEM Degrees Awarded to Hispanics

Omar S. López

Abstract Sponsored by the National Science Foundation, this article showcases the findings of a two-year study using Data Envelopment Analysis (DEA) with data from the Integrated Postsecondary Education Data System (IPEDS) at the National Center for Education Statistics, to effectively measure the efficiency of Hispanic Serving Institutions (HSIs) graduating Hispanics with STEM degrees. The guiding research question was: *On what input measures do we select efficient HSIs to serve other HSI institutions as possible benchmarks for program improvement in graduating Hispanics with STEM degrees*? An outcome of this work is a framework for investigating and organizing best practices at HSIs for increasing Hispanics earning STEM degrees. The project notably advances knowledge within the field of HSI research. Of the HSI studies published between 1978 and 2021, none used DEA and therefore, the proposed project has the potential to sprout a new branch of HSI research. The proposed project also has the potential to broadly impact desired American societal outcomes. Scientific literacy is one of the benefits from earning a STEM degree. Increasing Hispanics graduating with STEM degrees will here too, result in more voices that can purposefully engage in informed public discourse on science and technology policy, and especially on issues relevant to the Hispanic community.

Keywords Hispanic serving institutions · STEM degrees · Best practices

---

O. S. López (✉)
Texas State University, San Marcos, TX 78665, USA
e-mail: OL14@txstate.edu

A. Emrouznejad et al. (eds.), *Advances in the Theory and Applications of Performance Measurement and Management*, Lecture Notes in Operations Research,
https://doi.org/10.1007/978-3-031-61597-9_22

1 Introduction

1.1 Statement of the Problem

No single strategy has been shown to contribute exclusively to universities and colleges designated as Hispanic Serving Institutions (HSIs) graduating Hispanics with STEM degrees. Rather, the literature identifies a waterfront of practices researchers suggest are broadly relevant to student academic success. Two problems limit such studies. While they describe what practices HSIs should do, they do not prescribe how HSIs should do them. Moreover, one cannot determine if the researchers identified the practices in HSIs that efficiently graduate Hispanics with STEM degrees, casting doubt on the efficacy of the practices to transform less efficient HSIs into efficient ones.

To solve these problems requires a different type of research, one that is purposeful and built on field-based studies of best practices at institutions that are empirically proven efficient at graduating Hispanics with STEM degrees. Sponsored by the National Science Foundation, this article showcases the findings of a two-year study using DEA with data from the Integrated Postsecondary Education Data System (IPEDS) at the National Center for Education Statistics, to effectively measure the efficiency of HSIs graduating Hispanics with STEM degrees. The guiding research question was: *On what input measures do we select efficient HSIs to serve other institutions as possible benchmarks for program improvement in graduating Hispanics with STEM degrees*?

1.2 Background and Significance

Colleges and universities have played an increasingly important role in providing access and degree attainment for Hispanics in science, technology, engineering, and mathematics (STEM). At the forefront in this effort have been High Hispanic Enrollment (HHE) institutions, also known as Hispanic Serving Institutions (HSIs). According to the National Science Board [11], HHE institutions are degree-granting, nonprofit colleges and universities where full-time equivalent Hispanics represent 25 percent or more of the undergraduate enrollment. The literature commonly refers to the terms "HHE" and "HSI" interchangeably, but there are differences [6]. An HHE institution becomes HSI-eligible when its low-income undergraduate Hispanic student enrollment in need of financial aid meets or exceeds the proportional threshold of 50 percent. If so, an HHE institution qualifies to apply for federal grants as an HSI under select federal Title statutes [7].

Unfortunately, the U.S. Congress has not kept up with funding HSI-designated institutions under these federal Title programs [8]. At first, federal funding grew from 1998 to 2004 under Title V, which defined and authorized the federal programs for HSIs. Then funding leveled off until 2007 while the number of HSIs and Hispanic college students continued to grow. Federal funding did increase dramatically in 2008 with the legislation of the HSI STEM program, but with the recession soon after in 2009, funds declined and never recovered while HSIs and Hispanic enrollments increased even more rapidly.

In response, Congress increased recently HSI funding to $212.12 million for fiscal year 2020 by adding 14.8 percent ($27.32 million) to the $184.8 million budget level in 2019 [9], however, whether this closes any near-future HSI-funding gaps is not certain. As of 2019, there were 523 HSIs and 328 "emerging" HSIs with 10 to 24 percent Hispanic enrollments and growing, and two-thirds of all Hispanic undergraduate students in the nation were enrolled at HSIs [4].

Within the context of these historical events and trends, the study's aim was to use the project results to guide researchers to investigate efficient HSIs for best practices that less efficient HSIs could adopt and thereby, increase collectively Hispanics graduating with STEM degrees for the US academic, technical, and professional workforce.

2 Research Plan

2.1 Data Source: IPEDS

The Integrated Postsecondary Education Data System (IPEDS) from the National Center for Education Statistics (NCES) was the primary data source for the proposed project. Located in the U.S. Department of Education's Institute of Education Sciences, the NCES is the federal government's entity for collecting and analyzing data related to education in the United States [10]. Other data sources included the STEM Classification of Instructional Programs (CIP) codes maintained by the U.S. Veterans Benefits Administration [14] for educational benefits.

2.2 Conduct IPEDS Quality Data Audit

The following steps were taken to address the project's research question.

Step 1: Download the 2019 IPEDS SAS data files and format programs for the following survey areas available on the NCES data website.

- Institutional Characteristics (4 files; 376 variables)
- Admissions and Test Scores (1 file; 39 variables)
- Fall Enrollment (6 files; 117 variables)
- 12-Month Enrollment (2 files; 43 variables)
- Completions (4 files; 110 variables)
- Graduation Rates (4 files; 110 variables)
- Outcome Measures (1 file; 28 variables)

- Student Financial Aid and Net Price (2 files; 343 variables)
- Finance (3 files; 332 variables)
- Instructional Staff/Salaries (2 files; 85 variables)
- Fall Staff (4 files; 113 variables)
- Employees by Assigned Position (1 file; 13 variables)
- Academic Libraries (1 file; 31 variables)

The 2019 IPEDS data collection was selected because it was the latest that had all 13 survey data sets available at the start of the proposed project in fall 2021. The number of files and variables in parenthesis were based on the complete 2018 IPEDS data collection.

Step 2: Identify institutions in the IPEDS data that have 1,800 or more undergraduate enrollment and qualify as HSIs or High Hispanic Enrollment institutions namely, degree-granting institutions with 25 percent or more undergraduate Hispanic enrollment [6]. Among these HSIs, those with five or more Hispanics graduating with a STEM degree in the 2019 collection year. This last process used the IPEDS Completion data files and the STEM Classification of Instructional Programs (CIP) codes maintained by the U.S. Veterans Benefits Administration [14] for educational benefits. The output from this activity was a SAS HSI-ID data file containing the institutions unique identifier and name.

Step 3: Perform IPEDS data audit only for institutions listed in the SAS HSI-ID data file [12]. To conduct the data audit, SAS procedures were used to examine the values for IPEDS data variables using SAS descriptive statistics (PROC FREQ, PROC UNIVARIATE). Based on the results, removed were duplicate or irrelevant observations; structural errors (e.g., strange naming conventions, typos, or incorrect capitalization, etc.); unwanted outliers; and/or missing data. At the end of the data audit process, the findings were able to answer these questions in assessing HSI-select IPEDS data quality:

- Does the data make sense?
- Does the data follow the appropriate rules for its field as defined by NCES?
- Does the data quality support further statistical analysis?

3 Results

3.1 Selected IPEDS Data

For 467 public HSIs collected consisting of 202 4-year institutions and 265 2-year community colleges, the quality data audit resulted in the following variables from the IPEDS data files, for use in DEA modeling.

17 Input (Resource) Variables:

- Average Cost of Books and Supplies
- Faculty Index (Rank Strength)

- Full-time retention rate, 2019
- Hispanic Average SCH per STEM Program
- Hispanic Faculty Index (Rank Strength)
- Hispanic UG Graduation Rate at 150% Normal Time
- %FT/FT Undergraduates Awarded Pell Grant
- %Full-Time Faculty
- %Hispanic Employed at Institution (All Positions)
- %Hispanic Faculty–All Ranks
- %Hispanic Undergraduates
- %Library eCirculations of Total Circulations
- %Undergraduates Awarded Pell Grant
- Pell UG Graduation Rate at 150% Normal Time
- Number of STEM Programs at Institution
- Student-to-faculty ratio
- Student Services Cost per Student

3 Outcome Measures:

- %STEM Hispanic Degrees of Total Hispanic Degrees
- %STEM Hispanic Degrees of Total STEM Degrees
- Number of Hispanic UG STEM Degrees.

4 Future Research

There are two immediate needs for future research based on the project's results.

4.1 Selection of Inputs and Outputs and Number of HSIs

The number of inputs and outputs relative to the quantity of HSIs in a DEA model determines how good of a discrimination exists between efficient and inefficient HSIs [12]. Several guidelines for setting these relative levels exist in the literature. Boussofiane et al. [1] recommends that the number of HSIs in the model be at least the number of inputs multiplied by the number of outputs. In comparison, Golany and Roll [5] recommend that the number of HSIs in the model be at least twice the number of inputs and outputs. On the other hand, Bowlin [2] suggests there should be at least three times the number of HSIs as there are input and output variables. Lastly, Dyson et al. [3] recommends the number of HSIs in the model be at least two times the product of the number of input and output variables.

Table 1 below shows the optimal inputs recommended by each of these guidelines, assuming that all three output variables are kept for the DEA model. Golany and Roll [5] allow for the largest number of input/output variables in the model, even

Table 1 Number of optimal input variables relative to HSIs in the DEA model

Size	n	Outputs	Boussofiane et al. [1]	Golany and Roll [5]	Bowlin [2]	Dyson et al. [3]
			Optimal Inputs Recommended			
HSIs—Public Universities						
1,800–4,499	28	3	9	11	6	4
5,000–9,999	44	3	14	19	11	7
10,000–17,999	47	3	15	20	12	7
18,000–29,999	42	3	14	18	11	7
30,000+	41	3	13	17	10	6
HSIs—Community Colleges						
1,800–4,499	38	3	12	16	9	6
5,000–9,999	67	3	22	30	19	11
10,000–17,999	75	3	25	34	22	12
18,00–29,999	54	3	18	24	15	9
30,000+	31	3	10	12	7	5

beyond the 17 input variables available from the IPEDS findings. The Dyson et al. [3] guidelines limit the number of input/output variables the most for the model.

Based on these guidelines, future research will examine which of the 17 input variables HSI stakeholders (e.g., faculty, administrators, staff) consider most important to define comparability between their HSI and those identified as benchmark HSIs, that is, high performing. This is important because HSIs seeking to improve efficiency in graduating more Hispanics with STEM degrees must believe its inputs (i.e., resources) are similar enough to a benchmark HSI to consider adopting its STEM best practices for improving performance.

4.2 Validation of Most Highly Effective STEM Best Practices

One of the products that emerged during the last 3 months of the NSF grant was the identification of 107 STEM best practices structured into a graphic organizer, as shown in Fig. 1.

Based on the STEM best practices framework, future research will also investigate which of the 107 STEM best practices HSIs stakeholders (e.g., faculty, administrators, staff) consider most important in producing higher rates of Hispanics graduating with STEM degrees. Having done so, site-visits will then proceed to high performing benchmark HSIs to identify and document how these STEM best practices are implemented that result in higher levels of Hispanics graduating with undergraduate STEM degrees.

College/University Level (9)		
Program (4)	Monitor (1)	Community (4)

Division/College/School Level (22)				
Program (8)	Faculty (3)	Monitor (2)	Social Capital (2)	Community (7)

Department Level (54)				
Program (6)	Faculty (5)	Monitor (5)	Student Engagement (33): Recruiting (3), Advising (7), Retention (5) Academic Support (2), Mentoring (8), Social Capital (8)	Industry Relations (5)

Classroom Level (22)			
Curriculum (7)	Instruction (6)	Monitor (5)	Adjustments (4)

Fig. 1 STEM Best Practices Graphic Organizer. The number in parenthesis indicates how many STEM best practices were identified by level (e.g., department) and area (e.g., Faculty). The total of STEM best practices across all levels is 107

5 Broader Impact and Intellectual Merit

An outcome of this work is a framework for investigating and organizing best practices at HSIs for increasing Hispanics earning STEM degrees. In so doing, the project has the potential to advance knowledge within the field of HSI research. Of the HSI studies published between 1978 and 2021, none used DEA and therefore, the project may soon sprout a new branch of HSI research. The proposed project also has the potential to broadly impact desired American societal outcomes. Scientific literacy is one of the benefits from earning a STEM degree. Increasing Hispanics graduating with STEM degrees will here too, result in more voices that can purposefully engage in informed public discourse on science and technology policy, and especially on issues relevant to the Hispanic community.

References

1. Boussofiane, A., Dyson, R. G., & Thanassoulis, E. (1991). Applied data envelopment analysis. *European Journal of Operational Research, 52*, 1–15.
2. Bowlin, W. F. (1998). Measuring performance: An introduction to data envelopmentanalysis (DEA). *Journal of Cost Analysis, 7*, 3–27.
3. Dyson, R. G., Allen, R., Camanho, A. S., Podinovski, V. V., Sarrico, C. S., & Shale, E. A. (2001). Pitfalls and protocols in DEA. *European Journal of Operational Research, 132*, 245–259.

4. Excelencia in Education. (2019). Hispanic serving institutions (HSIs): 2017–18. Washington, D.C.: Excelencia in Education. Retrieved from https://www.edexcelencia.org/research/data/hispanic-serving-institutions-hsis-2017-2018.
5. Golany, B., & Roll, Y. (1989). *An application procedure for DEA. Omega, 17*, 237–250.
6. Hegji, A. (2018). The higher education act (HEA): A primer (Publication No. R43351). Washington, DC: Congressional Research Service. Retrieved from https://fas.org/sgp/crs/misc/R43351.pdf.
7. Hispanic Association of Colleges and Universities. (2020a). Hispanic-serving institution definitions. Retrieved from https://www.hacu.net/hacu/HSI_Definition.asp.
8. Hispanic Association of Colleges and Universities. (2020b). About Hispanic serving institutions (HSIs). Retrieved from https://hacuadvocates.net/abouthsis?0.
9. Hispanic Association of Colleges and Universities. (2020c). Fiscal year 2020 appropriations for HSIs and applications for new awards. Retrieved from https://www.hacu.net/NewsBot.asp?MODE=VIEW&ID=3146.
10. National Center for Education Statistics. (2020). IPEDS compare institutions. Washington, DC: NCES. Retrieved from https://nces.ed.gov/ipeds/use-the-data.
11. National Science Board. (2018). High-Hispanic-enrollment institutions: A typology. In Science and Engineering Indicators 2018 (Publication No. NSB-2018–1, Sidebar, pp. 2–15). Alexandria, VA: National Science Foundation. Retrieved from https://nsf.gov/statistics/2018/nsb20181/assets/nsb20181.pdf.
12. Sarkis, J. (2007). Preparing your data for DEA. In J. Zhu & W. D. Cook (Eds.), *Modeling data irregularities and structural complexities in data envelopment analysis* (pp. 305–320). Springer.
13. SAS Institute. (2020). Efficiency analysis: How to use data envelopment analysis to compare efficiencies of garages. Retrieved from https://documentation.sas.com/?docsetId=ormpex&docsetTarget=ormpex_ex22_sect001.htm&docsetVersion=15.2&locale=en.
14. U.S. Veterans Benefits Administration. (2020). STEM designated degree program list effective January 1, 2020. Washington, DC: U.S. Department of Veterans Affairs. Retrieved from https://benefits.va.gov/gibill/docs/fgib/STEM_Program_List.pdf.

DEA as a Tool for Policy Formulation in the Public Basic Education: An Application in Brazilian Schools

M. A. A. Sousa Júnior, A. P. Salgado Junior, P. C. P. W. Rebehy, Y. Marinzeck-Santos, and L. G. Medeiros

Abstract This article discusses the use of Data Envelopment Analysis (DEA) as a tool for evaluating the performance of educational institutions in Brazil. DEA is commonly employed in various sectors, including education, to assess efficiency and identify areas for improvement. Brazil faces an educational deficit compared to developed countries and its peers, particularly in the public education system. The importance of considering access conditions alongside student outcomes in school impact evaluations is emphasized due to the country's inequality. Public education policies in Brazil are striving to enhance instruction. The study proposes the use of DEA as a means to improve the educational effectiveness of basic public education. Methodologically, it involves a literature review of DEA BCC (Banker, Charnes, and Cooper) is suggested for data collection and calculating scores of Decision-Making Units (DMUs) due to its provision of confidence intervals for efficiency scores and bias-corrected estimates. The results highlight DEA as a valuable tool for measuring educational effectiveness. It operates primarily in the early stages of the public policy cycle, aiding in diagnosing needs and assisting policy formulation. Applied to school evaluation, DEA enables the comparison of individual school performance with benchmark groups and facilitates the identification of areas requiring improvement. By employing DEA, school managers and policymakers can develop effective programs to enhance educational performance and address the

M. A. A. S. Júnior · A. P. S. Junior · P. C. P. W. Rebehy · Y. Marinzeck-Santos ·
L. G. Medeiros (✉)
Universidade de São Paulo, Ribeirão Preto—SP, Brazil
e-mail: leonardomedeiros@usp.br

M. A. A. S. Júnior
e-mail: msouzajr@usp.br

A. P. S. Junior
e-mail: asalgado@usp.br

P. C. P. W. Rebehy
e-mail: perla@usp.br

Y. Marinzeck-Santos
e-mail: marinzeck@alumni.usp.br

A. Emrouznejad et al. (eds.), *Advances in the Theory and Applications of Performance Measurement and Management*, Lecture Notes in Operations Research,
https://doi.org/10.1007/978-3-031-61597-9_23

educational deficit in Brazil. Overall, DEA proves to be an important instrument for performance evaluation and policymaking in the education sector.

Keywords DEA (Data Envelopment Analysis) · Basic education · Cycle of public policy · Efficiency

1 Introduction

In recent years, Brazil has invested a lot in higher education, expanding the number of posts, courses and programs offered. However, this expansion had few qualitative aspects evaluated, not showing satisfactory quality [1, 2]. In addition, investments in basic education did not follow this expansion, so students start the cycle in higher education without the necessary knowledge.

In basic education, which is public and free, there are many gaps to be filled, such as pedagogy, training, management, and even investment [1]. Given the inequality in the country, the importance of school impact evaluations that take into account not only student outcomes but also their access conditions are highlighted [3].

In turn, Brazilian public education policies are struggling to improve instruction [1]. One of the main causes is the failure to use the full cycle of public policies. Several programs are formulated out of the interests of small groups or implemented without proper impact assessment, which leads to a waste of resources and delays educational development.

In addition, there is often a distance between the formulators (government level) and the implementers (municipal level), leading to implementation [3]. The municipalization of elementary education in Brazil has further highlighted the complexity of formulating effective public policies that meet local needs [4, 5].

The division of responsibility for education among municipalities [6] has placed demands on schools and municipal secretariats for which they were not prepared. For example, there are difficulties in the use of data for decision-making in the educational and administrative fields [1, 7]

Municipalities, represented by education secretariats, lack metrics to guide performance and the actions needed to improve instruction. Evidence-based practice, similar to health sector, has been shown to be useful in streamlining school management and directing decision making toward educational improvement [8]. In this sense, the application of the public policy cycle also becomes a way to systematize and organize ideas and design and implement improvement programs.

The cycle of public policy, a theoretical-methodological current, can be considered dynamic and constantly learning. Howlett and Ramesh [9] divide public policy into agenda building, policy formulation, decision making, policy implementation, and policy evaluation.

Based on the assumption that the decision-making process is guided by rationality, certain criteria are used to select the best solution to a given problem when there are alternatives [10]. In this sense, this study aims to propose the use of DEA as a tool in Public Policies to improve the educational effect of basic public education. Two methodological pillars were used for this study: a literature review of the main research topics and the use of previous works from GREFIC (Efficiency Study Group), such as Grounded Theory.

2 Education in Brazil

Despite recent progress, Brazil still has an educational deficit, not only compared to developed countries, but also compared to its peers [5, 11, 12]. The scenario becomes even worse when considering only the public education system.

In 2020, the Brazilian primary education system included a total of 47.3 million enrollments, more than 300,000 schools, of which more than two-thirds are public [13]. The concentration on the municipal public network is mainly due to the decentralization of education and the division of responsibility for education that has taken place in recent years [6].

The transfer and shift of responsibility, especially when there is no adequate preparation [1, 7], tends to create problems or exacerbate existing ones [5].

The improvement of Brazilian educational quality can be achieved by improving management factors and efficient use of resources [6, 11, 12]. In order for this outcome to be achieved, it is necessary to formulate public education policies that respond to the different Brazilian contexts and needs.

3 Public Policy Theory

The cycle of public policy, a theoretical-methodological current, can be considered dynamic and constantly learning. The public policy is divided into agenda building, policy formulation, decision making, policy implementation, and policy evaluation [9]. Also along these lines, are considers five phases in the public policy cycle: problem perception and definition, decision agenda formation, program formulation, policy implementation, and policy monitoring and evaluation [14].

However, more important than the sequence of phases is understanding the specific characteristics of each phase and their interrelationships [14]. Analyzing the public policy cycle through one or more phases is a methodological option and serves to streamline the process and systematize some elements, facilitating study and analysis. Based on the assumption that the decision-making process is guided by rationality, certain criteria are used to select the best solution to a given problem when there are alternatives [10].

Problems can be identified based on indicators (technical criteria) or result from emergency situations (epidemics, disasters, etc.). The formation of the decision agenda results from the perception of the relevance of the problems by the actors involved in the policy. These two phases often go hand in hand and are difficult to separate. The issue is on the systemic agenda at the moment when part of the population or the political community begins to perceive the importance of an issue and it is addressed by the government [15].

The formulation of the program or policy will guide the entire implementation and can have specific and detailed parameters or be more general (depending on the knowledge and complexity of the problem) [14]. It is in the formulation and decision-making phase that attention is focused on analysts, policy makers or officials. This phase must be guided by rationality, objectivity and delimitation of objectives and strategies [16]. Nevertheless, was noted a lack of cognitive rationality in this type of decision making, which is often made in the interest of public officials or interest groups and not in the interest of the general public [17].

The process of implementing public policies is considered to be one of the main causes, if not the main cause, of the failure of governments in relation to the proposed objectives set in advance when the policies are formulated [18]. It is not uncommon to find cases where public policies are made without implementing or achieving the expected objectives.

Implementing public policies in a complex environment (as in education) requires very close proximity between the formulators and the actors implementing the policies. This situation is highlighted in Brazil, where policies developed at the federal level were implemented only symbolically at the state level and offered little opportunity for action [4]. Implementation is based on a set of objectives, administrative controls and operational routines (rational-bureaucratic model), a strategy to establish the participation of implementation agents (human resources), internal and external relationships based on consensus, conflict and agreement (political), and a set of myths and symbols that help individuals justify their existence and feel like members of the community (symbolic) [19].

The implementation of public policy allows for the identification of possible errors in formulation as well as technical aspects or operationalization errors. This learning can inform policy formulation and improve its effectiveness [10]. Evaluating the degree of implementation of a policy is a fundamental aspect in education. The interpretation of the effectiveness of the results depends on which aspects have been implemented and how well this implementation has been [20].

Finally, monitoring and evaluation verifies whether the formulated program is implemented as planned and whether the policy has achieved satisfactory results after its implementation. It is important that the evaluation of the policy uses clear and well-defined indicators that can verify whether the objectives have been achieved [16].

In planning public policies in education, problems arise that are related to two extreme groups of planners: technocrats with an economistic model of education and educators who forgo effective planning because they lack the techniques to do so [21]. Public education policy must establish priorities and channel efforts and resources for their implementation, through a joint approach between government agencies, managers of the educational network and the active participation of society [21].

It is pointed out the lack of work that uses appropriate methods to evaluate education given the complexity of its processes, resulting in studies that limit the expansion of knowledge [22].

4 Public Policies in Education

In several countries in a situation of budget deficit, education policymakers have been concerned not only with the expected outcome, but also with the public expenditure estimated in the formulated policies [23]. In Brazil, in particular, the governance and coordination of education policies is very complex due to its large territorial extension [4]. This problem is compounded by the lack of evidence use in the cycle of these policies.

Unlike the health sector, which has been using data and evidence-based practices for many years, the education sector still struggles in this regard [8]. Decisions in education, whether at the strategic or instructional level, must be guided by research and evaluation information rather than by policy preferences or activities of the stakeholders involved [24].

Evidence-based education is conceptualized as "a set of principles and practices that can change the way people think about education, the way they approach educational policy and practice, and the basis on which they make professional judgments and use their knowledge" [25].

There are several sources of evidence, such as results from randomized trials and correlational research. Qualitative methods such as interviews and discourse analysis are also considered evidence. Each type of evidence better meets the existing need [25]. It is important to consider other types of data such as perceptions, motivation, processes, and behaviors in decision making [8].

Data should be aligned with the goals involved, e.g., grade-level data collection for student learning goals, school-level data collection for school goals. Goals that are quantitative and measurable can also be compared to local, regional, or national benchmarks [26]. Anecdotal data, considered less formal, such as observations, opinions, intuitions, and personal qualitative reports, are still a reality in the educational scenario [27].

Evidence-based practices are not limited to the public policy cycle, but are also applied at the school level [28–31]. Schools in several countries around the world have used data to monitor performance, diagnose areas for improvement, and make decisions that improve education for efficiency and effectiveness [26].

To this end, several countries have adopted models referred to as "evidence-based governance" or "evidence-based management." This strategy is considered the most important model for modernizing education and promises to make it more "rational" at all levels [24].

Also in this sense, evidence-based policy clearly and objectively assigns performance expectations and outputs to the education system. Stakeholders are involved in developing expectations and improvement measures by being consulted for their opinions and encouraged to respond to individual school performance [24].

5 Data Envelopment Analysis in Policy Cycle

This study aims to propose an evidence-based methodology for the identification of needs and agenda formation in public policy cycle that improves the school effect of public basic education. The proposal aims to develop a replicable methodology capable of satisfying a Brazilian demand by developing an educational policy cycle based on evidence management, capable of meeting the individual needs of each school and making planning more rational, efficient, and assertive (Fig. 1).

The first phase of the public policy cycle involves problem identification and agenda setting. This phase aims to establish technical indicators [14] in education, both from the point of view of efficiency and explanatory factors, in order to optimize the evaluation and definition of priorities. The objective is to establish a diagnosis of school performance from the point of view of efficiency, both at the macro level (country, states, regions or other stratifications) and at the micro level (schools and communities).

The formation of the political agenda, elaborated in a systematic way and aimed at the results of the best alternatives (model of rationality) [9, 16], is the goal of public policy, sometimes delegated to institutions without political dependence [32].

In this sense, this phase has as subproducts the ranking of schools by performance and the identification of explanatory factors for the school effect, related to the stratification of the sample as well as the benchmark in question. The sub-products are described below.

The evaluation of the quality of education must be done according to the objective of the study, the researcher, and society [33]. For the elaboration of public policies in the current Brazilian context, it is proposed to measure school performance from the perspective of efficiency, since it is important [26] and mandatory.

In this sense, the first step is to define the concept of efficiency in education. From this perspective, the use of school effect is also proposed, defined as "how much a particular school institution contributes to student learning through its internal

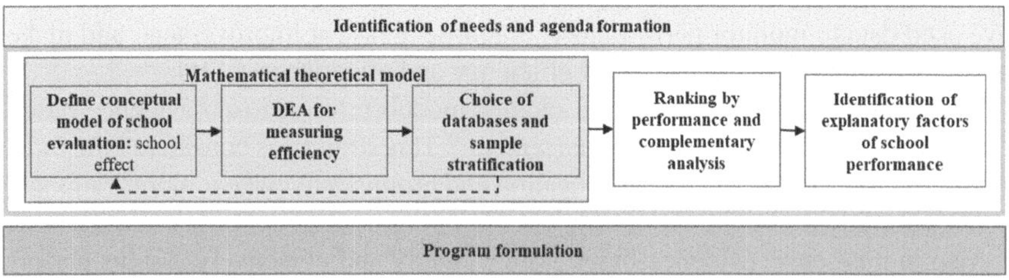

Fig. 1 Representation of the application of DEA in the cycle of public policies

policies and practices" [34]. This perspective takes into account the conditions of the student's entry and incorporates the student's social and educational conditions up to that point. It is important that this concept be thoroughly discussed and evaluated by researchers and policy makers as it will guide the next steps.

The choice of an appropriate technique allows the reduction of bias in the evaluation and more reliable results with respect to the expected objective. In this sense, Data Envelopment Analysis (DEA) has proven to be a useful tool for measuring educational effectiveness [35]. It is also worth noting that other techniques can be chosen depending on the interest of the researcher.

Although the definition of quality in education is complex and difficult to specify, educational planning can use the results of large-scale assessments as a measure of outcomes [21]. The use of such assessments is fundamental to measuring school impact through standardized comparisons from the perspective of different schools and contexts.

At this stage, it is important to use a quantitative database that contains school-level information, with the largest number of schools available, preferably at the national level. Such a definition accommodates the use of evidence-based policy and makes the process more streamlined.

The selection variables that represent the initial conditions for the transformation carried out by education and the resulting outcomes takes place. It is necessary to identify the variables as well as the availability of data that quantitatively represent this concept in order to perform a comparative analysis. Once the variables that can quantitatively represent the school effect are selected, the schools must be homogenized, that is, the sample must be selected based on criteria that make them comparable.

The comparative analysis must be conducted with schools that have a similar context and can be segmented, for example, by socioeconomic level, by public or private character, or by socioeconomic level.

In this step, it is possible to determine which schools have high performance and which have low performance. At this stage, it is still important to perform descriptive statistical analyses, such as Quartile Analysis and Descriptive Statistics.

Student achievement is the result of a complex interaction of factors, both within and outside the school environment [36], that can increase or decrease academic performance. Therefore, identifying these explanatory factors allows for more targeted intervention to make necessary changes and optimize resources and efforts. An appropriate technique must be chosen based on the quantity and quality of data available for analysis. In combination with DEA, multiple linear regression has proven to be a useful and widely used tool, especially in the educational field [37]. Using the explanatory factors, it is possible to define the benchmarks and goals of the schools that want to improve in the evaluation, either in the performance of the main model or in each explanatory factor.

6 Conclusions

This study focuses on the challenges and intricacies confronting the Brazilian education system, particularly concerning public basic education. Despite substantial investments in higher education and an expanded array of courses and programs, there has been insufficient attention paid to qualitative aspects, raising concerns about overall education quality. Additionally, the disparity in investments between basic education and higher education has resulted in students entering higher education without adequate foundational knowledge.

The decentralization of education to municipalities has presented its own set of challenges, as many local authorities were ill-equipped to handle the responsibilities and demands placed upon them. The formulation and implementation of public education policies have frequently been fragmented and lacked thorough impact assessments, leading to ineffective resource allocation and delayed progress in educational advancement.

To tackle these issues and enhance the efficacy of public policies in education, we propose the application of evidence-based practices and the adoption of the public policy cycle. By leveraging data and evidence, policymakers can make more informed decisions and prioritize areas in need of improvement. The public policy cycle, comprising distinct phases such as problem identification, agenda setting, policy formulation, implementation, and evaluation, offers a systematic and organized approach to the policy-making process.

At the core of our proposed methodology lies the use of Data Envelopment Analysis (DEA) to gauge the efficiency and effectiveness of schools concerning educational outcomes. DEA has proven to be a valuable tool in evaluating educational effectiveness, particularly when combined with other techniques like multiple linear regression, which identifies explanatory factors contributing to school performance.

Through the adoption of evidence-based governance and management strategies, education policymakers can align their objectives with measurable performance indicators and involve stakeholders in the decision-making process. This approach fosters rational, efficient, and effective planning, with a focus on achieving improved educational outcomes.

In summary, the integration of evidence-based practices, the public policy cycle, and methodologies like DEA can play a crucial role in transforming the Brazilian education system. As stakeholders at all levels collaborate to address challenges and seize opportunities, it is our hope that this proposed approach will pave the way for a more equitable, efficient, and quality-oriented education system in Brazil.

References

1. Bauer, A., Cassettari, N., & De Oliveira, R. P. (2017). Políticas docentes e qualidade da educação: Uma revisão da literatura e indicações de política. *Ensaio, 25*(97), 943–970.
2. Louzano, P., Rocha, V., Moriconi, G. M., & De Oliveira, R. P. (2010). Quem quer ser professor? Atratividade, seleção e formação docente no Brasil. *Estudos em Avaliação Educacional, 21*(47), 543–568.
3. de Andrade, R. J., & Soares, J. F. (2008). O efeito da escola básica brasileira. *Estudos em Avaliação Educacional, 19*(41), 379.
4. Lotta, G. S., Bauer, M., Jobim, R., & Merchan, C. R. (2021). Effects of changes in a multi-level implementation context: Analysis of the brazilian high school reform. *Revista de Administracao Publica, 55*(2), 395–413.
5. Bernardo, J. S., de Almeida, F. M., & Nascimento, A. C. C. (2020). Qualidade geral da educação municipal e as influências dos gastos públicos. *Education Policy Analysis Archives, 28*, 23.
6. Cardoso Gomes, M. H. S., Bresciani, L. P., & Amorim, W. A. (2016). Políticas sociais, educação e desenvolvimento econômico: Busca por evidências correlacionais a partir das avaliações do Ideb de três municípios paulistas. *Gestão & Regionalidade, 32*(94), 36–51.
7. Rosistolato, R., Do Prado, A. P., & Martins, L. R. A. (2018). "realidade" de cada escola e a recepção de políticas educacionais. *Ensaio, 26*(98), 112–132.
8. Mandinach, E. B., & Gummer, E. S. (2013). A systemic view of implementing data literacy in educator preparation. *Educational Researcher, 42*(1), 30–37.
9. Howlett, M., & Ramesh, M. (1995). *Studying public policy: Policy cycles and policy subsystems*. Oxford University.
10. Parente, C. D. M. D. (2017). Políticas de Educação Integral em Tempo Integral à Luz da Análise do Ciclo da Política Pública. *Educação & Realidade, 43*(2), 415–434.
11. Silva, P. A., & Giareta, P. F. (2020). O Financiamento Da Educação E O Pne: Implicações E Desafios Para a Meta 20 Do PNE. *Brazilian Journal of Development, 6*(12), 377–392.
12. Soares, J. F., & Alves, M. T. G. (2013). Efeitos de escolas e municípios na qualidade do ensino fundamental. *Cadernos de Pesquisa, 43*(149), 493–517.
13. Brasil. (2021). Instituto Nacional de Estudos e Pesquisas Educacionais Anísio Teixeira. Censo da Educação Básica 2020: notas estatísticas. INEP, Brasília, DF.
14. Raeder, S. (2014). Ciclo de Políticas: Uma abordagem integradora dos modelos para análise de políticas públicas. *Perspectivas em Políticas Públicas, 7*(13), 121–146.
15. Capella, A. C. N., Brasil, F. G., & Sudano, A. D. C. G. P. (2015). O estudo da agenda governamental: Reflexões metodológicas e indicativos para pesquisas. *Encontro Anual da ANPOCS, 1*(1), 1–43.
16. Estevão, R. B., & Ferreira, M. D. M. (2018). Análise de políticas públicas: Uma breve revisão de aspectos metodológicos para formulação de políticas. *Holos, 3*(1), 168–185.
17. Silva, T. L. B., Cissé Bá, S. A., Da Silva, A. V., Silva, G. C., Da Silva, L. A. (2020). A tomada de decisão na administração pública: Revisão integrativa. *Humanidades E Tecnologia, 19*(1), 239–247.
18. Silva, P. L. B., Melo, M. A. B. (2000). O processo de implementação de políticas públicas no Brasil: características e determinantes da avaliação de programas e projetos. Unicamp—Núcleo de Estudos e Políticas Públicas—NEPP, Campinas.
19. Najberg, E., & Barbosa, N. B. (2006). Abordagens sobre o Processo de Implementação de Políticas Públicas. *Interface, 3*(2), 31–43.
20. Van Kuijk, M., Mullender-Wijnsma, M., & Bosker, R. (2021). A Systematic review of studies addressing the implementation of the evidence-based whole-school reform "Success for All." *ECNU Review of Education, 4*(1), 128–163.
21. Najjar, J., Mocarzel, M., & Morgan, K. (2020). The municipal plans of education of the state of Rio de Janeiro: A mapping of goals, strategies and challenges. *Ensaio, 28*(109), 1033–1053.
22. Martins, A. M. (2013). O campo das políticas públicas de educação: Uma revisão da literatura. *Estudos em avaliação educacional, 24*(56), 276–299.

23. López-Torres, L., Prior, D., & Santín, D. (2019). Assessing the effect of educational programs on public schools' performance. *Applied Economics, 51*(48), 5205–5226.
24. Altrichter, H., Gamsjäger, M., Plaimauer, C., Prammer-Semmler, E., Steiner, R., Zuber, J., Handschuc, L. (2020). Making sense of evidence-based governance reforms: an exploratory analysis of teachers coping with the Austrian performance standard policy. *Leadership and Policy in Schools*, 1–21.
25. Davies, P. (1999). What is evidence-based education? *British Journal of Educacional Studies, 7*(2), 108–121.
26. Schildkamp, K. (2019). Data-based decision-making for school improvement: Research insights and gaps. *Educational Research, 61*(3), 257–273.
27. Maras, M. A., et al. (2014). School practitioners' perspectives on planning, implementing, and evaluating evidence-based practices. *Children and Youth Services Review, 47*(3), 314–322.
28. Flood, J., & Brown, C. (2018). Does a theory of action approach help teachers engage in evidence-informed self-improvement? *Research for All, 2*(2), 347–358.
29. Brown, C., Schildkamp, K., & Hubers, M. D. (2017). Combining the best of two worlds: A conceptual proposal for evidence-informed school improvement. *Educational Research, 59*(2), 154–172.
30. Sheard, M. K., & Sharples, J. (2016). School leaders' engagement with the concept of evidence-based practice as a management tool for school improvement. *Educational Management Administration and Leadership, 44*(4), 668–687.
31. Altrichter, H., & Gamsjäger, M. (2017). A conceptual model for research in performance standard policies. *Nordic Journal of Studies in Educational Policy, 3*(1), 6–20.
32. Souza, C. (2006). Políticas públicas: Uma revisão da literatura. *Sociologias, 16*, 20–45.
33. Cheng, Y. C., & Tam, W. M. (1997). Multi-models of quality in education. *Quality Assurance in Education, 5*(1), 22–31.
34. Brooke, N., & Soares, J. F. (2008). Pesquisa em Eficácia Escolar: Origem e Trajetórias. 1. ed. Editora UFMG, Belo Horizonte.
35. De Witte, K., & López-Torres, L. (2017). The interaction between dropout, graduation rates and quality ratings in universities. *Journal of the Operational Research Society, 68*(4), 339–363.
36. Povedano, R., Salgado Junior, A. P., Souza Junior, M. A. A., & Rebehy, P. C. P. W. (2021). Indicators and goals for school performance evaluation: A two-stage DEA analysis of the Ideb of municipal public schools. *Ensaio, 1*, 1–36.
37. Marinzeck, Y. S. S. (2021). Boas práticas administrativo-pedagógicas que impactam no desempenho dos alunos no ensino fundamental: um estudo quali-quantitativo em escolas públicas no Brasil e em Portugal. Dissertação de Mestrado—Faculdade de Economia Administração e Contabilidade de Ribeirão Preto—USP, Ribeirão Preto.

DEA-Viz: A Software for Visualization of Data Envelopment Analysis Problems

Shahin Ashkiani

Abstract The Visualization of data envelopment analysis (DEA) problems can be a very informative step in order to get insight into the problems and their characteristics. Through the various visualization methods, one can literally look at the DEA problem and the related data from different perspectives. Doing so would help to efficiently identify possible regularities and patterns, as well as irregularities and anomalies in the data. Despite its importance and benefits, visualization is vastly neglected in DEA applications. Even though there are several suggested methods for DEA visualization in the literature, the graphical investigation step is absent in the majority of the studies. The reason may lie in the fact that the DEA software packages severely lack any high-dimensional visualization features. This paper is an introduction to DEA-Viz, a new DEA software that is mainly focused on the visualization of DEA problems. The software is a free-to-use cloud-based applet that can shed light on a DEA problem using various visualization methods and thus from various aspects. The intuitive design of the DEA-Viz helps researchers generate diverse plots of their datasets in order to gain insight into the problems and augment the quantitative analysis. The software can be found here: https://ashkiani.shinyapps.io/dea-viz/.

Keywords Data envelopment analysis · Data visualization · Software

1 Introduction

The goal of this paper is the introduction of DEA-Viz, a data envelopment analysis (DEA) software specialized in visualization. In order to do so, it seems necessary to begin with an introduction to data visualization and its necessity in general. Therefore, this section reviews data visualization and tries to address its necessity, benefits, and purposes.

S. Ashkiani (✉)
Bank of England, Threadneedle St, London EC2R 8AH, London, UK
e-mail: contact@shahin-ashkiani.com

A. Emrouznejad et al. (eds.), *Advances in the Theory and Applications of Performance Measurement and Management*, Lecture Notes in Operations Research,
https://doi.org/10.1007/978-3-031-61597-9_24

Ward [1] defines visualization as "the communication of information using graphical representations". In this sense, visualization has been with humans for millennia. Even if we confine visualization to quantitative data visualization and define it as a graphical representation of the numerical data, it has been around for centuries and passed through several stages of boom and bust, with a recent rebirth [2].

While some of the findings of a proper visualization are also possible to discover through quantitative analytic methods, there may be some emergent properties" that appear through visualization that otherwise couldn't be found using conventional analytical approaches. One of such situations is famously portrayed by Anscombe's quartet [3]. Nevertheless, the emphasis on visualization should not overshadow the importance of quantitative analysis, since these are complementary approaches to the reality of the data.

Overall, Ware [4] finds the possibility of representation of large amounts of data, revealing emergent properties, shedding light on data patterns and peculiarities, as well as the generation of new hypotheses as some of the benefits of visualization.

Based on its purpose, visualization can be categorized into the following three categories: exploration, confirmation, and presentation [1]. DEA-Viz is a mixture of the first and last categories. DEA-Viz is for exploration of the data, to get insight into it, look at it from various perspectives, and find subjectively interesting points in it. DEA-Viz is an interactive applet, such that the users can zoom in, get further information about visualization components, and explore the data interactively. Finally, the findings can be downloaded and presented to a probable audience, although the main purpose of the applet is exploration.

All being said, it is of great importance to emphasize that visualization is a complementary tool to quantitative methods and not a rival to them. In the words of John Tukey [5]: "Exploratory data analysis can never be the whole story, but nothing else can serve as the foundation stone—as the first step."

2 Visualization of DEA Problems

While visualization is an indispensable step in quantitative studies nowadays, its limited presence in data envelopment analysis studies is conspicuous. For instance, there is no visualization chapter or section in the DEA handbook [6], despite the handbook's comprehensiveness. The reasons for such absence are perhaps threefold:

1. Multidimensionality of DEA problems: In order to use visualization in DEA problems, one has to decide which data set is intended to be visualized. The data sets of DEA problems, such as the inputs and outputs dataset or the dual multiplier profiles of the DMUs, are oftentimes high-dimensional. Adler [7] argues that the reason for the absence of visualization in DEA problems is their high dimensionality, which causes difficulty in the graphical presentation of the problems.

2. Doubt about the benefits of visualization: Another possible reason for not using data visualization in DEA is considering the visualization a superfluous step since the wide variety of DEA quantitative methods generate the expected results.
3. Lack of high-dimensional visualization features in DEA software packages: even if a researcher wants to use visualization methods, none of the current DEA software packages can be helpful in the case of comprehensive high-dimensional data visualization methods, so the researcher must code the visualization methods from scratch in a programming language.

This paper and the DEA-Viz applet intend to address all three issues. Firstly, we are not helpless in the face of the multidimensionality of DEA problems since several methods are already available in the DEA literature to cope with this difficulty, using dimension-reduction techniques such as principal component analysis (PCA), multidimensional scaling (MDS), and Self-Organizing maps (SOM) or techniques such as parallel coordinates.

The relative under-usage of visualization in DEA may lie in the fact that the current DEA packages, to the best of the author's search and knowledge, are limited to uni-variate or bi-variate visualizations. In other words, the visualization features of the current software packages lack most of the high-dimensional data visualization techniques, while DEA problems with more than two DMUs or more than two input or output variables are inevitably high-dimensional. High-dimensional visualization methods tend to retain the relations among variables and "holistically" visualize a given dataset.

In DEA, it is very common to plot the efficient frontier in a graph of two chosen variables, inputs or outputs, from the set of all variables. Doing so is a reductionist approach to a high-dimensional dataset since it does not retain all the components and the relations among them.

Despite the crucial importance of high-dimensional visualization, none of the current DEA software packages have proper high-dimensional visualization features. Hence, if a researcher believes in the power of high-dimensional visualization, none of the current software packages can help them do so, and the only available option would be to code the visualization process from scratch. Doing so is not only time-consuming but also demands coding knowledge. Possibly this obstacle, along with the other stated reasons, dissuades researchers from visualizing the DEA problems.

DEA-Viz intends to address all three possible causes of visualization neglect. The multidimensionality of DEA problems is not an obstacle anymore since the DEA-Viz software includes several data visualizations for high-dimensional DEA data. Additionally, DEA-Viz does not require its users to know any programming language, and its intuitive graphical user interface makes working with the package easy.

Through the rest of this section as well as the illustration of the software with a real dataset, the author is hopeful to motivate DEA researchers to benefit from data visualization in their studies. Thus, promotion of data visualization in the DEA field is another goal of this paper and DEA-Viz.

3 DEA-Viz

In this section, DEA-Viz and its capabilities are introduced using a real dataset. The chosen dataset will be visualized using various methods, but the methods are not introduced here extensively due to a lack of space. Interested audiences can learn about the available DEA visualization methods through [8].

3.1 *Exploration of a Real Dataset*

In order to illustrate the features of DEA-Viz, a real dataset is visualized and explored in this section. The dataset is the well-known dataset of 35 Chinese cities with 3 inputs and 3 outputs, which have been used in some DEA visualization studies, e.g. [7, 9], and some non-visualization DEA studies, e.g. [10]. Since this dataset has been used in previous DEA visualization studies, it connects this study to the previous ones and makes evaluation, comparison, and validation easier.

3.2 *Data Importation*

DEA-Viz has two main sections: Upload Data and Plots. The former is dedicated to data importation, and it has some limited data manipulation features. The data must be in the "comma separated values (CSV)" format. In addition, the values must be all numerical and non-negative. Missing values are not allowed as well.

3.3 *Distributions*

The first subsection of the "Plots" section is "distributions" (Fig. 1). Distributions include input and output distribution dotplots as well as the dotplots of various efficiency scores such as constant return to scale (CRS), variable return to scale (VRS), and free disposal hull (FDH). While these plots are univariate, they are very good starting points to get insight into the data. The efficiency scores are computed using the benchmarking package [11], and the plots are generated using the ggplot2 package [12].

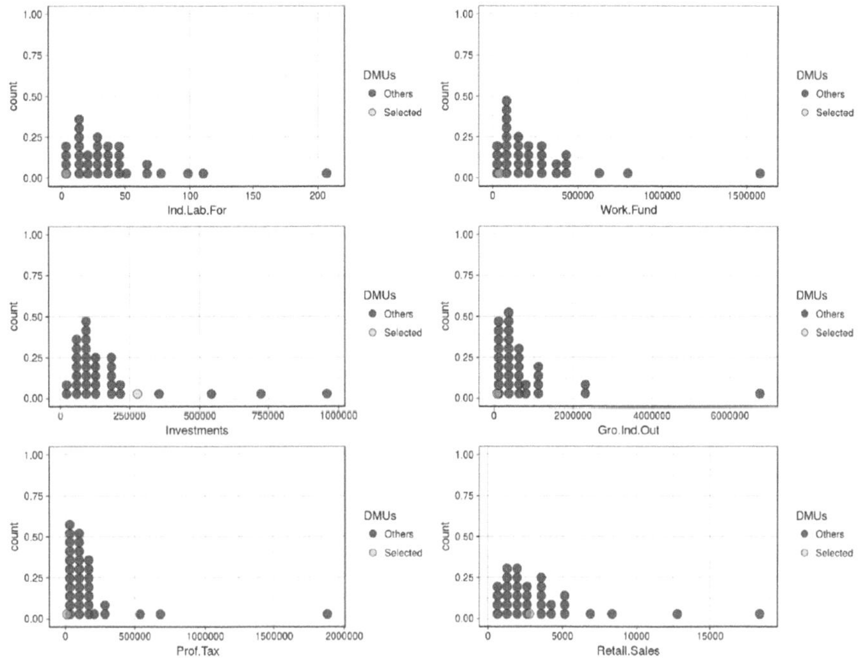

Fig. 1 Distributions of inputs and outputs with highlighted DMU 24

3.4 Correlations

The second preliminary set of plots is the Pearson's correlation plot. The correlations between each pair of inputs and/or outputs, as well as the efficiency scores of a specific model such as CRS, There are different approaches to visualization of correlations, but for low numbers of inputs and outputs, the comprehensive plot of the "Performance Analytics" package [13] (Fig. 2).

3.5 Cross-Efficiency Unfolding

The first high-dimensional DEA visualization method that we use to visualize the Chinese city dataset is Cross-efficiency unfolding. Suggested by Ashkiani and Mar Molinero [14], this method visualizes the cross-efficiency matrix (CEM) of the dataset.

Figure 3 is the unfolding map of CEM for 35 Chinese cities. The rating and rated objects are shown in blue dots and red triangles, respectively. Ideally, the distance between every pair of same-class objects is a reflection of the dissimilarity of the pair. Thus, two close-row objects probably have very low dissimilarity, i.e., high similarity.

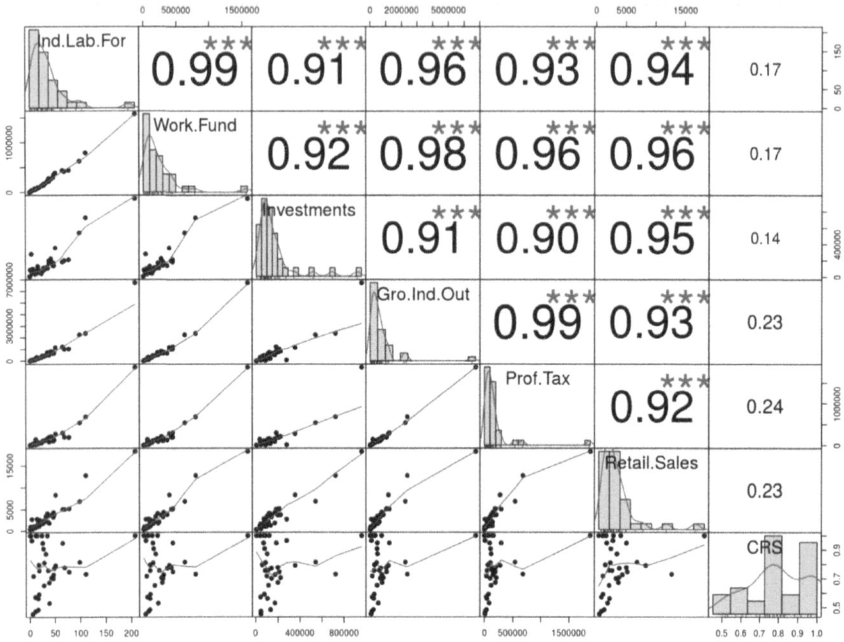

Fig. 2 Pearson's correlation coefficients and their statistical significance

The distance between a row object and a column object is ideally determined by its corresponding cross-efficiency score, and it is a measure of "preference".

3.6 Porembski Graph

Porembski et al. [15] suggest the first high-dimensional data visualization method for DEA. Their method is based on Sammon's mapping [16], which is a non-linear dimension reduction technique from the class of multidimensional scaling (MDS). Using this dimension reduction technique, the dimensionality of the dataset of inputs and outputs is reduced to two dimensions, i.e., the coordinates of DMUs in the final map. The authors suggest augmenting the bi-dimensional map using multiplier weights, i.e., lambda values. The lambda values are added to the map as edges or links between efficient units and inefficient units, such that the intensity of the edges is proportional to the magnitude of the lambdas. The dataset visualization using Porembski's method is presented in Fig. 4.

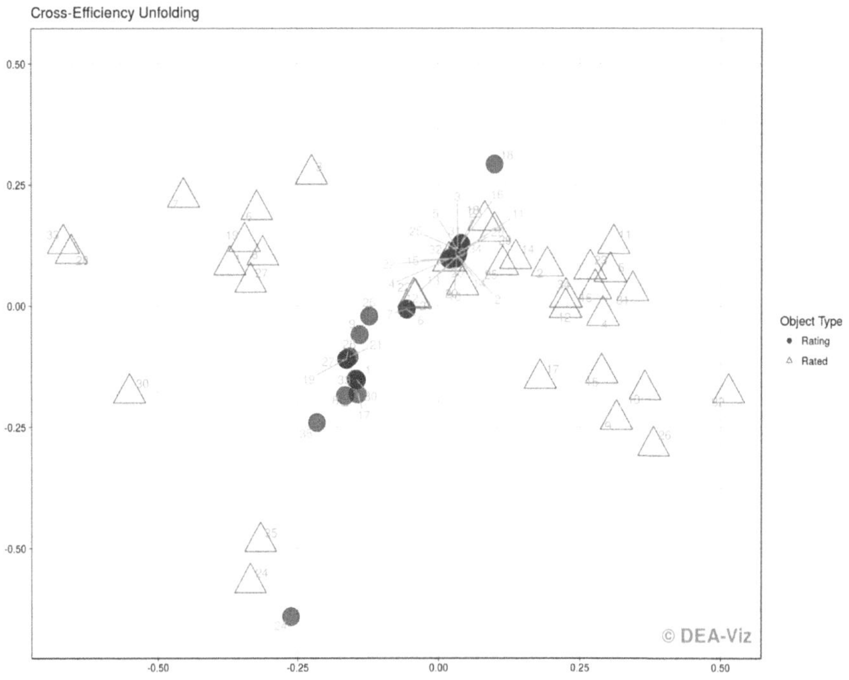

Fig. 3 Unfolding map of the benevolent CEM. Unit 24 seems an anomaly

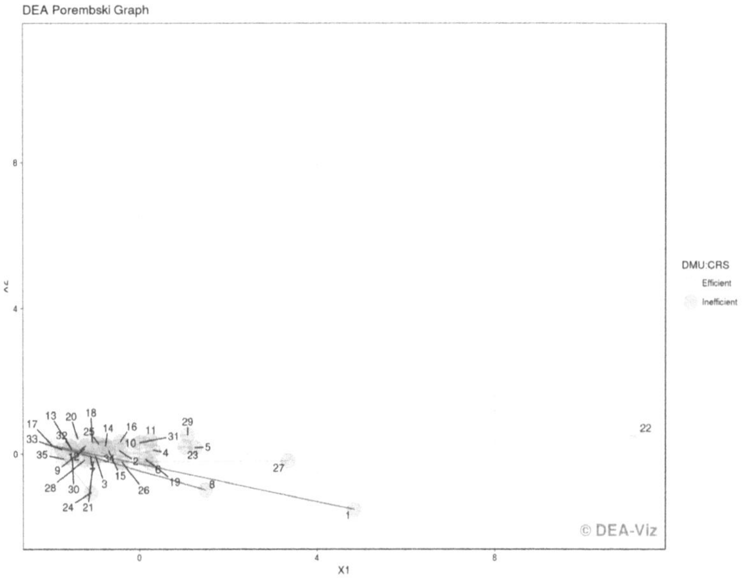

Fig. 4 Porembski's visualization based on Sammon's mapping. Unit 22 is an efficient DMU different from others

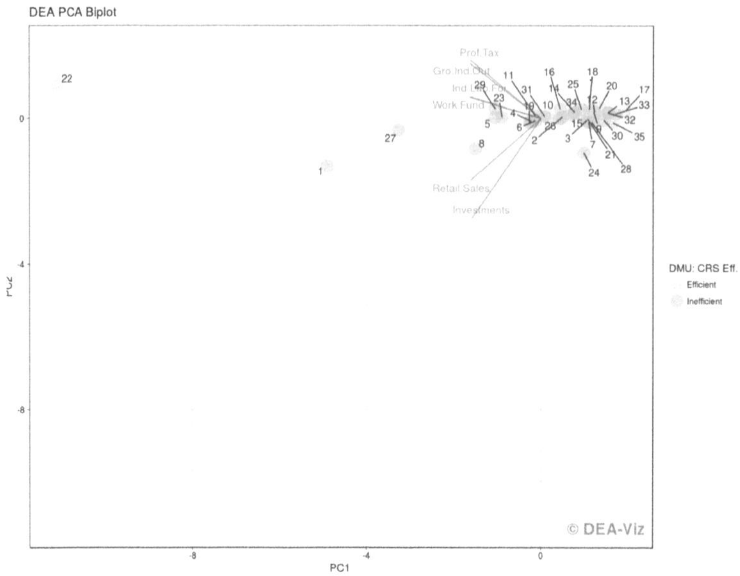

Fig. 5 PCA is used for dimension reduction and the variables are imposed on the map. Unit 1 is uniquely high in retail sales and investment, yet inefficient

3.7 PCA-Biplot

In PCA-Biplots, high-dimensional data is reduced using PCA, and then vectors of the original variables are superimposed on the PCA map. Using the direction of the vectors, one can investigate the differences between objects or clusters of objects on the map. The Biplot of the dataset is presented in Fig. 5.

3.8 Color-Coded Multidimensional Scaling

Multidimensional Scaling (MDS) is a class of non-linear dimension-reduction techniques. Compared to PCA, it makes fewer assumptions about data. For instance, PCA assumes that the topology of the data can be reduced in a linear fashion, or that the higher the variance of a variable, the higher the load of information it bears and the higher its importance. However, these assumptions in practice may be problematic. MDS tries to retain the original inter-object distances and represent them in low-dimensional space. For detailed information about MDS, excellent sources such as [18] are available. DEA-Viz's MDS plot is inspired by the DEA co-plot [7], except for the vectors. The map is presented in Fig. 6.

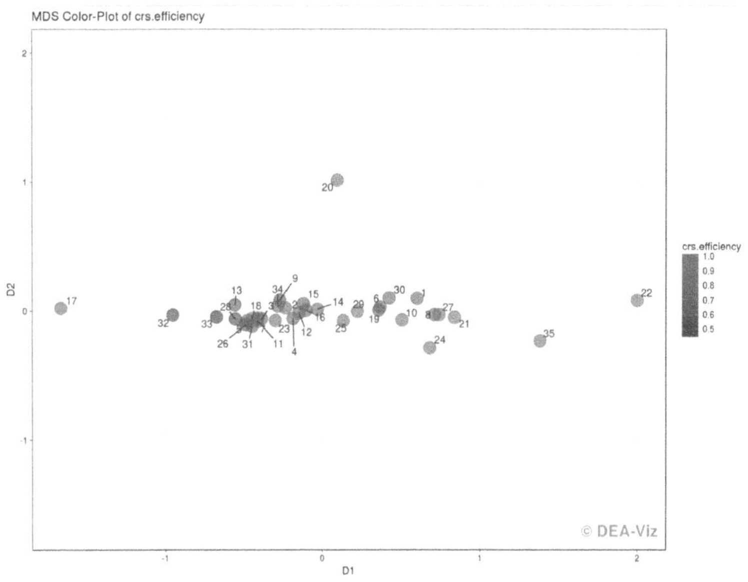

Fig. 6 Units 22, 35, 20, and 17 all seem efficient but in their own unique way

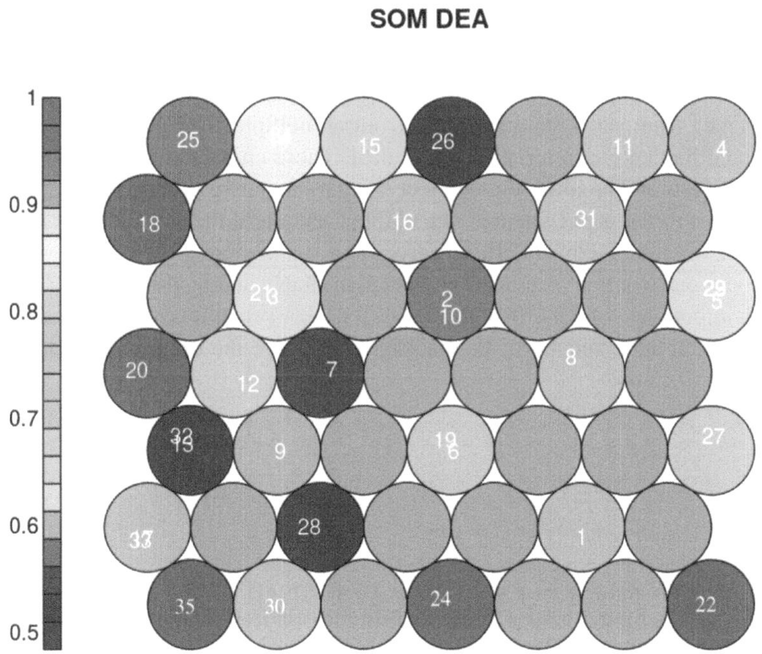

Fig. 7 Self-organizing map of DMUs based on their input and output profile. Unit 24 is CRS efficient with low similarity to other units

3.9 *Self-Organizing Map*

Self-Organizing Map (SOM), suggested by Kohonen [19], is an unsupervised learning method to represent a high-dimensional dataset into a two-dimensional grid of units such that the topology of the dataset in the original space is preserved. The main goal of the SOM is the division of the original data space into a pre-defined number of disjoint sub-spaces, and therefore the segmentation of the space rather than the observations.

4 Conclusion

DEA data-sets oftentimes are high-dimensional, and while several high-dimensional data visualization methods are suggested in the literature, the visualization features of DEA software packages are limited to uni-variate or bi-variate data. Therefore, these features are not very helpful to DEA practitioners in order to gain insight into DEA problems. Beside DEA software packages, DEA textbooks and handbooks are also vastly neglected for DEA visualization.

DEA-Viz intends to fill the shortage of visualization features in the current DEA packages.[1] Its specific focus is on high-dimensional visualization, and it benefits from the main suggested DEA visualization methods in the literature. Moreover, DEA-Viz is a cloud-based applet, available at https://ashkiani.shinyapps.io/dea-viz/, and can be run on any device with an internet connection. Through approaching the DEA data from various angles and providing multiple plots, DEA-Viz can help reveal some hidden characteristics of the data. These characteristics may make or mar the analysis and are difficult to detect using quantitative methods.

In contrast to the novel features, the current version of the software has its own limitations, mainly in terms of efficient frontier visualization. In the current version, there is only one frontier visualization method available among the outputs. Considering the importance of the efficient frontier for DEA practitioners and researchers, the focus of the next version of the package will be on the implementation of the frontier visualization.

[1] While some DEA softwear packages, such as DEAP, DEAFrontiner, OpenSourceDEA and DEA Toolbox, have no visualization features, others, such as Frontier Analyst, DEA-Solver Pro, MaxDEA, the R Benchmarking Package, and PIM-DEAsoft, are focused on low-dimensional visualizations such as barcharts of efficiency scores, scatterplots of efficient frontiers based on two selected variables, scatterplots of transformation curves, and Malmquist index time-series. The only high-dimensional visualization among all is the radar chart of unit variables in the Frontier Analyst package.

References

1. Ward, M. O., Grinstein, G., & Keim, D. (2010). *Interactive data visualization: Foundations, techniques, and applications*. CRC Press.
2. Chen, C. –H., et al. (2008). A brief history of data visualization. In: *Handbook of data visualization* (pp 15–56).
3. Anscombe, F. J. (1973). Graphs in statistical analysis. *The American Statistician, 27*(1), 17–21.
4. Ware, C. (2012). *Information visualization: Perception for design*. Waltham, MA.
5. Tukey, J. W. (1977). *Exploratory data analysis*, Vol. 2.
6. Cooper, W. W., et al. (2011). Avoiding large differences in weights in cross-efficiency evaluations: application to the ranking of basketball players. *Journal of CENTRUM Cathedra: The Business and Economics Research Journal*, 4(2), 197–215.
7. Adler, N., & Raveh, A. (2008). Presenting DEA graphically. *Omega, 36*(5), 715–729.
8. Ashkiani, S. (2019). *Four essays on data visualization and anomaly detection of data envelopment analysis problems* (pp. 29–64). https://www.tesisenred.net/bitstream/handle/10803/669593/shas1de1.pdf?sequence=1&isAllowed=y
9. e Costa, C. A. B., de Mello, J. C., & Meza, L. A. (2016). A new approach to the bi-dimensional representation of the DEA efficient frontier with multiple inputs and outputs. *European Journal of Operational Research*, 255(1), 175–186.
10. Sueyoshi, T. (1992). Measuring the industrial performance of Chinese cities by data envelopment analysis. *Socio-Economic Planning Sciences, 26*(2), 75–88.
11. Bogetoft, P., & Otto, L. (2018). Benchmarking with DEA and SFA, R package version 0.27.
12. Wickham, H., Chang, W., & Wickham, M. H. (2016). "Package 'ggplot2'." *Create elegant data visualisations using the grammar of graphics. Version,* 2(1), 1–189.
13. Peterson, B. G., Carl, P., et al. (2018). Package 'PerformanceAnalytics'. https://cran.r-project.org/web/packages/PerformanceAnalytics/index.html
14. Ashkiani, S., & Mar Molinero, C. (2017). Visualization of cross-efficiency matrices using multidimensional unfolding. In *Proceedings of the 15th International Conference of DEA* (pp. 242–249). ISBN 9781854494337.
15. Porembski, M., Breitenstein, K., & Alpar, P. (2005). Visualizing efficiency and reference relations in data envelopment analysis with an application to the branches of a German bank. *Journal of Productivity Analysis, 23*, 203–221.
16. Sammon, J. W. (1969). A nonlinear mapping for data structure analysis. *IEEE Transactions on Computers, 100*(5), 401–409.
17. Young, F. W. (1999). Principal components: BiPlots.
18. Borg, I., Groenen, P. J. F., & Mair, P. (2012). *Applied multidimensional scaling*. Springer Science & Business Media.
19. Kohonen, T. (1982). Self-organized formation of topologically correct feature maps. *Biological cybernetics, 43*(1), 59–69.

The manufacturer's authorised representative in the EU is Springer Nature Customer Service Centre GmbH, Europaplatz 3, 69115 Heidelberg, Germany. If you have any concerns regarding our products, please contact ProductSafety@springernature.com

Printed and bound by CPI Group (UK) Ltd, Croydon, CR0 4YY
01/10/2026
02225849-0009